Second Edition

Real and Functional Analysis

PART A
REAL ANALYSIS

MATHEMATICAL CONCEPTS AND METHODS IN SCIENCE AND ENGINEERING

Series Editor: Angelo Miele
Mechanical Engineering and Mathematical Sciences
Rice University

Recent volumes in this series:

17 **MATHEMATICAL PRINCIPLES IN MECHANICS AND ELECTROMAGNETISM, Part B: Electromagnetism and Gravitation** • *C. C. Wang*

18 **SOLUTION METHODS FOR INTEGRAL EQUATIONS: Theory and Applications** • *Edited by Michael A. Golberg*

19 **DYNAMIC OPTIMIZATION AND MATHEMATICAL ECONOMICS** • *Edited by Pan-Tai Liu*

20 **DYNAMICAL SYSTEMS AND EVOLUTION EQUATIONS: Theory and Applications** • *J. A. Walker*

21 **ADVANCES IN GEOMETRIC PROGRAMMING** • *Edited by Mordecai Avriel*

22 **APPLICATIONS OF FUNCTIONAL ANALYSIS IN ENGINEERING** • *J. L. Nowinski*

23 **APPLIED PROBABILITY** • *Frank A. Haight*

24 **THE CALCULUS OF VARIATIONS AND OPTIMAL CONTROL: An Introduction** • *George Leitmann*

25 **CONTROL, IDENTIFICATION, AND INPUT OPTIMIZATION** • *Robert Kalaba and Karl Spingarn*

26 **PROBLEMS AND METHODS OF OPTIMAL STRUCTURAL DESIGN** • *N. V. Banichuk*

27 **REAL AND FUNCTIONAL ANALYSIS, Second Edition Part A: Real Analysis** • *A. Mukherjea and K. Pothoven*

28 **REAL AND FUNCTIONAL ANALYSIS, Second Edition Part B: Functional Analysis** • *A. Mukherjea and K. Pothoven*

29 **AN INTRODUCTION TO PROBABILITY THEORY WITH STATISTICAL APPLICATIONS** • *Michael A. Golberg*

Second Edition

Real and Functional Analysis

PART A
REAL ANALYSIS

A. Mukherjea and K. Pothoven

University of South Florida
Tampa, Florida

Plenum Press · New York and London

Library of Congress Cataloging in Publication Data

Mukherjea, Arunava, 1941–
Real and functional analysis.

(Mathematical concepts and methods in science and engineering; 27–)
Bibliography: pt. A, p.
Includes index.
Contents: pt. A. Real analysis.
1. Functions of real variables. 2. Functional analysis. I. Pothoven, K. II. Title. III. Series: Mathematical concepts and methods in science and engineering; 27, etc.
QA331.5.M84 1984 515.8 84-8363
ISBN 0-306-41557-7 (v. 1)

A Division of Plenum Publishing Corporation
233 Spring Street, New York, N.Y. 10013

Printed in the United States of America

Preface to the Second Edition

The second edition is composed of two volumes. The first volume, Part A, is entitled *Real Analysis* and contains Chapters 1, 2, 3, 4, parts of Chapter 5, and the first five sections of Chapter 7 of the first edition together with additional material added to each of these chapters. The second volume, Part B, is entitled *Functional Analysis* and is based on Chapters 5 and 6 and the Appendix of the first edition together with additional topics in functional analysis. Included in this volume are many new problems, new proofs of theorems, and additional material. Our goal has been, as before, to present the essentials of real analysis as well as to include in the book many interesting, useful, and relevant results (usually not available in other books) so that the book can be used as a reference for the student of analysis.

Although there is more than sufficient material in the text for a full year of study in measure and integration theory, often only a semester is available. Clearly only the essentials can then be presented. For such a short course in measure and integration theory, we suggest the following sections:

Semester course

Section 1.3 through Theorem 1.4
Section 2.1 up to Proposition 2.3
Sections 2.2 and 2.3
Theorem 2.5 of Section 2.4
Sections 3.1 and 3.2
Sections 3.4 through Theorem 3.10 (omitting Theorem 3.9)
Sections 4.1, 4.2, and 4.3.

As in the first edition, certain portions of the text designated by (•) can be omitted. In this volume, problems which are designated by (×) are an integral part of the text and should be worked by the student. Problems which are difficult are starred (*). Finally, for the convenience of the reader, an index specifying page numbers for various theorems, propositions, lemmas, etc., has been provided.

We are again grateful to friends and colleagues who have pointed out errors in the first edition and given suggestions to us for improving the text. Particularly, we thank Professors D. Basu, G. Högnäs, and R. A. Johnson.

Tampa, Florida

A. Mukherjea
K. Pothoven

Contents

1. Preliminaries on Set Theory and Topology **1**
1.1. Basic Notions of Sets and Functions 1
1.2. Relations, Orderings, Zorn's Lemma, and the Axiom of Choice . 9
1.3. Algebras, σ-Algebras, and Monotone Classes of Sets 16
1.4. Topological Spaces 26
1.5. Connected Spaces, Metric Spaces, and Fixed Point Theorems . . 45
1.6. The Stone–Weierstrass Theorem and the Ascoli Theorem . . 78

2. Measure **85**
2.1. Measure on an Algebra 87
2.2. Lebesgue Measure on Intervals 93
2.3. Construction of Measures: Outer Measures and Measurable Sets 97
2.4. Non-Lebesgue-Measurable Sets and Inner Measure 117

3. Integration **131**
3.1. Measurable Functions 133
3.2. Definition and Properties of the Integral 143
3.3. Lebesgue–Stieltjes Measure and the Riemann–Stieltjes Integral . 159
3.4. Product Measures and Fubini's Theorem 168
3.5. The L_p Spaces 186

4. Differentiation **201**
4.1. Differentiation of Real-Valued Functions 202
4.2. Integration versus Differentiation I: Absolutely Continuous Functions 214
4.3. Integration versus Differentiation II: Absolutely Continuous Measures, Signed Measures, the Radon–Nikodym Theorem . . 233
4.4. Change of Variables in Integration 261

5. Measure and Topology **267**

5.1. The Daniell Integral 268
5.2. Topological Preliminaries. Borel and Baire Sets 278
5.3. Measures on Topological Spaces; Regularity 291
5.4. Riesz Representation Theorems 307
5.5. Product Measures and Integration 319

Appendix **333**

A. Differentiation of Borel Measures; The Change-of-Variable Formula 333
B. Product Measures 343

BIBLIOGRAPHY 349
DEFINITION, THEOREM, PROPOSITION, LEMMA, AND COROLLARY INDEX 353
SYMBOL AND NOTATION INDEX 357
SUBJECT INDEX 359

1

Preliminaries on Set Theory and Topology

One of the aims of this chapter is to introduce the reader to those parts of set theory and topology that are used frequently in the main theme of this text. While providing these preliminaries, for completeness we also consider here almost all basic concepts in point-set topology. The sections covering this material are presented in a manner somewhat different from that used for the other sections. Thus readers with very little or no background in topology may have to do a little (but not much) extra work while studying these sections.

We also consider fixed point theorems in Section 1.5. In recent years, the study of fixed point theorems has been found to be very fascinating and extremely useful in various problems in analysis, and it is now a subject in its own right. Here we have attempted to present a somewhat complete account of those fixed point theorems that depend only on very elementary concepts in metric spaces. A part of this account is given in the form of problems.

Finally, we present in this chapter two very important results in analysis that have been used frequently in the rest of this text: the Stone–Weierstrass Theorem and the Ascoli Theorem.

1.1. Basic Notions of Sets and Functions

It is not our intention to give a rigorous treatment of the theory of sets or even to give an elaborate discourse on this theory. Readers interested in an axiomatic treatment of the theory of sets should consult texts such as

Axiomatic Set Theory by P. Suppes† or *Naive Set Theory* by P. Halmos.‡ It is our intention, rather, in this brief section to establish notation and terminology to be used in the text as well as to explicitly list needed relationships and theorems involving sets and functions, many of which the reader is probably already familiar with.

Sets will generally be denoted by capital letters as A, B, C, X, Y, or Z and elements of sets by small letters a, b, c, x, y, or z. The small Greek letters $\alpha, \beta, \gamma, \ldots$ will generally be used to represent real or complex numbers when used as scalars in vector spaces. Classes or families of sets will be denoted by capital script letters as $\mathscr{A}, \mathscr{B}, \mathscr{C}, \ldots$.

As is standard, if $P(x)$ denotes some property about x, the notation

$$\{x: P(x)\}$$

will be used to mean "the set of all elements x (from some universal set) for which $P(x)$ is true."

In this text the following notations will be used:

$\varnothing$ is the empty set.
N or Z^+ is the set of natural numbers $\{1, 2, 3, \ldots\}$.
Z is the set of integers $\{\ldots, -3, -2, -1, 0, 1, 2, 3, \ldots\}$.
Q is the set of rational numbers.
R is the set of real numbers.

Definition 1.1. If A and B are sets, *A is a subset of B*, written $A \subset B$, if each element of A is an element of B. If in addition there is an element of B not in A we write $A \subsetneqq B$. Set A equals set B, written $A = B$, if $A \subset B$ and $B \subset A$.

The *union* $A \cup B$ of A and B is defined as

$$A \cup B \equiv \{x: x \in A \text{ or } x \in B\};$$

the *intersection* $A \cap B$ by

$$A \cap B \equiv \{x: x \in A \text{ and } x \in B\};$$

the *complement* of B in A by

$$A - B \equiv \{x: x \in A \text{ and } x \notin B\};$$

† P. Suppes, *Axiomatic Set Theory*, Dover, New York (1972).
‡ P. Halmos, *Naive Set Theory*, Springer-Verlag, New York (1970).

and the *symmetric difference* $A \triangle B$ by

$$A \triangle B \equiv (A - B) \cup (B - A).$$

If U is the universal set, the complement A^c of A is the set $U - A$.

If $\mathscr{A}$ is some family of sets, then

$$\bigcup_{A \in \mathscr{A}} A \equiv \{x\colon x \in A \text{ for some } A \text{ in } \mathscr{A}\},$$

and

$$\bigcap_{A \in \mathscr{A}} A \equiv \{x\colon x \in A \text{ for each } A \text{ in } \mathscr{A}\}.$$

If I is a nonempty set and for each $i \in I$ there is associated a set A_i (an indexed family of sets), then

$$\bigcup_{i \in I} A_i \equiv \{x\colon x \in A_i \text{ for some } i \in I\},$$

and

$$\bigcap_{i \in I} A_i \equiv \{x\colon x \in A_i \text{ for each } i \in I\}.$$

A collection of sets $\mathscr{A}$ is said to be *pairwise disjoint* if $A \cap B = \varnothing$ whenever A and B are in $\mathscr{A}$ and $A \neq B$. ▮

The following theorem is easy to verify.

Theorem 1.1. If A, B, and C are sets and $\mathscr{A}$ is a family of sets,

(i) $B \cap \left(\bigcup_{A \in \mathscr{A}} A\right) = \bigcup_{A \in \mathscr{A}} (B \cap A)$

(ii) $C \cup \left(\bigcap_{A \in \mathscr{A}} A\right) = \bigcap_{A \in \mathscr{A}} (C \cup A)$

distributive laws,

(iii) $B - \left(\bigcup_{A \in \mathscr{A}} A\right) = \bigcap_{A \in \mathscr{A}} (B - A)$

(iv) $C - \left(\bigcap_{A \in \mathscr{A}} A\right) = \bigcup_{A \in \mathscr{A}} (C - A)$

De Morgan's laws. ▮

Definition 1.2. Let X and Y be sets. The *Cartesian product* of X and Y is the set $X \times Y$ given by

$$X \times Y \equiv \{(x, y)\colon x \in X \text{ and } y \in Y\}.$$

Two elements (x, y) and (x', y') of $X \times Y$ are equal if and only if $x = x'$ and $y = y'$. [Note that if x and y are elements of $X \cap Y$, $(x, y) = (y, x)$ if and only if $x = y$.] ▮

Definition 1.3. Let X and Y be sets. A *relation f from X into Y* is any subset of $X \times Y$. The *domain* of a relation f is the set

$$D_f = \{x \in X: (x, y) \in f \text{ for some } y \in Y\}.$$

The *range* R_f of a relation f is the set

$$R_f = \{y \in Y: (x, y) \in f \text{ for some } x \in X\}.$$

If f is a relation from X into Y, the *inverse relation* f^{-1} is given by

$$f^{-1} = \{(y, x): (x, y) \in f\}.$$

(Note that $D_{f^{-1}} = R_f$ and $R_{f^{-1}} = D_f$.) If f and g are relations from X into Y and from Y into Z, respectively, the *composition* $g \circ f$ is the relation from X into Z given by

$$g \circ f = \{(x, z) \in X \times Z: (x, y) \in f \text{ and } (y, z) \in g \text{ for some } y \text{ in } Y\}.$$

(Note $g \circ f \neq \varnothing$ if and only if $R_f \cap D_g \neq \varnothing$.) If f and g are both relations from X into Y and $f \subset g$, then f is said to be a *restriction* of g or g is said to be an *extension* of f. ∎

As is customary, we write $y = f(x)$ if $(x, y) \in f$, where f is a relation. More generally we have the following definition.

Definition 1.4. If f is a relation from X into Y and A is a subset of X, the *image* of A under f is the set

$$f(A) \equiv \{y \in Y: y = f(x) \text{ for some } x \in A\}.$$

If B is a set in Y, the *inverse image* of B under f is the set

$$f^{-1}(B) = \{x \in X: y = f(x) \text{ for some } y \in B\}. \qquad ∎$$

The reader may prove the following.

Proposition 1.1. Let f be a relation from X into Y and $\mathscr{A}$ a collection of subsets from X. Then

(i) $f\big(\bigcup_{A \in \mathscr{A}} A\big) = \bigcup_{A \in \mathscr{A}} f(A)$,

(ii) $f\big(\bigcap_{A \in \mathscr{A}} A\big) \subset \bigcap_{A \in \mathscr{A}} f(A)$. ∎

Definition 1.5. A *function f from X into Y*, denoted $f: X \to Y$, is a relation from X into Y such that

(i) $D_f = X$,
(ii) if $(x, y) \in f$ and $(x, y') \in f$, then $y = y'$.

[Thus if $y = f(x)$ and $y' = f(x)$, then $y = y'$. We can thereby speak of y as *the image* of x under f, denoted by $f(x)$.] If $R_f = Y$ then f is called a *surjection* or an *onto* function from X into Y. If $f(x) = f(x')$ always implies $x = x'$, then f is called *injective* or a *one-to-one* function. If f is surjective and injective, it is called *bijective* or a *one-to-one correspondence*. A function is also called a *mapping* or *transformation*. ∎

Proposition 1.2. If $f: X \to Y$ and $\mathscr{B}$ is a collection of subsets of Y, then

(i) $f^{-1}\left(\bigcup_{B \in \mathscr{B}} B\right) = \bigcup_{B \in \mathscr{B}} f^{-1}(B)$,
(ii) $f^{-1}\left(\bigcap_{B \in \mathscr{B}} B\right) = \bigcap_{B \in \mathscr{B}} f^{-1}(B)$,
(iii) $f^{-1}(B^c) = [f^{-1}(B)]^c$,
(iv) $f(f^{-1}(B)) \subset B$,
(v) $f^{-1}(f(A)) \supset A$. ∎

The proofs of these easily verified statements are left for the reader.

There are special types of functions that will be encountered frequently. Corresponding to each set A is the identity function $1_A: A \to A$ given by

$$1_A(a) = a, \qquad a \in A.$$

Corresponding to every subset A of some universal set U is the characteristic function $\chi_A: U \to \{0, 1\}$ given by

$$\chi_A(x) = \begin{cases} 1, & \text{if } x \in A; \\ 0, & \text{if } x \notin A. \end{cases}$$

If $D = \{(x, y) \in X \times X: x = y\}$ for some set X, then *Kronecker's δ function* corresponding to X is the function $\delta: X \times X \to \{0, 1\}$ given by

$$\delta_{xy} \equiv \delta(x, y) = \chi_D(x, y).$$

A function $f: N \to X$, where X is any set, is called a *sequence* in X. Because a sequence is uniquely and completely determined by the values x_i equal to $f(i)$ for $i \in N$, a sequence is usually denoted by $(x_i)_{i \in N}$ without explicit reference to f. The value x_i is called the ith term of the sequence $(x_i)_N$. A *finite sequence* in X or an *n-tuple in* X is a function from $\{1, 2, \ldots, n\}$ into X, denoted by $(x_i)_{i=1}^n$, where $x_i = f(i)$.

Definition 1.6. A *finite set* is any set that is empty or the range of a one-to-one correspondence on the set $\{1, 2, \ldots, n\}$ for some $n \in N$. (See Problem 1.1.6.) Any set not finite is called *infinite*. A *countably infinite* set is any set that is the range of a one-to-one correspondence on N. A *countable set* is any set that is finite or countably infinite. An *uncountable* set is any set not countable.

Two sets A and B have the same *cardinality* (written card A = card B) if there is a bijection from one set onto the other. It is easy to show that for any positive integers n and m, card $\{1, 2, \ldots, m\}$ = card $\{1, 2, \ldots, n\}$ if and only if $n = m$. By virtue of this fact we can define the *cardinality* of a finite set A to be n if there is a bijection from $\{1, 2, \ldots, n\}$ onto A. The cardinality of $\varnothing$ is zero. The cardinality of a countably infinite set is $\aleph_0$, and the cardinality of R is c. Also card $A \leq$ card B if there is an injection from A to B. ∎

We have already spoken of an "indexed" family of sets in Definition 1.1. Generally we say that a nonempty set of elements A is an *indexed set* if there is a nonempty set I, called the *index set*, and an onto function f: $I \to A$. A can then be denoted by $(a_i)_{i \in I}$, where $a_i = f(i)$ for $i \in I$. Clearly an indexed family of sets is nothing more than an indexed subset of the *power set* 2^U, the set of all subsets of some universal set U.

Clearly any countably infinite set can be indexed by N. Conversely any infinite set indexed by N is countably infinite. Indeed suppose $A = (a_i)_{i \in N}$ is an infinite set. Define f: $N \to A$ recursively as follows: Let $f(1) = a_1$; assuming $f(1), f(2), \ldots, f(k)$ are chosen, denote $f(1)$ by a_{n_1}, $f(2)$ by a_{n_2}, $\ldots$, and $f(k)$ by a_{n_k}; define $f(k + 1)$ to be $a_{n_{k+1}}$ where n_{k+1} is the smallest positive integer n greater than $n_1, \ldots, n_k$ such that $a_n \notin \{a_{n_1}, a_{n_2}, \ldots, a_{n_k}\}$. That n_{k+1} exists follows from the fact that every nonempty subset of N has a least element and the fact that A is an infinite set. The reader may check that f: $N \to A$ is a bijection.

The facts recorded in the following propositions are important.

Proposition 1.3. Any subset B of a countable set A is countable. ∎

Proof. Assume B is not empty. Since A is countable, A can be written as $(a_n)_N$. Let n_1 be the smallest positive integer n such that $a_n \in B$. Inductively, assuming $B - \{a_{n_1}, \ldots, a_{n_k}\}$ is not empty, let n_{k+1} be the smallest positive integer n such that $a_n \in B - \{a_{n_1}, \ldots, a_{n_k}\}$. If for some k, $B - \{a_{n_1}, a_{n_2}, \ldots, a_{n_k}\} = \varnothing$, then B is finite. If not, we have defined recursively a function $f: N \to B$ given by $f(k) = a_{n_k}$. f is easily seen to be bijective. ∎

Proposition 1.4. The Cartesian product $N \times N$ is countable. ∎

Proof. Define $f: N \to N \times N$ by $f(n) = (p, q)$, where $n = 2^{p-1}(2q - 1)$ and p is the smallest positive integer such that 2^p does not divide n. Since for any q in N, $2q - 1$ is odd, f is clearly onto and one-to-one. ∎

Proposition 1.5. The union of a countable family of countable sets is a countable set. ∎

Proof. The countable family can be written as $(A_n)_{n \in N}$, where each A_n is countable. Let $A = \bigcup_{n \in N} A_n$. Since an empty set adds nothing to the union, we may assume $A_n \neq \varnothing$ for each n. For each n there is a surjection $g_n: N \to A_n$. Define $g: N \times N \to A$ by $g(n, m) = g_n(m)$, an element of A_n. It is obvious that g is a surjection. Letting h be the composition $g \circ f$, where f is a bijection from N onto $N \times N$ as in Proposition 1.4, h is a surjection from N onto A. According to our remarks preceding Proposition 1.3, A is countable. ∎

An immediate consequence of Proposition 1.5 is the fact that the sets Z and Q are countable sets. Indeed,

$$Z = \{-n: n \in N\} \cup \{0\} \cup N$$

and

$$Q = \bigcup_{n=1}^{\infty} \{m/n: m \in Z\}.$$

In contrast the set R is uncountable. The proof of this is realized by showing that the subset (0, 1) of positive real numbers less than 1 is uncountable. Since each real number in (0, 1) can be written as an infinite decimal, if (0, 1) were a countable set $(r_n)_{n \in N}$, then each element in (0, 1) could be expressed as an infinite decimal

$$r_n = 0.s_{n_1}s_{n_2}\cdots$$

for some n and for some sequence $(s_{n_i})_{i \in N}$, where $s_{n_i} = 0, 1, \ldots,$ or 9 for each i. To see that this is impossible let

$$x = 0.x_1x_2\cdots,$$

where

$$x_n = \begin{cases} 1, & \text{if } s_{n_n} \neq 1; \\ 2, & \text{if } s_{n_n} = 1. \end{cases}$$

Then x is an infinite decimal representing a real number in $(0, 1)$ but not in $(r_n)_{n\in N}$ since $x_1 \neq s_{1_1}$, $x_2 \neq s_{2_2}$, ... Hence $(0, 1) \neq (r_n)_{n\in N}$.

As a final consideration in this section, let $(A_n)_N$ be any countable collection of subsets of a set X. The *limit superior* of $(A_n)_N$, also denoted by $\overline{\lim}\, A_n$, is given by

$$\lim_n \sup A_n \equiv \bigcap_{n=1}^{\infty} \left(\bigcup_{m=n}^{\infty} A_m \right).$$

The *limit inferior*, also denoted by $\underline{\lim}\, A_n$, is given by

$$\lim_n \inf A_n \equiv \bigcup_{n=1}^{\infty} \left(\bigcap_{m=n}^{\infty} A_m \right).$$

The reader should convince her- or himself that $\lim_n \sup A_n$ is the set of all points from X that belong to A_n for infinitely many n, whereas $\lim_n \inf A_n$ consists of all points from X belonging to A_n for all but finitely many n. The *limit of* A_n is defined as the common value of $\lim_n \sup A_n$ and $\lim_n \inf A_n$, if it exists.

Problems

1.1.1. Prove the following:

(i) $A \triangle B = A \cup B - A \cap B$.

(ii) $A = A \triangle B$ if and only if $B = \varnothing$.

(iii) $A \cap (B \triangle C) = (A \cap B) \triangle (A \cap C)$.

(iv) $\bigcup_{i=1}^{\infty} (A_i \cup B_i) = \left(\bigcup_{i=1}^{\infty} A_i\right) \triangle \left[\left(\bigcup_{i=1}^{\infty} A_i\right)^c \cap \left(\bigcup_{i=1}^{\infty} B_i\right)\right]$.

1.1.2. Suppose $f: X \to Y$ and $g: Y \to X$ are functions.

(i) If $g \circ f = 1_X$, prove f is injective.

(ii) If $f \circ g = 1_Y$, prove f is surjective.

(iii) If $g \circ f = 1_X$ and $f \circ g = 1_Y$, prove $g = f^{-1}$.

1.1.3. Show that the inclusion in part (ii) of Proposition 1.1. may be proper.

1.1.4. Show that the inclusions in parts (iv) and (v) of Proposition 1.2 may be proper and that parts (i), (ii), and (iii) of Proposition 1.2 may fail for arbitrary relations.

1.1.5. Show that if X is a set, then 2^X, the set of all subsets of X, cannot be indexed by X. [Hint: if $f\colon X \to 2^X$ consider $\{x \in X: x \notin f(x)\}$.]

1.1.6. (i) Prove that nonempty set A is finite if and only if there is a bijection $f\colon A \to \{1, \ldots, n\}$ for some n.

(ii) Prove that set A is countably infinite if and only if there is a bijection $f\colon A \to N$.

1.1.7. If $(A_n)_N$ is a monotone sequence of sets (that is, $A_{n+1} \subset A_n$ for all $n \in N$ or $A_{n+1} \supset A_n$ for all $n \in N$), then $\lim_n A_n$ is $\bigcup_n A_n$ or $\bigcap_n A_n$.

1.1.8. If B is a set and $(A_n)_N$ is a sequence of sets, prove

$$B - \lim_n \sup A_n = \lim_n \inf(B - A_n),$$
$$B - \lim_n \inf A_n = \lim_n \sup(B - A_n),$$
$$\lim_n \sup A_n - B = \lim_n \sup(A_n - B),$$
$$\lim_n \inf A_n - B = \lim_n \inf(A_n - B).$$

1.1.9. Let X be a nonempty set. Let $f\colon 2^X \to 2^X$ be such that $f(A) \subset f(B)$ whenever $A \subset B$. Prove that there exists E such that $f(E) = E$. [Hint: Let $E = \cup \{A: A \subset f(A)\}$.]

1.1.10. (Application of 1.1.9: The Schroeder–Bernstein Theorem). Let X and Y be arbitrary nonempty sets. Let $f\colon X \to Y$ and $g\colon Y \to X$ be one-to-one maps. Then there exists a bijection $h\colon X \to Y$. [Hint: Consider $F\colon 2^X \to 2^X$ defined by $F(A) = X - g(Y - f(A))$. Then $F(E) = E$ for some E by 1.1.9. Note that $X - E = g(Y - f(E))$.]

1.1.11. (Another Proof of the Schroeder–Bernstein Theorem). Let X, Y, f, and g be as in 1.1.10. Let $h(x) = g(f(x))$. Then $h\colon X \to g(Y)$ is one-to-one. Let $g(Y) \neq X$. Let $A = \{x \in X: h^n(y) = x$ for some $y \in X - g(Y)$ and some nonnegative integer $n\}$, where $h^n(y) = h(h^{n-1}(y))$ and $h^0(y) = y$. Define F by $F(x) = h(x)$ if $x \in A$, and $= x$ if $x \in X - A$. Show that $g^{-1} \circ F$ is a bijection from X to Y.

1.2. Relations, Orderings, Zorn's Lemma, and the Axiom of Choice

Given two sets X and Y, we have defined a relation from X to Y as a subset of $X \times Y$. We now wish to consider relations *in* X, that is relations from X to X. Our first definition gives the characteristics of a special type of relation in a set X.

Definition 1.7. Let X be a set. An *equivalence relation* $\mathscr{r}$ in X is a relation $\mathscr{r}$ in X ($\mathscr{r} \subset X \times X$) such that

(i) $x \mathscr{r} x$ for all x in X (reflexivity);

(ii) if $x \mathscr{r} y$ for x and y in X, then $y \mathscr{r} x$ (symmetry);

(iii) if $x \mathscr{r} y$ and $y \mathscr{r} z$ for x, y, and z in X, then $x \mathscr{r} z$ (transitivity),

where $x \mathscr{r} y$ is notation meaning $(x, y) \in \mathscr{r}$.

If $\mathscr{r}$ is an equivalence relation in X and $x \in X$, let

$$[x] = \{y \in X : x \mathscr{r} y\}.$$

$[x]$ is called the *equivalence class* of x with respect to $\mathscr{r}$. ∎

It is easy to prove the statement $[x] = [y]$ if and only if $x \mathscr{r} y$. The nonequal equivalence classes partition X in the sense that X can be written as the union of disjoint equivalence classes. In fact if a *partition* of X is defined as a collection of pairwise disjoint nonempty subsets of X whose union is X, there is a one-to-one correspondence between equivalence relations on X and partitions of X. The reader is referred to Problem 1.2.1.

Another special type of relation in X is given in the next definition.

Definition 1.8. Let X be a set. A *partial ordering* $\mathscr{p}$ on X is a relation in X such that

(i) $x \mathscr{p} x$ for each x in X (reflexivity);

(ii) if $x \mathscr{p} y$ and $y \mathscr{p} x$, then $x = y$ (antisymmetry);

(iii) if $x \mathscr{p} y$ and $y \mathscr{p} z$, then $x \mathscr{p} z$ (transitivity);

here again $x \mathscr{p} y$ is notation meaning $(x, y) \in \mathscr{p}$.

A *total ordering* on X is a partial ordering $\mathscr{p}$ on X such that

(iv) if x and y are in X, then $x \mathscr{p} y$ or $y \mathscr{p} x$.

A *well ordering* on X is a total ordering $\mathscr{p}$ on X such that

(v) if A is any nonempty subset of X then there exists an element a in A such that $a \mathscr{p} x$ for all x in A.

A *partially ordered* (or, respectively, *totally ordered* or *well ordered*) set is a pair $(X, \mathscr{p})$, where X is a set and $\mathscr{p}$ is a partial (or total or well) ordering on X. ∎

Examples

1.1. Probably the most familiar example of a totally ordered set is the set of real numbers with the usual "less than or equal to" relation denoted by $\leq$. This is not a well ordering.

1.2. Likewise a very familiar example of a well-ordered set is the set of natural numbers N with the relation $\leq$. The fact that $(N, \leq)$ is well ordered is equivalent to the *principle of mathematical induction*: If A is a subset of N such that $1 \in A$ and $n + 1 \in A$ whenever $n \in A$ for any n in N, then $A = N$. (See Problem 1.2.2.)

Usually an arbitrary partial ordering on an arbitrary set is denoted by $\leq$, notation which is obviously derived from the "less than or equal to" partial ordering on R. In this case $x < y$ means $x \leq y$ but $x \neq y$.

1.3. A simple example of a partial ordering that is not a total ordering is the inclusion ordering on the set of subsets 2^X of a set X with two or more elements given by

$$A \leq B \text{ if and only if } A \subset B \text{ for } A, B \in 2^X.$$

Definition 1.9. Suppose $(X, \leq)$ is a partially ordered set. If $A \subset X$, then $x \in X$ is an *upper bound* of A if $a \leq x$ for all $a \in A$. If x is an upper bound of A and $x \leq y$ whenever y is an upper bound of A, then x is the *least upper bound* of A, written $x = \sup A$. (By antisymmetry there can be at most one.) Likewise a *lower bound* and a *greatest lower bound* of A (written $\inf A$) are defined. $(X, \leq)$ is *order complete* if every nonempty set A in X with an upper bound has a least upper bound. (See Problem 1.2.3.) ▮

Definition 1.10. Let $(X, \leq)$ be a partially ordered set. $x \in X$ is a *maximal element* in X if whenever $x \leq y$ for y in X, then $x = y$. Similarly a *minimal element* is defined. A *chain* in $(X, \leq)$ is any subset C of X that is totally ordered by $\leq$. ▮

Examples

1.4. The real numbers R with the usual partial ordering $\leq$ form an order-complete partial ordered set whereas the rational numbers Q with the same ordering do not.

1.5. If $\mathscr{A}$ is the family of sets $\{\varnothing, \{0\}, \{1\}\}$ with the inclusion relation, then $\{0\}$ and $\{1\}$ are maximal elements, $\varnothing$ is a minimal element and a lower bound for $\mathscr{A}$. A chain $\mathscr{C}$ in $\mathscr{A}$ is the class $\{\varnothing, \{0\}\}$.

Let $(X_i)_{i \in I}$ be any indexed collection of sets indexed by set I. A function $c: I \to \bigcup_{i \in I} X_i$ such that $c(i) \in X_i$ for each $i \in I$ is called a *choice function*. If we write $c_i = c(i)$ for each $i \in I$ the indexed set $(c_i)_I$ completely describes the action of c on set I. The set of all choice functions has a special name.

Definition 1.11. If $(X_i)_I$ is an indexed family of sets then the *Cartesian product* $\times_{i \in I} X_i$ is the set of all choice functions $c: I \to \bigcup_I X_i$. ∎

If $\{A_1, A_2, \ldots, A_n\}$ is a finite collection of sets then $\times_{i \in I} A_i$, where $I = \{1, 2, \ldots, n\}$, is also denoted by $A_1 \times A_2 \times \cdots \times A_n$ and identified with the set

$$\{(a_1, a_2, \ldots, a_n): a_i \in A_i \text{ for } i = 1, 2, \ldots, n\}$$

when $(a_1, a_2, \ldots, a_n)$ is identified with the choice function

$$a: \{1, 2, \ldots, n\} \to \bigcup_{i=1}^{n} A_i$$

given by $a(i) = a_i$.

An axiom that seems entirely reasonable is the following Axiom of Choice. It was proposed by Zermelo early in the 1900's and shown to be independent of—neither derivable from nor contradictory to—other axioms of set theory by P.J. Cohen around 1963 (see [13]). Actually the axiom is very strong and has many important consequences in mathematics. We will use the axiom both implicitly and explicitly in this book.

Axiom 1.1. *Axiom of Choice.* If $I \neq \varnothing$ and $A_i \neq \varnothing$ for each $i \in I$, then $\times_I A_i \neq \varnothing$. ∎

If $\mathscr{A}$ is any nonempty family of sets, then $\mathscr{A}$ can be considered as an indexed family of sets indexed by set $\mathscr{A}$ by virtue of the function i: $\mathscr{A} \to \mathscr{A}$ given by $i(A) = A$. The Axiom of Choice thereby says that if $\mathscr{A}$ is any nonempty family of nonempty sets an element can be "chosen" from each set in $\mathscr{A}$.

Two important consequences of the Axiom of Choice are Zorn's Lemma and The Well-Ordering Principle. In fact the following is true.

Theorem 1.2. The following statements are equivalent (each is a consequence of the other):

(i) Axiom of Choice.

(ii) Zorn's Lemma: Each nonempty partially ordered set in which each chain has an upper bound has a maximal element.

(iii) Well-Ordering Principle: If X is a set, then there exists a well-ordering for X.

(iv) Hausdorff Maximal Principle: If $\mathscr{A}$ is a family of sets and $\mathscr{C}$ is a chain in $\mathscr{A}$, then there is a maximal chain $\mathscr{C}_0$ that contains $\mathscr{C}$. ∎

We do not prove the equivalence of these statements in this book. The interested reader may consult other set theoretic books such as [21], [55], or the appendix of [29] for a discussion of the Axiom of Choice and related results. An immediate consequence of the Well-Ordering Principle is the existence of an uncountable set as described in the next proposition.

Proposition 1.6. There is an uncountable set $\bar{X}$ and a well-ordering $\leq$ on $\bar{X}$ such that

(i) there is an element Ω in $\bar{X}$ such that $\alpha \leq \Omega$ for all α in $\bar{X}$, and
(ii) for any β in $\bar{X}$, $\beta \neq \Omega$, the set $\{\alpha\colon \alpha \leq \beta\}$ is countable. ∎

(We will use $\bar{X}$ in constructing examples in later chapters. Ω will be called the first *uncountable ordinal*, $X = \bar{X} - \{\Omega\}$ the set of *countable ordinals*, and $\bar{X}$ the set of *ordinals* less than or equal to the first uncountable ordinal.)

Proof. Let Y be any uncountable set and well-order Y with some well-ordering $\leq$. If Y does not have an element y_0 such that $y \leq y_0$ for all y in Y, adjoin an element y_0 to Y forming $Y_0 = Y \cup \{y_0\}$ and extend the partial ordering $\leq$ to Y_0 by requiring that $y \leq y_0$ for all y in Y_0. Y_0 is then uncountable and well ordered by $\leq$. Let $Z = \{y \in Y_0\colon \{x \in Y_0\colon x \leq y\}$ is uncountable$\}$. Since $y_0 \in Z$, $Z \neq \varnothing$ and has a least element, say Ω. Let $\bar{X} = \{x \in Y_0\colon x \leq \Omega\}$. $\bar{X}$ is the required set. ∎

Earlier (in Example 1.2) we mentioned that the Principle of Mathematical Induction follows from the fact that the positive integers N are well ordered. Analogously, any well-ordered set satisfies the following Principle of Transfinite Induction. We will make use of this fact later in particular for the set X just constructed.

Theorem 1.3. *Principle of Transfinite Induction.* Let $(W, \leq)$ be a well-ordered set. For any $a \in W$ let

$$I(a) = \{x \in W\colon x < a\}.$$

If A is a subset of W such that $a \in A$ whenever $I(a) \subset A$, then $A = W$. ∎

Proof. Assume $A^c \cap W \neq \varnothing$ and let a be the least element of $A^c \cap W$; that is, $a \in A^c \cap W$ and $a \leq w$ for all $w \in A^c \cap W$. Then $I(a) \subset A$, so that by hypotheses $a \in A$. This contradiction shows $A^c \cap W = \varnothing$ or $A = W$. ∎

A totally ordered set with both a maximal and a minimal element that we shall encounter frequently is the set $\bar{R}$ of *extended real numbers*. It is the set of real numbers with the usual partial ordering $\leq$ together with two additional elements $-\infty$ and ∞, which are required to satisfy $-\infty < x < \infty$ for each real number x. In addition we define the binary operations $+$ and $\cdot$ on $\bar{R}$ to be the usual $+$ and $\cdot$ on R and to additionally satisfy for all real numbers x the following equations:

$$x + \infty = \infty, \qquad x + (-\infty) = -\infty,$$
$$x \cdot \infty = \infty \quad \text{if } x > 0, \qquad 0 \cdot \infty = 0, \qquad x \cdot (-\infty) = \infty \quad \text{if } x < 0,$$
$$\infty + \infty = \infty, \qquad -\infty - \infty = -\infty,$$
$$\infty \cdot (\pm\infty) = \pm\infty, \qquad -\infty \cdot (\pm\infty) = \mp\infty.$$

It is obvious that each nonempty subset of $\bar{R}$ has an upper bound and a lower bound. The least upper bound of $\varnothing$ is defined to be $-\infty$ and the greatest lower bound of $\varnothing$ to be ∞. If S is any nonempty set of real numbers with no upper bound in R, then we define $\sup S = \infty$. Similarly if S has no lower bound in R, then we define $\inf S \equiv -\infty$. Thus for each subset S of $\bar{R}$, $\sup S$ exists and $\inf S$ exists.

If $(x_n)_N$ is a sequence of real numbers we define the *limit superior* of $(x_n)_N$, denoted as $\lim \sup x_n$ or $\overline{\lim}_n x_n$, by

$$\lim_n \sup x_n \equiv \inf\{s_n \colon s_n = \sup_{k \geq n} x_k\}.$$

The *limit inferior* of $(x_n)_N$, also denoted by $\underline{\lim}_n x_n$, is defined by

$$\lim_n \inf x_n \equiv \sup\{i_n \colon i_n = \inf_{k \geq n} x_k\}.$$

The following criteria are useful.

Remarks

1.1. The real number x is the limit superior of $(x_n)_N$ if and only if

(a) given $\varepsilon > 0$, there exists n such that $x_k < x + \varepsilon$ for all $k \geq n$, and

(b) given $\varepsilon > 0$ and given $n \in N$, there exists $k \geq n$ such that $x_k > x - \varepsilon$.

1.2. The extended real number ∞ is the $\lim \sup x_n$ if and only if given any $M \in R$ and $n \in N$ there is a $k \geq n$ such that $x_k > M$.

1.3. $-\infty = \lim_n \sup x_n$ if and only if given $M \in R$, there exists $m \in N$ such that for $n \geq m$, $x_n < M$.

1.4. Similar statements can be formulated for the limit inferior of $(x_n)_N$.

The proofs of the statements in Remarks 1.1–1.4 as well as the formulation in Remark 1.4 are left as exercises for the reader (see Problem 1.2.5). Other properties are recorded in Problem 1.2.9.

Finally, a sequence $(x_n)_N$ of real numbers is said to *converge* to an extended real number x if $\lim_n \inf x_n = \lim_n \sup x_n = x$. The reader is asked to show (Problem 1.2.6) that $(x_n)_N$ converges to a real number x, written $x = \lim_n x_n$, if and only if, given $\varepsilon > 0$, there exists $n \in N$ such that $| x_m - x | < \varepsilon$ whenever $m \geq n$. The notation $x_n \to x$ is also used to mean that $x = \lim_n x_n$.

If $(x_n)_N$ is a *nondecreasing sequence* of real numbers—that is, $x_n \leq x_{n+1}$ for all $n \in N$—then $\lim x_n$ exists in the extended real numbers and equals $\sup x_n$. Similarly if $(x_n)_N$ is a nonincreasing sequence, $\lim x_n = \inf x_n$. The notations $x_n \uparrow x$ or $x_n \downarrow x$ are used sometimes to mean that $(x_n)_N$ is a nondecreasing or nonincreasing sequence, respectively, converging to x.

Problems

1.2.1. Prove that there is a one-to-one correspondence between partitions of a nonempty set X and equivalence relations defined in X.

1.2.2. Prove that the fact that $(N, \leq)$ is a well-ordered set where $\leq$ is the usual "less than or equal to" relation is equivalent to the fact that the Principle of Mathematical Induction holds.

1.2.3. Prove that a partially ordered set $(X, \leq)$ is order complete if and only if every nonempty set A in X with a lower bound has a greatest lower bound.

1.2.4. Let $X = \bar{X} - \{\Omega\}$, the set of ordinals less than the first uncountable ordinal Ω. Show that every countable subset E of X has a least upper bound in X.

1.2.5. (i) Verify Remarks 1.1–1.3.

(ii) Formulate statements similar to Remarks 1.1–1.3 for $\lim \inf x_n$.

1.2.6. (i) Prove that the real number $x = \lim x_n$ if and only if, given $\varepsilon > 0$, there exists $n \in N$ such that $| x_m - x | < \varepsilon$ whenever $m \geq n$.

(ii) Prove that the real sequence $(x_n)_N$ converges to ∞ (also termed "diverging to ∞") if and only if, for any $M \in R$, there exists an $n \in N$ for $m \geq n$, $x_m \geq M$.

(iii) Prove that the real sequence $(x_n)_N$ converges to $-\infty$ (also termed "diverging to $-\infty$") if for any $M \in R$ there exists $n \in N$ such that $x_m < M$ whenever $m \geq n$.

1.2.7. Prove that if $(x_n)_N$ is a nondecreasing sequence of real numbers then (x_n) converges to $\sup x_n$ in the extended real numbers.

1.2.8. A sequence (x_n) of real numbers is a *Cauchy sequence* if for each $\varepsilon > 0$, there exists a natural number n_0 such that for $n, m \geq n_0$, $| x_n - x_m | < \varepsilon$. Using the order completeness of $(R, \leq)$ show that every Cauchy sequence in R converges to some real number x.

1.2.9. Let $(x_n)_N$ be a sequence of real numbers.

(i) Prove $\liminf x_n \leq \limsup x_n$.

(ii) Prove $\limsup (-x_n) = -\liminf x_n$.

(iii) If $(y_n)_N$ is another sequence of real numbers, prove

$$\begin{aligned}\liminf x_n + \liminf y_n &\leq \liminf (x_n + y_n)\\ &\leq \limsup x_n + \liminf y_n\\ &\leq \limsup (x_n + y_n)\\ &\leq \limsup x_n + \limsup y_n,\end{aligned}$$

provided that none of the sums here is of the form $\infty - \infty$.

1.3. Algebras, σ-Algebras, and Monotone Classes of Sets

Of primary importance in the definition and consideration of measures, which are essentially functions on certain classes of sets, is the domain of these functions. In Chapter 2 we will consider measures whose domains are algebras or σ-algebras of sets. In Chapter 5 it will be most instructive to consider measures on σ-rings as well as σ-algebras. The purpose of this section is to introduce these various classes of sets, compare them, and give essential properties.

Given a set X there are six types of classes of subsets of X we wish to examine. They are rings, algebras, σ-rings, σ-algebras, monotone classes, and Dynkin systems. The latter of these systems we will only encounter in Chapter 5. Let us begin with the definition of a ring and σ-ring.

Definition 1.12. A *ring of sets* $\mathscr{R}$ is a nonempty family of sets such that whenever A and B are in $\mathscr{R}$, then

$$A \cup B \text{ is in } \mathscr{R}$$

and

$$A - B \text{ is in } \mathscr{R}.$$

A *σ-ring of sets* is a ring $\mathscr{R}$ satisfying the additional condition that whenever $(A_i)_N$ is a countable class of sets in $\mathscr{R}$ then $\bigcup_{i=1}^{\infty} A_i$ is in $\mathscr{R}$. ▮

Remarks

1.5. If $\mathscr{R}$ is a ring, $\varnothing \in \mathscr{R}$ since $\mathscr{R}$ is a nonempty family containing a set A and $\varnothing = A - A$.

1.6. If $\mathscr{R}$ is a ring and A and B are in $\mathscr{R}$, then $A \triangle B$ and $A \cap B$ are in $\mathscr{R}$ since

$$A \triangle B = (A - B) \cup (B - A),$$
$$A \cap B = (A \cup B) - (A \triangle B).$$

1.7. If $\mathscr{R}$ is a ring and $(A_i)_{i=1}^n$ is any finite collection in $\mathscr{R}$, then

$$\bigcup_{i=1}^n A_i \in \mathscr{R} \qquad \text{and} \qquad \bigcap_{i=1}^n A_i \in \mathscr{R}.$$

1.8. If $\mathscr{R}$ is a σ-ring and $(A_i)_N$ is a countable collection of sets in $\mathscr{R}$, then the identity

$$\bigcap_{i=1}^\infty A_i = A - \bigcup_{i=1}^\infty (A - A_i),$$

where $A = \bigcup_{i=1}^\infty A_i$, implies that $\bigcap_{i=1}^\infty A_i \in \mathscr{R}$.

1.9. If $\mathscr{R}$ is a σ-ring and $(A_i)_N$ is a countable collection of sets in $\mathscr{R}$, then

$$\lim_i \inf A_i \in \mathscr{R} \qquad \text{and} \qquad \lim_i \sup A_i \in \mathscr{R}.$$

Examples

1.6. $\mathscr{R} = \{\varnothing\}$ and $\mathscr{R} = 2^X$, where X is any set, are examples of σ-rings.

1.7. If X is any set, then the class of all finite subsets of X is a ring. This class is a σ-ring if and only if X is itself a finite set. The class of all countable subsets of X is a σ-ring.

1.8. The class $\mathscr{R}$ of finite unions of "half-open" intervals (a, b] for real numbers a and b is a ring.

1.9. If $\mathscr{C}$ is a collection of rings (or a collection of σ-rings) then $\bigcap_{\mathscr{R} \in \mathscr{C}} \mathscr{R}$ is a ring (or a σ-ring, respectively); that is, the intersection of a collection of rings (σ-rings) is a ring (σ-ring).

The definition of an algebra of sets has one essential difference from that of a ring of sets. More precisely, we have the following.

Definition 1.13. Let X be a set. An *algebra of sets* $\mathscr{A}$ *in* X is a nonempty family of subsets of X such that

$$A \cup B \text{ is in } \mathscr{A} \text{ whenever } A \text{ and } B \text{ are in } \mathscr{A}$$

and

$$A^c \text{ is in } \mathscr{A} \text{ whenever } A \text{ is in } \mathscr{A}.$$

A σ-algebra in X is an algebra of sets $\mathscr{A}$ in X such that $\bigcup_{n=1}^{\infty} A_n$ is in $\mathscr{A}$ whenever $(A_n)_N$ is a countable collection from $\mathscr{A}$. ∎

Remarks

1.10. Each algebra (σ-algebra) $\mathscr{A}$ is a ring (σ-ring) because if A and B are in $\mathscr{A}$, then $A - B \in \mathscr{A}$ since

$$A - B = (A^c \cup B)^c.$$

That the converse is false is evident from Example 1.7 if X is an infinite set and from Example 1.8.

1.11. If X is a set, then a ring $\mathscr{R}$ of subsets of X is an algebra in X if and only if $X \in \mathscr{R}$. Indeed, if $\mathscr{R}$ is an algebra then $X = A^c \cup A$, where A is some element of $\mathscr{R}$ so that $X \in \mathscr{R}$. Conversely, if $X \in \mathscr{R}$ then $A^c \in \mathscr{R}$ whenever $A \in \mathscr{R}$, since

$$A^c = X - A,$$

so that $\mathscr{R}$ is an algebra. Similarly a σ-ring $\mathscr{R}$ of subsets of X is a σ-algebra if and only if $X \in \mathscr{R}$.

1.12. The intersection of any collection of algebras (σ-algebras) of subsets of set X is again an algebra (σ-algebra).

Examples

1.10. As mentioned above, Example 1.8 is not an algebra. However, if we consider the collection of all finite unions of intervals of the type $(a, b]$ for $-\infty \leq a < b < \infty$ or (b, ∞) for $-\infty \leq b < \infty$, then the collection is an algebra.

1.11. If X is any set, the class of countable subsets of X is a σ-algebra if and only if X is countable. If X is any set, the class of all subsets A of X such that A or A^c is countable is a σ-algebra.

Of significance will be rings, σ-rings, algebras, and σ-algebras related to a given class of sets. More precisely, if X is a set and $\mathscr{E}$ is a class of subsets of X we will be interested in the ring, σ-ring, algebra, or σ-algebra *generated* by $\mathscr{E}$. By definition the *ring generated by* $\mathscr{E}$ is the "smallest" ring of subsets of X *containing* $\mathscr{E}$ and is denoted by $\mathscr{R}(\mathscr{E})$. (Ring $\mathscr{R}$ is smaller than ring $\mathscr{R}'$ if $\mathscr{R} \subset \mathscr{R}'$.) Similarly, *the σ-ring generated by* $\mathscr{E}$ is defined as the smallest σ-ring $\sigma_r(\mathscr{E})$ containing $\mathscr{E}$, the *algebra generated by* $\mathscr{E}$ as the smallest algebra $\mathscr{A}(\mathscr{E})$ containing $\mathscr{E}$, and the *σ-algebra generated by* $\mathscr{E}$ as the smallest σ-algebra $\sigma(\mathscr{E})$ containing $\mathscr{E}$.

Proposition 1.7. If $\mathscr{E}$ is any class of subsets of X, then $\mathscr{R}(\mathscr{E})$, $\sigma_r(\mathscr{E})$, $\mathscr{A}(\mathscr{E})$, and $\sigma(\mathscr{E})$ exist. ∎

Proof. Let

$$\mathscr{C} = \{\mathscr{R} : \mathscr{R} \text{ is a ring of subsets of } X \text{ and } \mathscr{E} \subset \mathscr{R}\}.$$

Since $2^X \in \mathscr{C}$, $\mathscr{C}$ is a nonempty collection, and by Example 1.9, $\bigcap_{\mathscr{R}\in\mathscr{C}} \mathscr{R}$ is a ring in $\mathscr{C}$. Clearly $\bigcap_{\mathscr{R}\in\mathscr{C}} \mathscr{R} = \mathscr{R}(\mathscr{E})$. Similarly $\sigma_r(\mathscr{E})$, $\mathscr{A}(\mathscr{E})$, and $\sigma(\mathscr{E})$ exist. ∎

If $\mathscr{E}$ is any class of subsets of X, then the following diagram is valid:

$$\begin{array}{ccc} \mathscr{R}(\mathscr{E}) & \subset & \mathscr{A}(\mathscr{E}) \\ \cap & & \cap \\ \sigma_r(\mathscr{E}) & \subset & \sigma(\mathscr{E}) \end{array}$$

If $\mathscr{E}$ is any class of subsets of set X and $A \subset X$, then $\mathscr{E} \cap A$ denotes the class of all sets of the form $E \cap A$, where $E \in \mathscr{E}$. The equalities

$$\left(\bigcup E_i\right) \cap A = \bigcup (E_i \cap A)$$

and

$$(E_1 - E_2) \cap A = (E_1 \cap A) - (E_2 \cap A)$$

for arbitrary sets show in particular that if $\mathscr{E}$ is a ring or σ-ring of subsets of X, then $\mathscr{E} \cap A$ is a ring or σ-ring, respectively, of subsets of A. Moreover, $\mathscr{E} \cap A$ is an algebra of subsets of A whenever $\mathscr{E}$ is an algebra in X or $\mathscr{E}$ is a ring and $A \subset E$ for some $E \in \mathscr{E}$. The following proposition says even more.

Proposition 1.8. If $\mathscr{E}$ is any class of subsets of a set X and $A \subset X$, then

(i) $\mathscr{R}(\mathscr{E}) \cap A = \mathscr{R}(\mathscr{E} \cap A)$,

(ii) $\sigma_r(\mathscr{E}) \cap A = \sigma_r(\mathscr{E} \cap A)$,

(iii) $\mathscr{A}(\mathscr{E}) \cap A = \mathscr{A}(\mathscr{E} \cap A)$ (as an algebra of subsets of A),

(iv) $\sigma(\mathscr{E}) \cap A = \sigma(\mathscr{E} \cap A)$ (as a σ-algebra of subsets of A). ∎

Proof. We prove (ii) and leave the other proofs as an exercise. Clearly

$$\sigma_r(\mathscr{E} \cap A) \subset \sigma_r(\mathscr{E}) \cap A,$$

since $\sigma_r(\mathscr{E}) \cap A$ is a σ-ring containing $\mathscr{E} \cap A$ and $\sigma_r(\mathscr{E} \cap A)$ is the smallest such σ-ring by definition. Now let $B \in \sigma_r(\mathscr{E} \cap A)$ and let $C \in \sigma_r(\mathscr{E})$. Since B is a subset of A, we have

$$B = B \cap A = [B \cup (C - A)] \cap A.$$

However, the class $\mathscr{C}$ of all sets of the form $B \cup (C - A)$, where $B \in \sigma_r(\mathscr{E} \cap A)$ and $C \in \sigma_r(\mathscr{E})$, is a σ-ring (verify!) that contains $\mathscr{E}$, since if $E \in \mathscr{E}$ then

$$E = (E \cap A) \cup (E - A)$$

with $E \cap A \in \sigma_r(\mathscr{E} \cap A)$ and $E \in \sigma_r(\mathscr{E})$. This means $\mathscr{C} \supset \sigma_r(\mathscr{E})$ and

$$\sigma_r(\mathscr{E} \cap A) = \mathscr{C} \cap A \supset \sigma_r(\mathscr{E}) \cap A. \qquad \blacksquare$$

The next two results will be of special interest in Chapter 5.

Proposition 1.9. If $\mathscr{E}$ is any class of sets and $A \in \sigma_r(\mathscr{E})$ $[A \in \sigma(\mathscr{E})]$, then there exists a countable subclass of $\mathscr{E}$, call it $\mathscr{C}$, such that $A \in \sigma_r(\mathscr{C})$ $[A \in \sigma(\mathscr{C})]$. ∎

Proof. The union of all σ-rings $\sigma_r(\mathscr{C})$ for every countable subclass $\mathscr{C}$ of $\mathscr{E}$ is a σ-ring. Since this union contains $\mathscr{E}$ and is contained in $\sigma_r(\mathscr{E})$, it must be equal to $\sigma_r(\mathscr{E})$. ∎

Proposition 1.10. Let $\mathscr{E}$ be a class of subsets of some set Y. The following statements are true:

(i) If $\mathscr{E}$ is a countable class, then $\mathscr{A}(\mathscr{E})$ is countable.

(ii) If card $\mathscr{E} \leq c$, then card $\sigma_r(\mathscr{E}) \leq c$ and card $\sigma(\mathscr{E}) \leq c$. ∎

Remark. To prove part (ii), there are additional facts from set theory we must use but do not prove. They are as follows:

(1) If $(A_i)_I$ is an indexed family of sets with card $A_i \leq c$ for each $i \in I$ and card $I \leq \aleph_0$, then card $(\times_I A_i) \leq c$.

(2) If $(A_i)_I$ is an indexed family of sets with card $A_i \leq c$ for each $i \in I$ and card $I \leq c$, then card $(\bigcup_I A_i) \leq c$.

In light of (2) the set X constructed in Proposition 1.6 can be seen to have cardinality less than or equal to c since Y in the proof of Proposition 1.6 can be chosen to be R so that Y_0 is the set $R \cup \{y_0\}$ and card $Y_0 \leq c$.

Proof. We prove (ii) and leave the analogous proof of (i) for the reader (Problem 1.3.3). For each class of sets $\mathscr{C}$ let $\mathscr{C}^*$ denote the class of countable unions of differences of sets of $\mathscr{C}$, that is,

$$\mathscr{C}^* = \{\bigcup_{i=1} A_i - B_i\colon A_i, B_i \in \mathscr{C}\}.$$

It is clear that if $\mathscr{C} \subset \sigma(\mathscr{E})$, then $\mathscr{C}^* \subset \sigma(\mathscr{E})$. Moreover if $\varnothing$ and Y are in $\mathscr{C}$, then obviously $\varnothing$ and Y are in $\mathscr{C}^*$ and in particular $\mathscr{C} \subset \mathscr{C}^*$. In addition, from the preceding remark it follows that if card $\mathscr{C} \leq c$, then card $\mathscr{C}^* \leq c$.

To prove (ii) it suffices to show card $\sigma(\mathscr{E}) \leq c$, and to this end we may assume that $\varnothing$ and Y are in $\mathscr{E}$. Letting X be the well-ordered set $\{\alpha\colon \alpha < \Omega\}$ of countable ordinals constructed in Proposition 1.6, we will show that for each element α in X there exists a class $\mathscr{E}_\alpha$ of subsets of Y such that

(a) if $\alpha < \beta$, then $\mathscr{E} \subset \mathscr{E}_\alpha \subset \mathscr{E}_\beta \subset \sigma(\mathscr{E})$,

(b) card $\mathscr{E}_\alpha \leq c$,

(c) $\sigma(\mathscr{E}) = \bigcup_{\alpha \in X} \mathscr{E}_\alpha$.

For the least element 1 of X, let $\mathscr{E}_1 = \mathscr{E}$. Assuming that for each element $\alpha < \delta$, $\mathscr{E}_\alpha$ has been defined satisfying (a) and (b), let

$$\mathscr{E}_\delta = (\bigcup_{\alpha<\delta} \mathscr{E}_\alpha)^*.$$

Clearly (a) and (b) are satisfied for all $\alpha \leq \delta$, and using the Principle of Transfinite Induction (Theorem 1.3), a class $\mathscr{E}_\alpha$ exists for each α in X so that (a) and (b) are satisfied. Statement (c) is proved by showing that $\bigcup_{\alpha \in X} \mathscr{E}_\alpha$ is a σ-algebra. To this end let $(E_i)_N$ be a countable collection of sets from $\bigcup_{\alpha \in X} \mathscr{E}_\alpha$. For each $i \in N$, there exists an $\mathscr{E}_{\alpha_i}$ such that $E_i \in \mathscr{E}_{\alpha_i}$. Let $\alpha = \sup \alpha_i$, an element of X by Problem 1.2.4. Since $\mathscr{E}_{\alpha_i} \subset \mathscr{E}_\alpha$ by (a), $E_i \in \mathscr{E}_\alpha$ for each $i \in N$. Since $\varnothing \in \mathscr{E}_\alpha$, this means that

$$\bigcup_{i=1}^{\infty} E_i = \bigcup_{i=1}^{\infty} (E_i - \varnothing) \in \mathscr{E}_\alpha^* \subset \bigcup_{\alpha \in X} \mathscr{E}_\alpha.$$

In a similar fashion, it can be shown that whenever E_1 and E_2 are in $\bigcup_{\alpha \in X} E_\alpha$, then $E_1 - E_2 \in \bigcup_{\alpha \in X} E_\alpha$.

Since card $\mathscr{E}_\alpha \leq c$ and card $X \leq c$, card $\sigma(\mathscr{E}) = \text{card}\,(\bigcup_{\alpha \in X} \mathscr{E}_\alpha) \leq c$ by our preceding remark. ∎

Having discussed σ-rings and σ-algebras, our next consideration is that of monotone classes and their relation to σ-rings.

Definition 1.14. A nonempty class $\mathscr{M}$ of sets is called a *monotone class* if for every monotone sequence (i.e., a nonincreasing or nondecreasing sequence) of sets (E_n),

$$\lim_n E_n \in \mathscr{M}. \qquad \blacksquare$$

Remarks

1.13. If X is a set, $\{\varnothing, X\}$ and 2^X are examples of monotone classes.

1.14. Any σ-ring $\mathscr{R}$ is a monotone class. A monotone class $\mathscr{M}$ is a σ-ring if and only if $\mathscr{M}$ is a ring (Problem 1.3.4).

1.15. The intersection of any collection of monotone classes is a monotone class.

1.16. If X is a set and $\mathscr{E}$ is a collection of subsets of X, then there exists a smallest monotone class $\mathscr{M}(\mathscr{E})$ containing $\mathscr{E}$, $\mathscr{M}(\mathscr{E}) \subset \sigma_r(\mathscr{E})$.

The next result compares monotone classes with σ-rings and σ-algebras.

Theorem 1.4. *Monotone Class Theorem.* If $\mathscr{R}$ is a ring, then $\sigma_r(\mathscr{R}) = \mathscr{M}(\mathscr{R})$. If $\mathscr{R}$ is an algebra, then $\sigma(\mathscr{R}) = \mathscr{M}(\mathscr{R})$. $\blacksquare$

Proof. Suppose $\mathscr{R}$ is a ring. Since $\sigma_r(\mathscr{R})$ is a monotone class, clearly $\mathscr{M}(\mathscr{R}) \subset \sigma_r(\mathscr{R})$. The proof is accomplished by showing $\mathscr{M}(\mathscr{R})$ is a σ-ring, for then $\mathscr{M}(\mathscr{R}) \supset \sigma_r(\mathscr{R})$. From Remark 1.14 it is sufficient to show that $\mathscr{M}(\mathscr{R})$ is a ring. For each A in $\mathscr{M}(\mathscr{R})$ let

$$\mathscr{M}_A = \{B \in \mathscr{M}(\mathscr{R}) \colon A \cup B,\ A - B, \text{ and } B - A \in \mathscr{M}(\mathscr{R})\}.$$

It is trivial to see (using Problem 1.1.8) that $\mathscr{M}_A$ is a monotone class contained in $\mathscr{M}(\mathscr{R})$. Note also that $B \in \mathscr{M}_A$ if and only if $A \in \mathscr{M}_B$. Inasmuch as $\mathscr{R} \subset \mathscr{M}_A$ whenever $A \in \mathscr{R}$, we have $\mathscr{M}(\mathscr{R}) \subset \mathscr{M}_A$ whenever $A \in \mathscr{R}$. Therefore if $B \in \mathscr{M}(\mathscr{R})$, then $B \in \mathscr{M}_A$ and hence $A \in \mathscr{M}_B$. Since A is an arbitrary element of $\mathscr{R}$, $\mathscr{R} \subset \mathscr{M}_B$ and hence $\mathscr{M}(\mathscr{R}) \subset \mathscr{M}_B$ for every B in $\mathscr{M}(\mathscr{R})$. The inclusion $\mathscr{M}(\mathscr{R}) \subset \mathscr{M}_B$ for every B in $\mathscr{M}(\mathscr{R})$ implies that $\mathscr{M}(\mathscr{R})$ is a ring.

The proof of the second statement follows immediately upon noting that if $\mathscr{R}$ is an algebra, $\mathscr{M}(\mathscr{R})$ is not only a σ-ring but also a σ-algebra. $\blacksquare$

Our final consideration in this section is that of a Dynkin system.

Definition 1.15. Let $\mathscr{D}$ be a class of subsets of set X. $\mathscr{D}$ is a *Dynkin system* if

(i) $X \in \mathscr{D}$;
(ii) whenever $A, B \in \mathscr{D}$ and $B \subset A$, then $A - B \in \mathscr{D}$;
(iii) for every sequence $(A_i)_N$ of pairwise disjoint sets in $\mathscr{D}$, $\bigcup_{i=1}^{\infty} A_i \in \mathscr{D}$. ∎

Remarks

1.17. Every σ-algebra of subsets of X is a Dynkin system.

1.18. Every Dynkin system is a monotone class (Problem 1.3.7).

1.19. For every collection $\mathscr{E}$ of subsets of X, there is a smallest Dynkin system $\mathscr{D}(\mathscr{E})$ containing $\mathscr{E}$ (Problem 1.3.5).

Proposition 1.11. A Dynkin system $\mathscr{D}$ is a σ-algebra if and only if $A \cap B \in \mathscr{D}$ whenever $A, B \in \mathscr{D}$. ∎

Proof. The necessity is obvious, so we prove the sufficiency. Clearly Definition 1.15(i) and (ii) imply that if $A \in \mathscr{D}$ then $A^c \in \mathscr{D}$. If A and B are in $\mathscr{D}$, then the equations

$$A \cup B = A \cup [B - (A \cap B)] \quad \text{and} \quad A \cap [B - (A \cap B)] = \varnothing$$

imply that $A \cup B \in \mathscr{D}$. By induction, for any finite sequence $(A_i)_{i=1}^n$ from $\mathscr{D}$, $\bigcup_{i=1}^n A_i \in \mathscr{D}$. Now let $(A_i)_N$ be any countable collection from $\mathscr{D}$. Letting $(B_i)_N$ be the collection in $\mathscr{D}$ given by

$$\begin{aligned}
B_1 &= A_1, \\
B_2 &= (A_1 \cup A_2) - A_1, \\
B_3 &= (A_1 \cup A_2 \cup A_3) - (A_1 \cup A_2), \\
&\vdots \\
B_n &= (A_1 \cup A_2 \cdots \cup A_n) - (A_1 \cup A_2 \cdots \cup A_{n-1}),
\end{aligned}$$

we have a pairwise disjoint sequence in $\mathscr{D}$ and $\bigcup_{i=1}^{\infty} A_i = \bigcup_{i=1}^{\infty} B_i$. ∎

Proposition 1.12. If $\mathscr{E}$ is a class of subsets of X for which $A \cap B \in \mathscr{E}$ whenever A and B are in $\mathscr{E}$, then

$$\mathscr{D}(\mathscr{E}) = \sigma(\mathscr{E}).$$ ∎

Proof. Clearly $\mathscr{D}(\mathscr{E}) \subset \sigma(\mathscr{E})$ as $\sigma(\mathscr{E})$ is a Dynkin system. To show $\sigma(\mathscr{E}) \subset \mathscr{D}(\mathscr{E})$ it suffices to show that $\mathscr{D}(\mathscr{E})$ is a σ-algebra. By Proposition

1.11 it suffices to show that $A \cap B \in \mathscr{D}(\mathscr{E})$ whenever A and B are in $\mathscr{D}(\mathscr{E})$. If $A \in \mathscr{D}(\mathscr{E})$, define

$$\mathscr{D}_A = \{B \in 2^X: A \cap B \in \mathscr{D}(\mathscr{E})\}.$$

It is easy to see that $\mathscr{D}_A$ is a Dynkin system. Inasmuch as $\mathscr{D}_A \supset \mathscr{E}$ whenever $A \in \mathscr{E}$ according to our hypothesis on $\mathscr{E}$, we have $\mathscr{D}_A \supset \mathscr{D}(\mathscr{E})$ whenever $A \in \mathscr{E}$. This means that for every B in $\mathscr{D}(\mathscr{E})$ and every $A \in \mathscr{E}$, $B \cap A \in \mathscr{D}(\mathscr{E})$ so that $A \in \mathscr{D}_B$. This means $\mathscr{E} \subset \mathscr{D}_B$ and hence $\mathscr{D}(\mathscr{E}) \subset \mathscr{D}_B$. However, $\mathscr{D}(\mathscr{E}) \subset \mathscr{D}_B$ for arbitrary B in $\mathscr{D}(\mathscr{E})$ means $A \cap B \in \mathscr{D}(\mathscr{E})$ whenever A and B are in $\mathscr{D}(\mathscr{E})$. ∎

Problems

1.3.1. If $\mathscr{A}$ is an algebra and $(A_i)_N$ is a countable collection of sets in $\mathscr{A}$, prove there is a sequence $(B_i)_N$ from $\mathscr{A}$ such that $B_i \cap B_j = \varnothing$ if $i \neq j$ and $\bigcup_{i=1}^{\infty} A_i = \bigcup_{i=1}^{\infty} B_i$.

× **1.3.2.** (i) Prove that if $\mathscr{E}$ is a collection of subsets of set X and $A \in \mathscr{R}(\mathscr{E})$, then there is a finite collection of sets in $\mathscr{E}$ that cover A, that is, their union contains A. (Hint: Show the class of sets that can be covered by a finite collection from $\mathscr{E}$ is a ring.)

(ii) Prove that if $A \in \sigma_r(\mathscr{E})$, then there is a countable collection of sets in $\mathscr{E}$ covering A.

1.3.3. Prove Proposition 1.10 (i).

1.3.4. Prove Remark 1.14.

1.3.5. If $\mathscr{E}$ is a collection of subsets of X, prove that a smallest Dynkin system $\mathscr{D}(\mathscr{E})$ containing $\mathscr{E}$ exists.

× **1.3.6.** Prove that a nonempty collection of subsets $\mathscr{A}$ of set X is an algebra if and only if (a) $A^c \in \mathscr{A}$ whenever $A \in \mathscr{A}$ and (b) $A \cap B \in \mathscr{A}$ whenever $A, B \in \mathscr{A}$.

1.3.7. Prove that a collection of subsets $\mathscr{D}$ of set X is a Dynkin system if and only if (a) $X \in \mathscr{D}$, (b) $A, B \in \mathscr{D}$ with $A \subset B$ implies $B - A \in \mathscr{D}$, (c) $A, B \in \mathscr{D}$ and $A \cap B = \varnothing$ implies $A \cup B \in \mathscr{D}$, and (d) $\bigcup_n A_n \in \mathscr{D}$ for every increasing sequence $(A_n)_N$ in $\mathscr{D}$.

× **1.3.8.** Suppose $\mathscr{E}$ and $\mathscr{F}$ are classes of subsets of X and Y, respectively. Let $\mathscr{G} = \{E \times F: E \in \mathscr{E}$ and $F \in \mathscr{F}\}$. Let $\sigma_r(\mathscr{E}) \times \sigma_r(\mathscr{F})$ be the smallest σ-ring containing sets of the form $A \times B$, where $A \in \sigma_r(\mathscr{E})$ and $B \in \sigma_r(\mathscr{F})$. Prove $\sigma_r(\mathscr{G}) = \sigma_r(\mathscr{E}) \times \sigma_r(\mathscr{F})$.

× **1.3.9.** (i) If $f\colon X \to Y$ is a function and $\mathscr{F}$ is a σ-ring of subsets of Y, then prove

$$\mathscr{E} = \{E \subset X\colon E = f^{-1}(F) \text{ for } F \in \mathscr{F}\}$$

is a σ-ring. What happens if $\mathscr{F}$ is a σ-algebra?

(ii) If $f\colon X \to Y$ and $\mathscr{E}$ is a σ-ring of subsets of X, prove

$$\mathscr{F} = \{F \subset Y\colon f^{-1}(F) \in \mathscr{E}\}$$

is a σ-ring in Y. What happens if $\mathscr{E}$ is a σ-algebra?

(iii) Prove that $\mathscr{F} \equiv \{E \subset X\colon A \in \mathscr{E} \Rightarrow A \cap E \in \mathscr{E}\}$ is a σ-algebra in X containing the σ-ring $\mathscr{E}$ in X.

× **1.3.10.** Let $\mathscr{F}$ be a σ-ring of subsets of X such that $X \notin \mathscr{F}$. Show that the smallest σ-algebra containing $\mathscr{F}$ is $\mathscr{A} = \{A \subset X\colon A \in \mathscr{F} \text{ or } X - A \in \mathscr{F}\}$.

★ **1.3.11.** *On σ-Classes* and *σ-Algebras*. A collection $\mathscr{F}$ of subsets of a nonempty set X containing X is called a *σ-class* if it is closed under countable *disjoint* unions and complementations. Suppose $\mathscr{A}$ is the σ-class generated by a collection $\mathscr{C}$ of subsets of X. Then prove the following result due to T. Neubrunn: $\mathscr{A}$ is a σ-algebra if and only if one of the following conditions holds:

(i) If $A, B \in \mathscr{C}$, then $A - B \in \mathscr{A}$.

(ii) If $A, B \in \mathscr{C}$, then $A \cap B \in \mathscr{A}$.

[Hint: Notice that (1) $A, B \in \mathscr{A}$ and $A \subset B$ imply $B - A = (A \cup B^c)^c \in \mathscr{A}$, and (2) if $A_1 \supset A_2 \supset \cdots$, $A_i \in \mathscr{A}$, then $\bigcap_{i=1}^{\infty} \in \mathscr{A}$. Now assume (i) above. It is sufficient to show that $A - B \in \mathscr{A}$ whenever $A, B \in \mathscr{A}$. Let $D \in \mathscr{C}$ and $\varphi(D) = \{E \in \mathscr{A}\colon D - E \in \mathscr{A}\}$. Show that $\varphi(D)$ is a σ-class. Then since $\mathscr{C} \subset \varphi(D)$, $\mathscr{A} = \varphi(D)$. Again for fixed $A \in \mathscr{A}$, let $\psi(A) = \{B \in \mathscr{A}\colon B - A \in \mathscr{A}\}$. As before, show that $\psi(A)$ is a σ-class containing $\mathscr{C}$ so that $\psi(A) = \mathscr{A}$. The proof of (i) is now clear.]

★ **1.3.12.** *Cardinalities of σ-Algebras*. Regarding cardinal numbers, the following are facts:

(i) For any cardinal number a, $2^a > a$.

(ii) $b \leq a$ and a infinite implies $a + b = a$.

(iii) $0 < b \leq a$ and a infinite implies $a \cdot b = a$.

Use these to prove that card $\sigma(\mathscr{F}) \leq a^{\aleph_0}$ whenever card $\mathscr{F} = a \geq 2$, where $\mathscr{F}$ is a family of subsets of a set X containing $\varnothing$. [Hint: An argument similar to the one used in the proof of Proposition 1.10 is helpful.]

Show also that a σ-algebra cannot be countably infinite—its cardinality is either finite or $\geq c$.

× **1.3.13.** *The Borel Subsets of R.* The sets in the smallest σ-algebra containing the open subsets of R are called the Borel subsets of R. (The definition for the Borel sets in a metric space is similar.) Show that the Borel subsets of R are also in the smallest σ-algebra containing the closed and bounded intervals of R, and that they have cardinality c.

1.3.14. *The Smallest σ-Algebra Containing a Given σ-Algebra $\mathscr{A}$ and a Given Set E* (all sets are here subsets of a set X) is the class of all sets of the form $(A \cap E) \cup (B \cap E^c)$, A and $B \in \mathscr{A}$. Verify this and then use this to prove that card $\sigma(\mathscr{F}) \leq 2^{2^n}$ if $\mathscr{F}$ is a class of subsets of X with cardinality at most n. Can the reader find an alternative proof?

1.3.15. A σ-algebra is called *separable* if it is generated by a countable class of sets. (a) Show that the σ-algebra of Borel subsets of R is separable. (b) Let X be an uncountable set and $\mathscr{A} = \{A \subset X\colon$ either A or $X - A$ is countable$\}$. Show that $\mathscr{A}$ is a σ-algebra that is not separable. (c) Let $\mathscr{A}_1 \subset \mathscr{A}_2$ be two σ-algebras of subsets of X. Suppose that $\mathscr{A}_2$ is separable. Is $\mathscr{A}_1$ then necessarily separable?

★ **1.3.16.** Let $\mathscr{A}_1 \subsetneqq \mathscr{A}_2 \subsetneqq \cdots$ be an infinite sequence of properly increasing σ-algebras. Show that $\bigcup_{n=1}^{\infty} \mathscr{A}_n$ is not a σ-algebra.

• 1.4. Topological Spaces

One of the primary reasons for including this section in the text is to make the book self-contained with respect to topological notions that are used frequently throughout the text. The student should treat this section as a synopsis of topology that is encountered in the text. For completeness, topological concepts closely related to normality, local compactness, compactness, and separability are also considered.

One of the main aims of this text is to present a comprehensive account of the basic theory of integration. In Chapter 5, it will be seen how this theory has an ultimate connection with some topological concepts. Though a detailed study of topology is not possible here, we present many of the basic results of the theory in a form resembling a typical "Texas-style" course: The reader is required to supply many of the proofs of theorems and remarks, which in most cases is easy and straightforward. However, propositions, theorems, and examples that are not straightforward and demand a little more than routine arguments have all been proved or given in detail. Readers interested in a detailed account of this discipline should consult [29] or [60].

Topological spaces are generalizations of the spaces R^n $(n \geq 1)$. One of the main reasons for considering such generalizations is to study the concept of limit of functions on spaces other than R^n. The definition of limit and the theory of convergence in R^n are based on the notion of distance between points, and use is made of only the following few properties of this distance:

(i) $d(x, y) = 0$ if and only if $x = y$,

(ii) $0 \leq d(x, y) = d(y, x) < \infty$,

(iii) $d(x, z) \leq d(x, y) + d(y, z)$,

for every $x, y, z \in R^n$. These properties alone provide the framework for a class of topological spaces called the metric spaces. Nevertheless, there are situations that demand an even more general approach—for instance, the study of pointwise convergence in a space of bounded real-valued functions. This is one motivation behind the consideration of general topological spaces.

Definition 1.16. A *topology* $\mathscr{T}$ for a set X is a collection of subsets of X, called *open sets*, such that the following conditions hold:

(i) $\varnothing \in \mathscr{T}$, $X \in \mathscr{T}$.

(ii) $A \cap B \in \mathscr{T}$ whenever $A \in \mathscr{T}$ and $B \in \mathscr{T}$.

(iii) $\cup A_\alpha \in \mathscr{T}$ whenever $A_\alpha \in \mathscr{T}$ for each α. ∎

Examples

1.12. If $\mathscr{T} = 2^X$ (the class of all subsets of X), then $\mathscr{T}$ is a topology, called the *discrete topology* for X.

1.13. If $\mathscr{T} = \{\varnothing, X\}$, then $\mathscr{T}$ is a topology called the *indiscrete topology* for X.

1.14. Let $X = R$ and $\mathscr{T}$ be the collection of all subsets G of R such that $x \in G$ implies that for some $\mu > 0$, $\{y \in R: |y - x| < \mu\} \subset G$. Then $\mathscr{T}$ is a topology for R, called its *usual topology*.

1.15. Let $A \subset X$ and $\mathscr{T}$ be a topology for X. Let $\mathscr{T}|_A = \{A \cap T: T \in \mathscr{T}\}$. Then $\mathscr{T}|_A$ is a topology for A, called the *relative topology* for A.

Definition 1.17. Let $\mathscr{T}$ be a topology for X. A point $x \in X$ is a *limit point* of $A \subset X$ if whenever $x \in T$ and $T \in \mathscr{T}$ then $T \cap (A - \{x\}) \neq \varnothing$. The union of A and the set of all its limit points is called its closure, $\bar{A}$. A set A is called closed if $A = \bar{A}$. (Note that $A = \bar{A}$ if and only if $X - A \in \mathscr{T}$.) ∎

Remarks. Let $A, B \subset X$ and $\mathscr{T}$ be a topology for X. Then

1.20. $A \subset B \Rightarrow \bar{A} \subset \bar{B}$.

1.21. $\overline{A \cup B} = \bar{A} \cup \bar{B}$.

1.22. $\overline{A \cap B} \subset \bar{A} \cap \bar{B}$.

Definition 1.18. $(X, \mathscr{E})$ is called a *topological space* if $\mathscr{E}$ is a topology for X. A point $x \in X$ is called an *interior point* of $A \subset X$ if there is $T \in \mathscr{E}$ such that $x \in T \subset A$. The set of all interior points of A is called its *interior* and denoted by A°. ▮

Remarks. Let $(X, \mathscr{E})$ be a topological space and A, $B \subset X$. Then

1.23. $A \subset B \Rightarrow A^\circ \subset B^\circ$.

1.24. $(A \cap B)^\circ = A^\circ \cap B^\circ$.

1.25. $A = A^\circ$ if and only if $A \in \mathscr{E}$.

Definition 1.19. In a topological space $(X, \mathscr{E})$, $x \in X$ is called a *boundary point* of $A \subset X$ if

$$x \in T \in \mathscr{E} \Rightarrow T \cap A \neq \varnothing$$

and

$$T \cap (X - A) \neq \varnothing.$$

The *boundary* A_b of A is the set of all its boundary points. ▮

Remarks

1.26. $A_b = \bar{A} - A^\circ$.

1.27. $(A \cup B)_b \subset A_b \cup B_b$.

1.28. $(\bar{A})_b = A_b$.

Definition 1.20. A mapping $f\colon X \to Y$ where $(X, \mathscr{E}_1)$ and $(Y, \mathscr{E}_2)$ are topological spaces, is called *continuous* (relative to $\mathscr{E}_1$ and $\mathscr{E}_2$) if $f^{-1}(V) \in \mathscr{E}_1$ for each $V \in \mathscr{E}_2$. The mapping f is called *open* (or *closed*, respectively) if $f(T)$ is open (closed) for each open (closed) T. ▮

Definition 1.21. Let $\mathscr{L}$, $\mathscr{B}$ be collections of sets such that $\varnothing \in \mathscr{L}$ and $\mathscr{B}$ consists of all the finite intersections of sets in $\mathscr{L}$. Let $\mathscr{E}$ consist of all sets that are arbitrary unions of sets in $\mathscr{B}$. Then $\mathscr{L}$ is called a *subbase* and $\mathscr{B}$ a *base* for $\mathscr{E}$. ▮

The reader can easily verify that $\mathscr{E}$ is a topology for the set

$X = \cup\{A: A \in \mathscr{B}\}$. Note that if $(X, \mathscr{C})$ is a topological space and $\mathscr{B}$ is a base for $\mathscr{C}$, then given $x \in V \in \mathscr{C}$, there exists $U \in \mathscr{B}$ with $x \in U \subset V$.

Example 1.16. Let $(X_\lambda, \mathscr{C}_\lambda)$ be a topological space for each $\lambda \in \Lambda$, an indexed set. Let $X = \times_{\lambda \in \Lambda} X_\lambda$ (see Definition 1.11 for definition) and $\pi_\lambda: X \to X_\lambda$ be defined by $\pi_\lambda(x) = x(\lambda)$. Consider the topology $\mathscr{C}$ generated by $\{\pi_\lambda^{-1}(V): V \in \mathscr{C}_\lambda, \lambda \in \Lambda\}$ as a subbase. Then $(X, \mathscr{C})$ is a topological space, called *the product of the spaces* $(X_\lambda, \mathscr{C}_\lambda)$, $\lambda \in \Lambda$. *The functions* π_λ *are all open and continuous.* It is also clear that if $(X_1, \mathscr{C}_1)$ and $(X_2, \mathscr{C}_2)$ are two topological spaces, then $\{U \times V: U \in \mathscr{C}_1, V \in \mathscr{C}_2\}$ is *a base for the product* topology for $X_1 \times X_2$.

Proposition 1.13. Let $(X_1, \mathscr{C}_1)$ and $(X_2, \mathscr{C}_2)$ be topological spaces and $f: X_1 \to X_2$. Then the following are equivalent:

(i) f is continuous.

(ii) If $x \in X_1$, and $f(x) \in V \in \mathscr{C}_2$, then there is $U \in \mathscr{C}_1$ such that $x \in U$ and $f(U) \subset V$.

(iii) $f(\bar{A}) \subset \overline{f(A)}$ for all $A \subset X_1$.

(iv) $f^{-1}(B)$ is closed whenever B is closed. ∎

Proposition 1.13 (ii) motivates us to define *continuity at a point of* X. We say f is continuous at $x \in X$ if for any open V containing $f(x)$, there is an open U containing x such that $f(U) \subset V$.

Definition 1.22. A class $\mathscr{A}$ of subsets of X is a cover of X if $X = \cup\{A: A \in \mathscr{A}\}$. A cover $\mathscr{A}$ of a topological space $(X, \mathscr{C})$ is *open* if $A \in \mathscr{A}$ implies $A \in \mathscr{C}$. $(X, \mathscr{C})$ is called *compact* if each open cover of X contains a finite subcover. If $A \subset X$, then A is called compact if A is compact in its relative topology. ∎

Remarks on Compactness

1.29. The continuous image of a compact set is compact.

1.30. A closed subset of a compact set is compact.

1.31. Any nonempty collection of closed subsets in a compact topological space X with *finite intersection property* (a collection of subsets has the finite intersection property if any finite number of sets in this collection has nonempty intersection) has a nonempty intersection. [To prove this, suppose $\mathscr{A}$ is the collection and $\bigcap_{A \in \mathscr{A}} A = \varnothing$. Then $X = \bigcup_{A \in \mathscr{A}} (X - A)$

and by compactness of X, there are $A_1, A_2, \ldots, A_n \in \mathscr{A}$ such that $X = \bigcup_{x=1}^{n}(X - A_i)$ and therefore, $\bigcap_{i=1}^{n} A_i = \varnothing$, a contradiction.]

1.32. A topological space $(X, \mathscr{T})$ is compact if and only if each nonempty collection of closed subsets of X having f.i.p. (finite intersection property) has nonempty intersection.

1.33. If $\mathscr{A}$ is a nonempty family of subsets of X having f.i.p., then by the Hausdorff's Maximal Principle, $\mathscr{A}$ is contained in $\mathscr{B}$ ($\subset 2^X$), which is maximal relative to having f.i.p. Because of maximality, $\mathscr{B}$ has the following properties:

(a) $A, B \in \mathscr{B} \Rightarrow A \cap B \in \mathscr{B}$,

(b) $B \cap A \neq \varnothing$ for each $A \in \mathscr{B} \Rightarrow B \in \mathscr{B}$.

1.34. X is compact if and only if every nonempty collection $\mathscr{A}$ of *subsets* of X with f.i.p. has the property $\cap \{\bar{A}: A \in \mathscr{A}\} \neq \varnothing$. This statement remains true if the "collection" above is replaced by a "collection" maximal relative to having f.i.p.

Theorem 1.5. (*Tychonoff*) The product of compact spaces is compact. (This theorem was first proved by Tychonoff (or Tihonov) in the case where the compact spaces are all $[0, 1]$.[†] Later on, E. Čech proved it in the general case.[‡]) ▌

Proof. Let $X = \times_{\lambda \in \Lambda} X_\lambda$ and each X_λ, $\lambda \in \Lambda$, be compact. Let $\mathscr{A}$ be a collection of subsets of X maximal relative to having f.i.p. Since each X_λ is compact, $\cap \{\overline{\pi_\lambda(A)}: A \in \mathscr{A}\}$ is nonempty and must contain some element $x(\lambda) \in X_\lambda$. Let $x \in X$ such that $\pi_\lambda(x) = x(\lambda)$. We claim that $x \in \bar{A}$, for each $A \in \mathscr{A}$. Let $\bigcap_{i=1}^{n} \{\pi_{\lambda_i}^{-1}(U_{\lambda_i}): U_{\lambda_i}$ open in $X_{\lambda_i}\}$ be an open set containing x. Then since $x_{\lambda_i} = \pi_{\lambda_i}(x) \in U_{\lambda_i}$, $U_{\lambda_i} \cap \pi_{\lambda_i}(A) \neq \varnothing$ for each $A \in \mathscr{A}$. This means that $\pi_{\lambda_i}^{-1}(U_{\lambda_i}) \cap A \neq \varnothing$ for each $A \in \mathscr{A}$. By Remark 1.33(b), $\pi_{\lambda_i}^{-1}(U_{\lambda_i}) \in \mathscr{A}$ and by Remark 1.33(a), $\bigcap_{i=1}^{n} \pi_{\lambda_i}^{-1}(U_{\lambda_i}) \in \mathscr{A}$, and therefore by f.i.p. of $\mathscr{A}$, $\bigcap_{i=1}^{n} \pi_{\lambda_i}^{-1}(U_{\lambda_i})$ has nonempty intersection with each $A \in \mathscr{A}$. The claim now follows and the theorem follows by Remark 1.34. ▌

The reader may very well recall now the following theorem, from his or her first course in "Real Analysis."

[†] A. Tychonoff, *Math. Ann.* **102**, 544 (1930).

[‡] E. Čech, *Ann. Math.* **38**(2), 823 (1937).

Theorem 1.6. (*Heine–Borel–Bolzano–Weierstrass*) A subset A of R^n is compact if and only if it is closed and bounded. ∎

Definition 1.23. Let $(X, \mathscr{T})$ be a topological space. Then we have the following:

(i) X is called T_1 if each singleton $\{x\}$ in X is closed.

(ii) X is T_2 (or Hausdorff) if $x \neq y$, $x, y \in X$ imply that there are open sets V, W such that $x \in V$, $y \in W$, and $V \cap W = \varnothing$.

(iii) X is called *regular* if for any closed set A and $x \notin A$, there are open sets V containing x and $W \supset A$ with $V \cap W = \varnothing$.

(iv) X is called *normal* if for any two disjoint closed sets A and B, there are open sets $V \supset A$ and $W \supset B$ with $V \cap W = \varnothing$. Equivalently, X is normal if given a closed subset A and an open subset V such that $A \subset V$, there exists an open subset U such that $A \subset U \subset \bar{U} \subset V$.

(v) A regular T_1 space is called T_3 and a normal T_1 space is called T_4. ∎

Remarks

1.35. $T_4 \Rightarrow T_3 \Rightarrow T_2 \Rightarrow T_1$. The set of real numbers with the usual topology (see Example 1.14) is a T_4 space.

1.36. $(X, \mathscr{T})$ is Hausdorff if and only if $\{(x, x): x \in X\}$ is closed in the product topology for $X \times X$.

1.37. A compact subset of a Hausdorff space is closed. A compact Hausdorff space is normal.

1.38. Let A be a compact subset of a topological space $(X, \mathscr{T})$ and B be a compact subset of a topological space $(Y, \mathscr{T}')$ such that $A \times B \subset G$, an open set in $X \times Y$. Then there are open sets $V \in \mathscr{T}$, $W \in \mathscr{T}'$ such that $A \times B \subset V \times W \subset G$.

1.39. Let $(X_\lambda, \mathscr{T}_\lambda)$, $\lambda \in \Lambda$ be a family of topological spaces and let $X = \times_{\lambda \in \Lambda} X_\lambda$, their product with the product topology. Then (a) if each X_λ is T_1, then X is T_1; (b) if each X_λ is T_2, then X is T_2; (c) if each X_λ is regular, then X is regular.

1.40. If $(X, \mathscr{T}_1)$ is a Hausdorff space and $(X, \mathscr{T}_2)$ is compact, then $\mathscr{T}_1 \subset \mathscr{T}_2$ implies $\mathscr{T}_1 = \mathscr{T}_2$. For, if i: $(X, \mathscr{T}_2) \to (X, \mathscr{T}_1)$ is the identity mapping, then i is one-to-one, onto, continuous, and closed; this means that i is open and therefore $\mathscr{T}_2 \subset \mathscr{T}_1$.

1.41. If $(X, \mathscr{T})$ is a compact Hausdorff space, then X is not compact with any topology $\mathscr{T}_1$ satisfying $\mathscr{T}_1 \supsetneqq \mathscr{T}$ and X is not Hausdorff with any topology $\mathscr{T}_2$ satisfying $\mathscr{T}_2 \subsetneqq \mathscr{T}$.

Examples

1.17. *A T_1 Space That Is Not T_2.* Let X be an uncountable set and $\mathscr{C} = \{\varnothing\} \cup \{U \subset X: X - U$ is at most countable$\}$. Then $\mathscr{C}$ is clearly a T_1 topology that is not T_2. [At most countable = finite or countable.]

1.18. *A T_2 Space That Is Not Regular.* Let X be the set of real numbers. Let all sets of the form $\{x\} \cup I_x$, where I_x contains only the rationals in an open interval around x, form a base for a topology $\mathscr{C}$. $\mathscr{C}$ is clearly Hausdorff. But $\mathscr{C}$ is *not* regular, since the irrationals in X form a closed set, and cannot be separated from a rational number by open sets.

1.19. *Regularity or Normality Does Not Imply T_2.* The indiscrete topology provides an example.

1.20. *Normality Need Not Imply Regularity.* A trivial example is the topology $\{\varnothing, \{x\}, \{y\}, \{x, y\}, \{x, y, z\}\}$ for the set $X = \{x, y, z\}$.

1.21. *T_3 Need Not Imply T_4.* The following example is due to Niemytzki. Let $X = \{(x, y) \in R^2: y \geq 0\}$. We consider that topology $\mathscr{C}$ on X which has as a base the set of all open disks in X along with all sets of the form $\{(x, 0)\} \cup D$, where D is an open disk in X touching the x axis at $(x, 0)$. Then $\mathscr{C}$ contains the relative topology of R^2 on X. Let $A \subset X$ be a closed set and $a \in X - A$. If a is not on the x axis, then there is an open disk U such that $a \in U \subset X - A$. Since the relative topology of R^2 is regular by Remarks 1.35 and 1.39, a and $X - U$ (and therefore, a and A) can be separated by sets open in this relative topology and therefore, in $\mathscr{C}$. If $a = (a_1, 0)$ and D be an open disk in X with radius r and tangent at a to the x axis such that $\{a\} \cup D \subset X - A$, then we can define the function f by $f(a) = 0$, $f(x) = 1$ if $x \notin \{a\} \cup D$ and $f(z, w) = [(z - a_1)^2 + w^2]/2rw$ if $(z, w) \in D$. It can be easily verified that f is a continuous mapping from X into R. Since $a \in f^{-1}((-\infty, \frac{1}{2}))$ and $A \subset f^{-1}((\frac{1}{2}, \infty))$, it follows that X is regular (in fact, *completely regular*†). But X is not normal. The reason is that the rationals and the irrationals on the x axis are disjoint and closed, but do not have any disjoint open sets containing them.

1.22. A compact Hausdorff topology on $[0, 1]$ different from the usual topology is the topology $\mathscr{C} = \{A: A \subset [0, 1)\} \cup \{B: B \subset [0, 1], 1 \in B$ and $[0, 1] - B$ is finite$\}$.

By definition, a sequence (x_n) in a topological space $(X, \mathscr{C})$ converges to x in X if for every open U with $x \in U$ there is an integer n_0 such that $x_n \in U$ whenever $n > n_0$.

† A topological space $(X, \mathscr{C})$ is called completely regular if for $x \notin A$ and A closed, there exists a continuous function $f: X \to [0, 1]$ with $f(x) = 0$ and $f(A) = 1$. A completely regular space is always regular.

Proposition 1.14. In a Hausdorff space, a convergent sequence has a unique limit. ∎

Proposition 1.15. Let X be the product of the topological spaces X_λ, $\lambda \in \Lambda$. Let f_λ, for each λ be a continuous mapping from a topological space Y into X_λ. If f: $Y \to X$ is defined by $[f(y)]\ (\lambda) = f_\lambda(y)$, then f is continuous. ∎

Proposition 1.16. Let X be a nonempty compact (or only countably compact, to be defined in Definition 1.28) space and f be a real-valued *lower-semicontinuous* function on X, i.e., the set $\{x \in X: f(x) \leq r\}$ is closed for every real r. Then there is $y \in X$ such that $f(y) \leq f(x)$ for each $x \in X$. ∎

Proof. Since $X = \bigcup_{n=1}^{\infty} \{x: f(x) > -n\}$ and X is compact, there is a k such that $X = \bigcup_{n=1}^{k} \{x: f(x) > -k\}$ and therefore f is bounded from below. Let $c = \text{g.l.b. } \{f(x): x \in X\}$. If $c \notin f(X)$, then $X = \bigcup_{n=1}^{\infty} \{x: f(x) > c + 1/n\}$ and by compactness, there is an integer m such that $X = \bigcup_{n=1}^{m} \{x: f(x) > c + 1/n\}$. This means that $c + 1/m \leq \text{g.l.b.} f(X) = c$, which is a contradiction. ∎

Definition 1.24. A topological space X is said to be *first countable* if for each $x \in X$ there exists a countable family $[V_n(x)]$ of open sets containing x such that whenever G is an open set containing x, $V_n(x) \subset G$ for some n. ∎

(Example 1.22 gives an example of a compact Hausdorff space, that is not first countable; the requirement for first countability fails at 1. Note that [0, 1], with the usual topology, is first countable.)

Remarks

1.42. A first countable topological space is Hausdorff if and only if every convergent sequence has a unique limit.

1.43. If X is first countable and T_1, then x is a limit point of E if and only if there exists a sequence of *distinct* points in E converging to x.

1.44. Let f be a mapping from a first countable topological space X into a topological space Y. Then f is continuous at $x \in X$ if and only if $f(x_n)$ converges to $f(x)$ whenever x_n converges to x.

Definition 1.25. A topological space is called *second countable* if there is a countable base for its topology. ∎

Remarks

1.45. Every second countable space is first countable, but not conversely. The discrete topology on an uncountable set illustrates this. The real numbers, with the usual topology, is second countable since all the open intervals with rational endpoints form a countable base.

1.46. In a second countable space, every open cover of a subset has a countable subcover.

To see this, let $\mathscr{A}$ be an open cover for $E \subset X$, which has (B_n) as a countable base. Let $N = \{n: B_n \subset A \text{ for some } A \in \mathscr{A}\}$. Then N is countable and $E \subset \bigcup_{n \in N} B_n$. For each $n \in N$, let $A_n \in \mathscr{A}$ such that $B_n \subset A_n$. Then it can be easily verified that $E \subset \bigcup_{n \in N} A_n$.

1.47. If $(X, \mathscr{C})$ is an infinite, second countable, and Hausdorff space, then card $\mathscr{C} = c$ and card $X \leq c$.

To see this, let (B_n) be a countable base for X. Then for $x \in X$, we have $\{x\} = \bigcap \{B_n: x \in B_n\}$. This means that the subset of all those n for which $x \in B_n$ determines the point x. Therefore, card $X \leq c$. By a similar argument (since each open set is a union of the B_n's), card $\mathscr{C} \leq c$. Now we observe that an infinite Hausdorff space contains an infinite sequence of pairwise disjoint nonempty open sets. [If X has no limit points, X has the discrete topology and it is easy to prove. If x is a limit point of X, then one can use an inductive argument and the Hausdorff property to prove this.] Now the union of each distinct subsequence (finite or infinite) of these open sets is a different open set. It follows that card $\mathscr{C} \geq c$. Hence, card $\mathscr{C} = c$.

Definition 1.26. A topological space is called *Lindelöf* if every open cover of the space has a countable subcover. ∎

Remarks

1.48. Every compact space is a Lindelöf space. A subspace of a compact Hausdorff space need not be Lindelöf; for instance, see Example 1.22, which also shows that "Lindelöf" need not imply "second countable."

1.49. Closed subspaces of Lindelöf spaces are Lindelöf.

1.50. Every regular Lindelöf space X is normal.

To see this, let A and B be disjoint closed sets. For $x \in A$, by regularity there is an open set $V(x)$ such that $x \in V(x) \subset \overline{V(x)} \subset X - B$. By the Lindelöf property, there exists a sequence of open sets (V_n) such that $A \subset \bigcup_{n=1}^{\infty} V_n$ and for each n, $\bar{V}_n \subset X - B$. Similarly, there exists a sequence of open sets (W_n) such that $B \subset \bigcup_{n=1}^{\infty} W_n$ and for each n, $\overline{W}_n \subset X - A$. Let

$$G = \bigcup_{n=1}^{\infty} \left(V_n - \bigcup_{i=1}^{n} \overline{W}_i \right)$$

and

$$H = \bigcup_{n=1}^{\infty} \left(W_n - \bigcup_{i=1}^{n} \bar{V}_i \right).$$

Then $G \cap H = \varnothing$, $A \subset G$, $B \subset H$ and G, H are open.

Definition 1.27. A subset A of a topological space X is *dense* if $\bar{A} = X$. If X has a countable dense subset, then X is called *separable*. ∎

Remarks

1.51. $R^n (n \geq 1)$ with its usual topology is separable. Any uncountable set with discrete topology, though first countable, is not separable or Lindelöf.

1.52. Every second countable space is separable. The converse need not be true. For instance, let X be an uncountable set and $\mathscr{T} = \varnothing \cup \{A \subset X: X - A \text{ is finite}\}$. Then $(X, \mathscr{T})$ is a *separable space*, since every infinite set is dense in X. But this space is *not even* first countable, since in that case each $x \in X$ is the intersection of a countable number of open sets, a contradiction.

1.53. Separability need not imply the Lindelöf property. Also, a closed subspace of a separable space need not be separable. (See Problem 1.4.10.)

We next present an important characterization of normality often useful in analysis.

Lemma 1.1. *Urysohn's Lemma.* A topological space X is normal if and only if given any two disjoint closed subsets A and B of X, there exists a continuous real-valued mapping $f: X \to [0, 1]$ such that $f(A) = \{0\}$ and $f(B) = \{1\}$. ∎

Proof. The "if" part is easy and left to the reader. For the "only if" part, let X be normal, and A and B two disjoint closed subsets. Let us first show by induction that we can associate with every rational number r of the form $k/2^n$, $0 \leq k \leq 2^n$, an open set $V(r)$ such that

$$r_1 < r_2 \Rightarrow A \subset V(r_1) \subset \overline{V(r_1)} \subset V(r_2) \subset \overline{V(r_2)} \subset B^c.$$

To do this, we construct, by induction on n, the classes

$$\mathscr{E}_n = \{V(k/2^n): 0 \leq k \leq 2^n\}.$$

The class $\mathscr{E}_0$ consists of an open set $V(0)$ and the open set $V(1) = B^c$ such that $A \subset V(0) \subset \overline{V(0)} \subset B^c$. Such a set $V(0)$ exists by the definition of normality. Assuming that $\mathscr{E}_n$ has been constructed, we construct $\mathscr{E}_{n+1}$ as follows. For a rational number of the form $(2k/2^{n+1})$, we simply take $V(k/2^n)$ in $\mathscr{E}_n$ as $V(2k/2^{n+1})$. For a rational number of the form $(2k + 1/2^{n+1})$, we notice that

$$k/2^n < (2k+1)/2^{n+1} < (k+1)/2^n;$$

also, since by induction hypothesis $\overline{V(k/2^n)} \subset V(k + 1/2^n)$, there exists by normality an open subset $V(2k + 1/2^{n+1})$ such that

$$\overline{V(k/2^n)} \subset V(2k + 1/2^{n+1}) \subset \overline{V(2k + 1/2^{n+1})} \subset V(k + 1/2^n).$$

This completes the induction argument. We now use the sets $V(r)$ constructed above to define the function f as follows:

$$f(x) = \begin{cases} 0, & \text{if } x \in V(r) \text{ for each } r; \\ \sup\{r: x \notin V(r)\}, & \text{otherwise.} \end{cases}$$

Then $f(A) = \{0\}$ and $f(B) = \{1\}$. Also, for $0 < a < 1$,

$$f(x) < a \Leftrightarrow x \in \bigcup_{r<a} V(r),$$

$$f(x) > a \Leftrightarrow x \in \bigcup_{r>a} (\overline{V(r)})^c.$$

It follows that the sets $f^{-1}([0, a))$ and $f^{-1}((a, 1])$ are both open. Thus, f is the desired continuous function. ∎

Theorem 1.7. *Tietze Extension Theorem.* A topological space X is normal if and only if every real-valued continuous function f from a closed subset A of X into $[-1, 1]$ can be extended to a continuous function $g: X \to [-1, 1]$. ∎

Proof. The "if" part follows easily using Urysohn's Lemma. For the "only if" part, let X be normal and f be a nonconstant continuous function from a closed subset A of X into $[-1, 1]$. Suppose $[a, b]$ is the smallest closed subinterval of $[-1, 1]$ containing $f(A)$. Then if

$$h(x) = \frac{2}{b-a}x - \frac{b+a}{b-a},$$

the smallest closed interval containing $h \circ f(A)$ is $[-1, 1]$. Since $h^{-1}: [-1, 1] \to [a, b]$ is continuous, the theorem will be proven for f if it is proven for $h \circ f$. Therefore, with no loss of generality, we assume that the smallest closed interval containing $f(A)$ is $[-1, 1]$.

Now let $C = \{x \in A: f(x) \leq -1/3\}$ and $D = \{x \in A: f(x) \geq 1/3\}$. Then C and D are nonempty disjoint closed subsets of X. Noting that Urysohn's Lemma holds easily for any two real numbers in place of the numbers 0 and 1, there exists a continuous $f_1: X \to [-1/3, 1/3]$ such that $f_1(C) = \{-1/3\}$ and $f_1(D) = \{1/3\}$. Then for each x in A, $|f(x) - f_1(x)| \leq 2/3$. By a similar argument applied to $f - f_1$, there is a continuous f_2: $X \to [-2/3^2, 2/3^2]$ such that $|f(x) - f_1(x) - f_2(x)| \leq (2/3)^2$ for each x in A. Continuing, we can find a sequence of continuous functions f_n: $X \to [-2/3^n, 2/3^n]$ such that $|f(x) - \sum_{i=1}^n f_i(x)| \leq (2/3)^n$ for each x in A. Since $\sum_{i=1}^\infty |f_i(x)| \leq 1$ for all x in X, it follows that $\sum_{i=1}^\infty f_i(x)$ is uniformly convergent to some continuous function $g(x)$. Then g is the desired extension of f. ∎

Next, we introduce some notions related to compactness.

Definition 1.28. A topological space X is called *countably compact* if every countable open cover of X has a finite subcover. ∎

Proposition 1.17. A topological space X is countably compact if and only if every sequence (x_n) in X has at least one *cluster point* (i.e., a point x such that every open set containing x contains x_n for infinitely many n.) ∎

Proof. It can be verified easily that X is countably compact if and only if every countable family of closed sets with f.i.p. has a nonempty intersection.

For the "if" part, let (A_i) be a sequence of closed sets with f.i.p. For each n, let $x_n \in \bigcap_{i=1}^n A_i$. Let x be a cluster point of (x_n). Then clearly $x \in \bigcap_{i=1}^\infty A_i$. For the "only if" part, let X be countably compact and (x_n) be a sequence in X. Let $A_n = \{x_n, x_{n+1}, \ldots\}$. Then (A_n) has f.i.p. and therefore, $\bigcap_{n=1}^\infty \bar{A}_n$ is nonempty and contains some point x. Then x is a cluster point of (x_n). ∎

Definition 1.29. A topological space X is called *sequentially compact* if every infinite sequence in X has a convergent subsequence. ∎

The following result is obvious.

Proposition 1.18. A sequentially compact space is countably compact. Also a first countable, countably compact space is sequentially compact. ∎

Examples

1.23. *A First Countable Sequentially Compact Space That Is Not Compact.* Let $(X, \leq)$ be the well-ordered set of all ordinals that are less then Ω, the first uncountable ordinal. (See Proposition 1.6.) Let all sets of the form $\{x: x < a\}$ or $\{x: a < x\}$ for some $a \in X$ be a subbase for a topology, called the *order topology* for X. Then this topology is *first countable*. For, if $y \in X$, then there is $z \in X$ such that $y < z < \Omega$ and $\{(x_1, x_2): x_1 < y < x_2 \leq z\}$ is countable by Proposition 1.6 (ii). A similar argument shows that this topology is *not second countable*.

Also, X has the following property. If $A \subset X$, A countable, then sup A (which exists by the well-ordering of $X \cup \{\Omega\}$) is less than Ω. This is because sup A has only a countable number of predecessors relative to the ordering and, therefore, sup $A < \Omega$. It is clear, therefore, that X is countably compact.

To prove that X is not compact, consider the open cover (for X) $\bigcup_{x\in X}\{y: y < x\}$. Since every countable union of these open sets is countable, it does not have a finite subcover and therefore X is *not compact*.

1.24. *A Compact Space That Is Not Sequentially Compact.* Consider $X = I^I$, the uncountable cartesian product of $[0, 1] = I$, with product topology. Since I is compact and Hausdorff (with usual topology), X is also compact and Hausdorff. Points of X are really functions $f: I \to I$. A sequence (f_n) converges to f if and only if for each $x \in I$, $f_n(x)$ converges to $f(x)$. This follows from the definition of product topology. To show that X is not sequentially compact, let $f_n(x)$ be the nth digit in the binary expansion of x. Suppose (f_{n_k}) is a convergent subsequence. There is $x \in I$ such that the sequence $f_{n_k}(x)$ is a sequence of infinitely many 0's and infinitely many 1's, so that the f_{n_k}'s cannot converge.

Definition 1.30. A topological space X is called *locally compact* if for each $x \in X$, there is an open set V such that $x \in V$ and $\bar{V}$ is compact. ▮

Remarks on Local Compactness

1.54. Every compact space is locally compact; $R^n (n \geq 1)$ (with usual topology) are locally compact, but not compact.

1.55. Let X be a locally compact Hausdorff space. We can compactify X by considering $X^* = X \cup \{\infty\}$ and taking a set in X^* to be open if and only if it is either an open subset of X or the complement of a compact subset of X. Clearly, the topological space X^* (thus formed) is a compact Hausdorff space. X^* is called the *one-point compactification* of X. Since every compact Hausdorff space is T_4 (Remark 1.37) and $T_4 \Rightarrow$ complete regularity (by Lemma 1.1), X^* is completely regular. Since any subspace of a completely regular space is also completely regular, X is completely regular. This means that in a locally compact Hausdorff space, given any open set V containing x, there exists an open set W such that $x \in W$, $\overline{W}$ compact and $\overline{W} \subset V$.

1.56. Every closed or open subspace of a locally compact space is locally compact.

1.57. Now that the reader is familiar with one-point compactifications of a topological space X, the question of n-point compactifications of X naturally comes up. An n-point compactification X^* of X (n a positive integer) is a compact T_2 space containing X as a dense subspace such that $X^* - X$ has cardinality n. Spaces X having an n-point compactification can be characterized as follows:

A T_2 space X has an n-point compactification X^* if and only if it is locally compact and contains a compact subset K whose complement is the union of n mutually disjoint open subsets G_i such that $K \cup G_i$, for each i, is not compact.

Let us prove the above result for $n = 2$. The same idea works for any finite n. (The reader is encouraged to work out the case of general n.) For the "only if" part, let $X^* = X \cup \{x_1, x_2\}$ be a two-point compactification of X. Since X^* is T_2, there are disjoint open subsets G_1' and G_2' containing x_1 and x_2, respectively. Let $G_1 = G_1' - \{x_1\}$ and $G_2 = G_2' - \{x_2\}$. Then the set $K = X - (G_1 \cup G_2) = X^* - (G_1' \cup G_2')$ is a compact subset of X (being closed and therefore compact in X^*). Notice that if $K \cup G_1$ is compact in X, then it is compact and therefore closed in X^* so that $\{x_1\} \cup G_2'$, its complement in X^*, is open; but, then $\{x_1\} = G_1' \cap (\{x_1\} \cup G_2')$ is open, which contradicts that X is dense in X^*. Thus, $K \cup G_1$ (and similarly, $K \cup G_2$) is not compact in X. Since $X = X^* - \{x_1, x_2\}$ is open in X^*, it is locally compact by 1.56. For the "if" part, let X be a locally compact T_2 space with its subsets $\{K, G_1, G_2\}$ forming a partition of X such that K is compact and G_1, G_2 are open. Choose two distinct elements x_1, x_2 not in X. Let $X^* = X \cup \{x_1, x_2\}$. Then the class of subsets defined by $\{A \subset X^*$: either A is an open subset of X or $A = B \cup \{x_i\}$, $i = 1$ or 2, where B is an open subset of X and $(K \cup G_i) \cap (X - B)$ is compact in $X\}$ can be verified to form a basis for a compact T_2 topology for X^*, containing X as a dense subspace.

1.58. The continuous open image of a locally compact space is locally compact.

1.59. $\times\{X_\lambda: \lambda \in \Lambda\}$ is locally compact if and only if all the X_λ are locally compact and all but finitely many X_λ are compact.

The "if" part is easy to prove. For the "only if" part, let the product space be locally compact. Then each projection π_λ is a continuous, open, onto mapping and therefore each X_λ is locally compact. Also, if V is open in $\times_{\lambda\in\Lambda}X_\lambda$ and $\bar{V}$ is compact, then $\pi_\lambda(\bar{V})$ is compact and $\pi_\lambda(\bar{V}) = X_\lambda$ for all but finitely many λ.

1.60. If X is Hausdorff and Y is a dense locally compact subspace, then Y is open.

To prove this, let $x \in Y$. Then there is open V such that $x \in V$ and $V \cap Y$ has its closure (in Y) compact. Suppose $V \cap (X - Y)$ is nonempty and contains z. Then if $\{W_\lambda(z)\}$ is the class of all open sets ($\subset V$) containing z, clearly $\{z\} = \cap\, W_\lambda(z)$. Now $\{\overline{W_\lambda(z)} \cap Y\}$ is a family of closed sets in Y with f.i.p. (since Y is dense). But it is not difficult to see that $\cap \{\overline{W_\lambda(z)} \cap Y\}$ is empty, contradicting the compactness of $\bar{V} \cap Y$ (= the closure of $V \cap Y$ in Y). Hence $V \subset Y$.

Proposition 1.19. (*Baire*) The intersection of any countable family of open dense sets in a locally compact Hausdorff space X is dense. ▮

Proof. Let (A_i) be a sequence of open dense sets. Let V be an open set. Then $V \cap A_1 \neq \varnothing$. By Remark 1.56 there is open B_1, such that $\bar{B}_1$ is compact and $\bar{B}_1 \subset V \cap A_1$. Similarly, working with B_1 and A_2, we see that there is open B_2 with $\bar{B}_2$ compact and $\bar{B}_2 \subset B_1 \cap A_2$. By induction, we can find a sequence of open sets (B_n), $\bar{B}_n \subset B_{n-1} \cap A_n$ for each n. Since $\bar{B}_1$ is compact and $(\bar{B}_n)$ have f.i.p., the proposition follows. ▮

Definition 1.31. A topological space X is called a *Baire* space if the intersection of any countable family of open dense sets is dense. ▮

Remarks on a Baire Space

1.61. Every locally compact Hausdorff space is a Baire space.

1.62. If X is a Baire space and $X = \bigcup_{n=1}^{\infty} A_n$, A_n closed, then for some n the interior of A_n is nonempty. The reason is that since $\bigcap_{n=1}^{\infty} X - A_n = \varnothing$, $X - A_n$ is not dense for some n and therefore $A_n{}^\circ \neq \varnothing$.

1.63. Because of Remark 1.62 the set of rationals with relative topology (in R) is *not* a Baire space.

1.64. A set A in a topological space is called *a set of the first category* if and only if it is a countable union of *nowhere dense sets* (i.e., sets whose closure have an empty interior). The interior of a set of the first category in a Baire space is empty. A Baire space is of the *second category*, i.e., a space that is not of the first category.

Problems

1.4.1. Let f be a mapping from a topological space X into a topological space Y. Then show that the following are equivalent:

(i) f is continuous.

(ii) If $B \subset Y$, then $\overline{f^{-1}(B)} \subset f^{-1}(\bar{B})$.

(iii) If $B \subset Y$, then $f^{-1}(B^\circ) \subset [f^{-1}(B)]^\circ$.

1.4.2. If Y is Hausdorff, X is any topological space and $f, g: X \to Y$ are continuous, then show that $\{x \in X: f(x) = g(x)\}$ is a closed set.

1.4.3. Show that a one-to-one, onto, and continuous mapping f from a compact space into a Hausdorff space is also open. [A continuous, open, one-to-one, and onto map from one topological space into another is called a *homeomorphism*.]

× **1.4.4.** Given A, a closed G_δ in a normal topological space X, show that there exists a continuous real-valued function f such that $A = f^{-1}\{0\}$. Note that a set $B \subset X$ is a G_δ if and only if $B = \bigcap_{n=1}^\infty G_n$, G_n open in X for each n. [Hint: Write $A = \bigcap_{n=1}^\infty G_n$, $G_n \supset G_{n+1}$ and G_n open for each n. By Urysohn's Lemma, there is a continuous function f_n with $f_n(A) = \{0\}$ and $f_n(X - G_n) = 1/n^2$. Then consider $f = \sum_{n=1}^\infty f_n$.]

1.4.5. Show that closures are preserved by homeomorphisms, but not necessarily by continuous mappings.

1.4.6. Show that the graph $G = \{(x, f(x)): x \in X\}$ (with relative product topology) of a continuous mapping $f: X \to Y$ is homeomorphic with X.

1.4.7. (a) A point x in a topological space X is called *isolated* if and only if $\{x\}$ is open. Show that there are an infinite number of isolated points in a compact Hausdorff space X if and only if there exists $x_0 \in X$ such that X and $X - \{x_0\}$ are homeomorphic.

(b) Let $L = (a_n)$ be a sequence of real numbers and $A(L)$ be the set of limit points of L. Suppose that $A(L)$ is nonempty and $p \in A(L)$. Is L then a disjoint union of subsequences S and T of L such that S converges to p and p is not a limit point of T? If not, then find a necessary and sufficient condition for L to be such a disjoint union. [Hint: Consider p to be an isolated point of $A(L)$.]

1.4.8. Show that normality is preserved by closed continuous mappings.

1.4.9. *Half-Open Interval Topology*. Show that the family of all sets of the form $[a, b)$ (a and b real numbers) form a basis for a topology in R (the reals). Show also that this topology is T_4, first countable, separable, *not* second countable, but Lindelöf. [Hint for T_4: For A and B closed,

$A \cap B = \emptyset$, let $[a, x_a) \subset X - B$ and $[b, x_b) \subset X - A$; then $(\bigcup_{a \in A}[a, x_a)) \cap (\bigcup_{b \in B}[b, x_b)) = \emptyset$.]

Furthermore, show that any compact set in this topology is a nowhere dense set in the usual topology of R. [Hint: If A is not nowhere dense, then $A \cap [a, b]$ is dense in some interval $[a, b]$; this means that if $a < b_i < b_{i+1} < b$ and $\lim b_i = b$, then $\{(-\infty, a), [a, b_1), [b_i, b_{i+1})$ for $1 \leq i < \infty$; $[b, \infty)\}$ is an open cover of A with no finite subcover.]

1.4.10. Let X be the reals with the half-open interval topology. Show the following:

(i) $S = \{(x, y): x = -y\}$ is a closed subset in $X \times X$ (with the product topology) and *discrete* in its relative topology.

(ii) S is neither separable nor Lindelöf; hence $X \times X$ is not Lindelöf.

(iii) $X \times X$ is not normal.

[Hint: Since $X \times X$ is separable, there are $2^{\aleph_0}$ continuous real-valued functions on $X \times X$. But there are 2^c continuous real-valued functions on S. Since S is closed, this contradicts normality for $X \times X$.]

1.4.11. A topological space X is called *pseudocompact* if and only if every continuous real-valued function on X is bounded. Show that every countably compact space is pseudocompact whereas a pseudocompact T_4 space is countably compact. [Hint: If X is not countably compact, then there is an infinite closed set $A \subset X$, discrete in the relative topology. If $(x_n) \subset A$ and $f(x_n) = n$ for each n, then the function f continuous on A cannot be extended to a bounded function on X.]

1.4.12. Consider an infinite set X and $\mathscr{T} = \emptyset \cup \{A \subset X: a \in A\}$, where a is some fixed point of X. Show that $(X, \mathscr{T})$ is a pseudocompact space that is *not* countably compact.

★ **1.4.13.** *Paracompact Spaces.* A cover (A_α) is called a *refinement* of a cover (B_β) if each $A_\alpha \subset B_\beta$ for some β. A cover is called *locally finite* if each point has an open set (containing it) intersecting only finitely many members of the cover. A topological space is called *paracompact* if every open cover has an open locally finite refinement. Show the following:

(i) Every paracompact Hausdorff (or regular) space is normal.

[Hint: Let X be paracompact regular, A and B closed disjoint subsets of X. For $x \in A$, let $V(x)$ be open, $x \in V(x)$, and $\overline{V(x)} \subset X - B$. Then if $\mathscr{F}$ is a locally finite refinement of $\{V(x): x \in A\} \cup \{X - A\}$, and if $G = \cup \{H: H \cap A \neq \emptyset$ and $H \in \mathscr{F}\}$, then G is open and $\supset A$. If $J = X - \cup \{\bar{H}: H \cap A \neq \emptyset$ and $H \in \mathscr{F}\}$, then J will be open and $\supset B$. Note that if $\mathscr{E}$ is a locally finite family of subsets of a topological space, then $\cup\{\bar{E}: E \in \mathscr{E}\} = \overline{\cup\{E: E \in \mathscr{E}\}}$.]

(ii) Every compact (or metric) topological space is paracompact. (The metric case will be considered in Section 1.5.)

1.4.14. Let X be a locally compact Hausdorff space and K a compact subset of X. Show that there is a continuous real-valued function f on X such that $f(x) = 1$, $x \in K$, and $\{x: f(x) \neq 0\}$ has compact closure.

1.4.15. Show that in a regular space, the closure of a compact set is compact.

1.4.16. *Separability of products.* Let $X_\beta = \times_{k \in A} Y_k$, where card $A = \beta$ and each Y_k is Hausdorff. Prove that X_β is separable if and only if each Y_k is separable and all but c of the Y_k's consist of a single point. [Hint: Let $A = [0, 1]$ in the "if" part. Let $(y_{k,n})_{n=1}^\infty$ be a countable dense subset of Y_k. For each finite mutually disjoint family of open intervals $I_1, I_2, \ldots, I_m$ with rational end points and each finite set $n_1, n_2, \ldots, n_m$ of positive integers, let $s(I_1, \ldots, I_m, n_1, \ldots, n_m)$ denote the element $(y_{k,s_k})_{k \in A}$ in the product, where $s_k = n_i$ if $k \in I_i$, $= 1$ otherwise. Then all such elements form a countable dense subset of X_β. For the "only if" part, let $A_0 = \{k \in A$: card $Y_k \geq 2\}$. For each $k \in A_0$, let G_k, H_k be nonempty disjoint open subsets of Y_k. Let D be a countable dense subset of the product X_β. For each $k \in A_0$, let D_k be those elements of D whose kth coordinate belongs to G_k. Then $k_1 \neq k_2 \Rightarrow D_{k_1} \neq D_{k_2}$. Thus, there is an injection from A_0 to the class of all subsets of D.]

1.4.17. *Product of Second (First) Countable Spaces.* Show that the product topology is second (first) countable if and only if the topology of each coordinate space is second (first) countable and all but a countable number of the coordinate spaces have the indiscrete topology.

1.4.18. *Functions with Compact Graphs.* Let f be a map from a topological space X into a topological space Y. Then show the following:

(i) Compactness of the graph of f implies that of X and $f(X)$.

(ii) The graph of f is compact whenever X is compact, Y is T_2, and f is continuous.

(iii) If the graph of f is compact and X is T_2, then f is continuous.

1.4.19. *Nets. A directed system* is a set A with a relation $<$ satisfying

(i) $a < b$ and $b < c$ imply $a < c$, and

(ii) for $a, b \in A$, there is $c \in A$ with $a < c$ and $b < c$.

[The family of all open sets containing a point x is a directed system, with $G_1 < G_2$, if $G_2 \subset G_1$.] *A net* (x_α) is a mapping of a directed system into a topological space X, where x_α is the value of the net at α. A net (x_α) is said to converge to $x \in X$ if given an open set G, $x \in G$, there is an $\alpha_0 \in A$ such that for all $\alpha > \alpha_0$, $x_\alpha \in G$. Prove the following.

(i) For $E \subset X$, $x \in \bar{E}$ if and only if it is the limit of a net (x_α) in E.

(ii) X is T_2 if and only if every net in X has at most one limit.

(iii) $f: X \to Y$ is continuous at x if and only if for each net (x_α) converging to x, the net $[f(x_\alpha)]$ converges to $f(x)$.

★ **1.4.20.** *The Stone–Čech Compactification.* Let X be a completely regular T_2 space and $F(X)$ be the family of continuous functions from X to the unit interval I. [Note that given any two distinct points x, y in X, there is $f \in F(X)$ such that $f(x) = 0$ and $f(y) = 1$.] Let $I^{F(X)}$ be the product space $\times_{f \in F(X)} I_f$, where each $I_f = I$. Let $t_X: X \to I^{F(X)}$ be defined by $(t_X(x))_f = f(x)$. Show that t_X is a homeomorphism (into). The compact T_2 space $t_X(X)$ is called the *Stone–Čech compactification* of X, and denoted by $\beta(X)$. Let g be a continuous map of X into a compact T_2 space Y. Show that (i) there is a unique continuous map $h: \beta(X) \to Y$ such that $g = h \circ t_X$; (ii) if W is a compactification of X, i.e., a compact T_2 space containing X as a dense subspace, with property (i), then W is homeomorphic to $\beta(X)$; (iii) $\beta(X)$ is the largest compactification of X in the sense that if W is a compactification of X, then there is a continuous surjection from $\beta(X)$ to W. Finally, consider the continuous function $\sin(1/x)$ on $(0, 1]$ to show that $[0, 1]$ is not the Stone–Čech compactification of $(0, 1]$. [Hint for (i): Note that $t_Y: Y \to \beta(Y)$ is a homeomorphism onto. Define $h_0: I^{F(X)} \to I^{F(Y)}$ by $h_0((x_f)_{f \in F(X)}) = (x_{f' \circ g})_{f' \in F(Y)}$. Show that h_0 is continuous and the map $h = t_Y^{-1} \circ h_0: \beta(X) \to Y$ is a continuous extension of g.]

1.4.21. Show that card $\beta(Z^+) =$ card $\beta(R) = 2^c$. [Hint: By Problem 1.4.16, I^c is separable where $I = [0, 1]$. Let D be a countable dense subset of I^c and g be a bijection from Z^+ to D. By Problem 1.4.20, there is a continuous extension $h: \beta(Z^+) \to I^c$, which must be onto.]

★ **1.4.22.** *n-Point Compactifications*:

(a) Prove with details the assertion made in Remark 1.57 for any positive integer n.

(b) Prove that the only n-point compactification of the complex plane (with usual topology) is its one-point compactification.

(c) Prove that the only n-point compactifications of the reals are the one- and two-point compactifications.

(d) Prove that an infinite discrete space has infinitely many n-point compactifications for each positive integer n.

(The above results are due to K. D. Magill, Jr. [32].)

★ **1.4.23.** Prove that R is not homeomorphic to $X \times X$ for any topological space X.

• 1.5. Connected Spaces, Metric Spaces, and Fixed Point Theorems

In this text, as will be found later, metric spaces are referred to often and are of utmost importance. They are studied in this section. Although connectedness does not appear explicitly in the theme of this text, for completeness we also study connectedness, an important topological concept. We also study in this section fixed point theorems specifying conditions when a mapping of a topological space into itself leaves one or more points invariant or fixed. Fixed point theorems are of immense importance in analysis; they are used often in the existence theory of differential and integral equations. In Part B, we present the well-known Kakutani Fixed Point Theorem and then use it to show the existence of a Haar measure in a compact topological group, a result of fundamental importance in analysis.

Definition 1.32. A topological space X is called *connected* if X and $\emptyset$ are the *only* subsets of X that are both open and closed. X is called *disconnected* if X is *not* connected. A subset A of X is called connected if A with its relative topology is connected. ∎

Remarks

1.65. A subset A of the reals with the usual topology, which contains at least two distinct points, is connected if and only if it is an interval.

To see this, suppose A is not an interval; then there are a, b in A and c not in A with $a < c < b$. This means $A \cap (-\infty, c)$ is open and closed in A, and A is not connected. Conversely, suppose A is an interval, and $B \subset A$ with B both open and closed in A, $B \neq \emptyset$ and $A - B \neq \emptyset$. Then there is $x \in A - B$ such that $x > y$ for some $y \in B$; otherwise for some $z \in A - B$, $B = (z, \infty) \cap A$ and is therefore not closed. Let $c = \text{l.u.b. } \{y \in B: y < x\}$. Then $c \in A$ (since A is an interval) and c is a limit point of B and so $c \in B$. Clearly, $c < x$ and $[c, x] \subset A$. But since B is open in A, there is d in B with $c < d < x$. This contradicts the l.u.b. property of c.

1.66. A topological space X is connected if and only if the only continuous mappings $f: X \to \{0, 1\}$ (with discrete topology) are the constant mappings. The reason is that if f is nonconstant, then $f^{-1}\{0\}$ is an open and closed subset different from both X and $\emptyset$; also if A is a nontrivial open and closed subset of X, then f defined by $f(A) = \{0\}$ and $f(X - A) = \{1\}$ is a continuous map.

1.67. Connectedness is preserved under continuous mappings. The intermediate value property of a continuous function follows easily from this fact.

1.68. If A is a connected set in a topological space and $A \subset B \subset \bar{A}$, then B is also connected. The reason is that if $D \subset B$ and D is open and closed in B, then $D \cap A$ is also open and closed in A.

1.69. The union of any family (A_λ) of connected sets, having a nonempty intersection, is connected. To see this, let $x \in \bigcap_{\lambda\in\Lambda} A_\lambda$ and H be a closed and open subset of $\bigcup_{\lambda\in\Lambda} A_\lambda = A$. If $x \in H$, then for each λ, $A_\lambda \cap H \neq \emptyset$ and is both open and closed in A_λ. Since A_λ is connected, $A_\lambda \cap H = A_\lambda$ or $A_\lambda \subset H$ for each λ. On the other hand, if $x \in A - H$, then $A_\lambda \cap H^c \neq \emptyset$ and is both open and closed in A_λ. Since A_λ is connected, $A_\lambda \cap H^c = A_\lambda$ or $A_\lambda \subset H^c$ for all λ, meaning $H = \emptyset$. In either case $H = A$ or $H = \emptyset$.

1.70. $X_1 \times X_2$ (with the product topology) is connected if X_1 and X_2 are connected. Notice that for $x_1 \in X_1$ and $x_2 \in X_2$, $\{x_1\}\times X_2$ and $X_1 \times \{x_2\}$ are both connected, $X_1 \times X_2 = \bigcup_{x_2 \in X_2}[\{x_1\}\times X_2 \cup X_1 \times \{x_2\}]$, which is connected by Remark 1.69 above.

1.71. If $n > 1$ and $A \subset R^n$ is countable, then $R^n - A$ is connected. The reason is simple: For simplicity, let $n = 2$ and $x, y \in R^2 - A$. For any line segment I_y meeting the segment $\overline{xy}$ in only one point (see Figure 1), if $z, z^1 \in I_y$, then $(\overline{xz} \cup \overline{zy}) \cap (\overline{xz^1} \cup \overline{z^1y}) = \{x, y\}$. Clearly, since I_y is uncountable and A is countable, $A \cap (\overline{xz_y} \cup \overline{z_y y}) = \emptyset$ for some $z_y \in I_y$. Hence, $R^2 - A = \cup\{(\overline{xz_y} \cup \overline{z_y y})\colon y \in R^2 - A\}$ is connected by Remark 1.69.

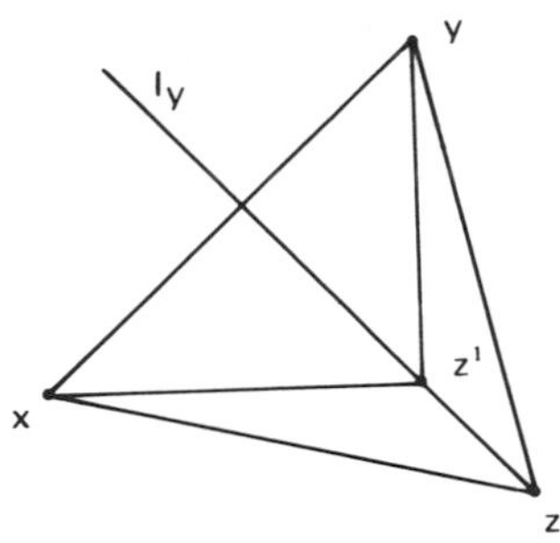

Fig. 1

1.72. R^1 and R^n $(n > 1)$ are not homeomorphic.[†] For if $f\colon R^1 \to R^n$ is a homeomorphism, then for $x \in R$, $R^1 - \{x\}$ is not connected, despite the

[†] Two topological spaces are called homeomorphic if there is a homeomorphism (i.e., a one-to-one, onto, continuous, and open map) from one into the other.

fact that $f(R^1 - \{x\}) = R^n - \{f(x)\}$, which is connected, by Remark 1.71. (It is relevant to point out here that for $n \neq m$, R^n and R^m are also *not* homeomorphic; the proof is more involved and is omitted.)

Theorem 1.8. *A Fixed Point Theorem.* Let f be a continuous function from $[0, 1]$ into itself. Then there is $x \in [0, 1]$ with $f(x) = x$. ▮

Proof. We may and do assume that $f(0) > 0$ and $f(1) < 1$. Let $g(x) = x - f(x)$. Then g is continuous, $g(0) < 0 < g(1)$. By the intermediate-value property of g, there is $x \in (0, 1)$ with $g(x) = 0$. ▮

This theorem is a special case of the famous Brouwer's Fixed Point Theorem (the earliest fixed point theorem, proven in 1912 by L. E. J. Brouwer).

Theorem 1.9. *The Brouwer Fixed Point Theorem.* Let $f: I^n \to I^n$ be a continuous map. Then there is $x \in I^n$ with $f(x) = x$. [Here $I^n = \{(x_1, x_2, \ldots, x_n): 0 \leq x_i \leq 1, 1 \leq i \leq n\}$.] ▮

An elementary proof of this theorem is recently given by Kiyoshi Kuga.[†] The proof is too lengthy to be included here.

Proposition 1.20. Let $\{X_\lambda\}$, $\lambda \in \Lambda$ be an indexed family of topological spaces. Then $X = \times_{\lambda \in \Lambda} X_\lambda$ is connected if and only if each X_λ is connected. ▮

Proof. The "only if" part follows by Remark 1.67, since the projection π_λ is a continuous map from X onto X_λ.

In the "if" part, let f be a continuous map from X into the discrete space $\{0, 1\}$. Suppose $x, y \in X$ and $f(x) = 0$. We claim that $f(y) = 0$. Let V be open in X, $x \in V$ and $f(V) = 0$. Let $\lambda_1, \lambda_2, \ldots, \lambda_n \in \Lambda$ be such that $\bigcap_{i=1}^n \pi_{\lambda_i}^{-1}(V_{\lambda_i}) \subset V$, V_{λ_i} open in X_{λ_i} and $x(\lambda_i) \in V_{\lambda_i}$. Let $z \in X$ be defined by $z(\lambda_i) = x(\lambda_i)$, $1 \leq i \leq n$ and $z(\lambda) = y(\lambda)$ for all other $\lambda \in \Lambda$. Then $z \in V$ and $f(z) = 0$. Now $W \equiv \{w \in X: w(\lambda) = y(\lambda)$ for all $\lambda \neq \lambda_i$, $1 \leq i \leq n\}$ is clearly homeomorphic to $X_{\lambda_1} \times X_{\lambda_2} \times \cdots \times X_{\lambda_n}$, which is connected by Remark 1.70. Since $z \in W$ and $y \in W$, $f(z) = f(y) = 0$. ▮

A disconnected space can be partitioned uniquely by maximal connected sets or components. The number of components gives a rough idea of how disconnected the space is.

[†] Kiyoshi Kuga, *SIAM J. Math. Anal.* **5**(6) (1974).

Definition 1.33. The *component* C_x of $x \in X$, a topological space, is the union of all connected subsets of X containing x. ▮

Clearly, C_x is a *maximal connected* set, which is closed, by Remark 1.68.

Examples

1.25. Let $X = \{1/n: n \text{ is a positive integer}\} \cup \{0\}$ with the relative topology from R. Then $\{0\}$ is a component, which is *not* open.

1.26. In $R - \{x\}$, with the relative topology, $(-\infty, x)$ and (x, ∞) are the components.

1.27. In R^2, the subspace $\{(x, y)$: either $xy = 1$ or $y = 0$ or $x = 0\}$ has three components. Note that the map f: $(0, \infty) \to R^2$ defined by $f(x) = (x, 1/x)$ is continuous and therefore $\{(x, 1/x): x > 0\}$ is a connected subset of R^2.

A topological concept called "*local connectedness*" provides examples of nonconnected spaces, where the *components are open*.

Definition 1.34. A topological space X is called *locally connected* at a point $x \in X$ if given V open with $x \in V$, there exists an open U, a connected set W such that $x \in U \subset W \subset V$. X is called locally connected if it is locally connected at each point in X. Equivalently, X is locally connected if components of open subsets of X are open in X. ▮

Clearly R and every interval of R is locally connected, whereas the set of rationals (with relative topology) is not locally connected at any of its points. The rationals are, in fact, *totally disconnected* (i.e., a space, where every singleton is a component). Note that a discrete topological space, though not connected, is locally connected.

Example 1.28. *A Connected Space That Is Not Locally Connected.* Consider f: $(0, 1] \to R$ defined by $f(x) = \sin(1/x)$. Let G be the graph of f. Then G, being the image of a continuous map of $(0, 1]$ into R^2, is a connected subset of R^2. Clearly, its closure in R^2, that is, $G \cup \{(0, y): -1 \leq y \leq 1\}$ is also connected. Since $\bar{G} \cap \{R \times (-\frac{1}{2}, \frac{1}{2})\}$ consists of separated line segments (verify by drawing a neat picture), the component of $(0, 0)$ in $\bar{G} \cap \{R \times (-\frac{1}{2}, \frac{1}{2})\}$ is the segment $0 \times (-\frac{1}{2}, \frac{1}{2})$, which is not open. Hence $\bar{G}$ is *not* locally connected.

One can easily verify that for any two points in R^2 there is a continuous function f: $[0, 1] \to R^2$ mapping 0 and 1 to these two points. In fact, in any open connected set in R^2, any two points can be joined† by such a function or *a path*. This motivates the following stronger concept of connectedness.

Definition 1.35. A topological space X is called *path connected* if for any $x_1, x_2 \in X$, there is a continuous map f: $[0, 1] \to X$ such that $f(0) = x_1$ and $f(1) = x_2$. ∎

Remarks

1.73. Path connectedness implies connectedness. The reason is that if A is a proper open and closed subset of X (path connected) then there is a continuous f: $[0, 1] \to X$ such that $f(0) \in A$ and $f(1) \in X - A$. This means that $f^{-1}(A)$ is a proper open and closed subset of $[0, 1]$, contradicting the connectedness of $[0, 1]$.

1.74. The set $\bar{G}$ in Example 1.28, though connected, is *not* path connected.

1.75. Every continuous image of a path-connected space is path connected.

1.76. For every family of path-connected sets (taken with the relative topology) having nonempty intersection, their union is also path connected.

Now we turn our attention to the most widely used topological space—metric spaces.

Definition 1.36. A metric space (X, d) is a nonempty set X with a nonnegative real-valued function d on $X \times X$ such that, for all x, y, and $z \in X$, the following properties hold:

(i) $d(x, y) = 0 \Leftrightarrow x = y$,
(ii) $d(x, y) = d(y, x)$,
(iii) $d(x, z) \leq d(x, y) + d(y, z)$.

The function d is called a *metric* (or *distance*). ∎

Examples

1.29. For $x = (x_1, x_2, \ldots, x_n)$ and $y = (y_1, y_2, \ldots, x_n) \in R^n$, let

$$d(x, y) = \left[\sum_{i=1}^{n} (x_i - y_i)^2 \right]^{1/2}.$$

† This will be clarified in Problem 1.5.4.

Then d defines a metric in R^n. To see this, it is sufficient to observe that

$$\text{(a)}\quad \Big(\sum_{i=1}^{n} a_i b_i\Big)^2 \leq \Big(\sum_{i=1}^{n} a_i^2\Big)\Big(\sum_{i=1}^{n} b_i^2\Big),$$

and

$$\text{(b)}\quad \Big[\sum_{i=1}^{n} (a_i + b_i)^2\Big]^{1/2} \leq \Big(\sum_{i=1}^{n} a_i^2\Big)^{1/2} + \Big(\sum_{i=1}^{n} b_i^2\Big)^{1/2}.$$

1.30. Let (X_n, d_n), $1 \leq n < \infty$ be an infinite sequence of metric spaces. For $x = (x_1, x_2, \ldots,)$ and $y = (y_1, y_2, \ldots)$ in $X = \times_{n=1}^{\infty} X_n$, let

$$d(x, y) = \sum_{n=1}^{\infty} \frac{1}{2^n} \cdot \frac{d_n(x_n, y_n)}{1 + d_n(x_n, y_n)}.$$

Then (X, d) is a metric space.

Proposition 1.21. In any metric space (X, d), the family of all sets $S_x(\varepsilon) = \{y\colon d(x, y) < \varepsilon\}$, where $x \in X$ and ε is a positive real number is a base for a topology (called *the topology induced by d*). ▮

Proof. If $z \in S_x(\varepsilon_1) \cap S_y(\varepsilon_2)$ and $\varepsilon = \min\{\varepsilon_1 - d(z, x), \varepsilon_2 - d(z, y)\}$, then $S_z(\varepsilon) \subset S_x(\varepsilon_1) \cap S_y(\varepsilon_2)$. This means that the family of all unions of sets of the forms $S_x(\varepsilon)$ is a topology. ▮

In what follows, a metric space will always be considered a topological space with topology induced by d.

Remarks

1.77. Two different metrics can, sometimes, induce the same topology on a set. For instance, in R^2, the metric d_1 defined by

$$d_1((x_1, y_1), (x_2, y_2)) = |x_1 - x_2| + |y_1 - y_2|$$

and the metric d_2 defined by

$$d_2((x_1, y_1), (x_2, y_2)) = \max\{|x_1 - x_2|, |y_1 - y_2|\}$$

induce the same topology as induced by the metric in Example 1.29.

Let us also mention at this point that in Example 1.30, the topology induced in the product space $X = \times_{n=1}^{\infty} X_n$ by the metric d is the product topology. Thus, a countably infinite product of metric spaces is a metric space. [See Problem 1.5.13(a).]

1.78. The metric d defined by

$$d(x, y) = \begin{cases} 0, & x = y, \\ 1, & x \neq y, \end{cases}$$

induces a discrete topology on any nonempty set X.

1.79. Every metric space is a first countable, T_4 space. To see this, it is sufficient to observe the following:

(a) For $x \neq y$ and $0 < 2\varepsilon < d(x, y)$, $S_x(\varepsilon) \cap S_y(\varepsilon) = \varnothing$.

(b) Given any open V containing x, there is a positive integer n such that $S_x(1/n) \subset V$.

(c) Given any two closed disjoint sets A and B, for $x \in A$ and $y \in B$ we can choose $\varepsilon_x > 0$ and $\varepsilon_y > 0$ such that $S_x(2\varepsilon_x) \cap B = \varnothing$ and $S_y(2\varepsilon_y) \cap A = \varnothing$.

If $G = \bigcup_{x \in A} S_x(\varepsilon_x)$ and $H = \bigcup_{y \in B} S_y(\varepsilon_y)$, then G and H are open disjoint sets containing A and B, respectively.

1.80. A metric space is separable if and only if it is second countable if and only if it is Lindelöf.

To see this, we need to observe only the following, since "second countable" implies easily "separable" and "Lindelöf."

(a) If (x_n) is a countable dense set, then the family of all sets of the form $S_{x_n}(1/k)$, where k is a positive integer, is a base for the topology.

(b) If the space is Lindelöf, then, for each positive integer k, the open cover $\{S_x(1/k): x \in X\}$ has a countable subcover, $\{S_{x_{n,k}}(1/k): x_{n,k} \in X\}$. Then the family $\{S_{x_{n,k}}(1/k): n \text{ and } k \text{ are positive integers}\}$ is a base for the topology.

Definition 1.37. A metric space (X, d) is called *bounded* if there exists a real number k such that $X = S_x(k)$ for each $x \in X$. It is called *totally bounded* if, given $\varepsilon > 0$, there is a finite set $A_\varepsilon \subset X$ such that $X = \bigcup_{x \in A_\varepsilon} S_x(\varepsilon)$. ∎

Remarks

1.81. A totally bounded metric space is separable. The reason is that the set $A = \bigcup_{n=1}^{\infty} A_{1/n}$ ($A_{1/n}$ is as in Definition 1.37) is a countable dense set.

1.82. Every countably compact metric space is totally bounded. For, if X is countably compact and not totally bounded, then there is $\varepsilon > 0$ and a sequence (x_n) in X such that for each n, $d(x_n, x_i) \geq \varepsilon$, $1 \leq i \leq n$; but this contradicts that (x_n) has a limit point in X.

Proposition 1.22. The following are equivalent for a metric space X:

(a) X is compact,

(b) X is countably compact,

(c) X is sequentially compact. ∎

Proof. By Remark 1.79, a metric space is first countable. Because of

Proposition 1.18, we only need to show that (b) $\Rightarrow$ (a). By Remarks 1.80–1.82 a countably compact metric space is Lindelöf and therefore, compact. ∎

Definition 1.38. A sequence (x_n) in a metric space (X, d) is called a *Cauchy sequence*, if given $\varepsilon > 0$ there is a positive integer $N(\varepsilon)$ such that for $m, n > N(\varepsilon)$, $d(x_m, x_n) < \varepsilon$. (X, d) is called *complete* if every Cauchy sequence in X converges to some point in X. ∎

Remarks

1.83. Completeness need not be preserved by homeomorphisms. For example, R and $(0, 1)$ are homeomorphic, whereas R is complete and $(0, 1)$ is not, with their usual distance metric.

1.84. The class $C_b(X)$ of bounded, real-valued continuous functions on an arbitrary topological space X is a complete metric space with the metric d_0 defined by $d_0(f, g) = \sup_{x\in X} |f(x) - g(x)|$. Completeness follows from the fact that for any Cauchy sequence (f_n) in $C_b(X)$ and for each $x \in X$, $(f_n(x))$ is a Cauchy sequence of real numbers; then if $f(x) = \lim_{n\to\infty} f_n(x)$, it follows easily that $f \in C_b(X)$ and $d_0(f_n, f) \to 0$ as $n \to \infty$.

Definition 1.39. Two metric spaces (X, d_1) and (Y, d_2) are called *isometric* if there is a mapping f from X onto Y such that $d_1(x, y) = d_2(f(x), f(y))$. Here f is called an *isometry*. (Clearly an isometry is a homeomorphism.) ∎

Proposition 1.23. Every metric space (X, d) is isometric to a dense subset of a complete metric space. ∎

Proof. Let $z \in X$. For each $x \in X$, let us define the function f_x by $f_x(t) = d(t, x) - d(t, z)$. Clearly $|f_x(t)| \leq d(x, z)$ and $f_x \in C(X)$. We will now consider the "sup" metric d_0 in $C(X)$ [as in Remark 1.84]. Then $d_0(f_x, f_y) = \sup_{t\in X} |d(t, x) - d(t, y)|$. By taking $t = x$ here, it follows that $d_0(f_x, f_y) \geq d(x, y)$. If $d_0(f_x, f_y) > d(x, y)$, then for some $t \in X$ we have either

$$d(t, x) - d(t, y) > d(x, y)$$

or

$$d(t, x) - d(t, y) < - d(x, y),$$

both of which contradict the triangular inequality property of d. Therefore $d_0(f_x, f_y) = d(x, y)$. Since the closure of $\{f_x: x \in X\}$, being a closed subset of $C(X)$, is a complete metric space, $f: x \to f_x$ is the desired isometry. ∎

Theorem 1.10. A metric space is compact if and only if it is both complete and totally bounded. ▮

Proof. The "only if" part follows by Remark 1.82 and the fact that a Cauchy sequence always converges to its limit point.

For the "if" part, suppose (X, d) is complete and totally bounded. Let (x_n) be an infinite sequence. By the total boundedness property, there exists y_1 in X such that $S_{y_1}(1)$ contains infinitely many terms of (x_n). For the same reason, there exists y_2 in X such that $S_{y_1}(1) \cap S_{y_2}(\frac{1}{2})$ contains infinitely many terms of the sequence (x_n). In this way, we can get a sequence of open sets $S_{y_n}(1/n)$ such that for every positive integer k, the set $B_k = \bigcap_{i=1}^{k} S_{y_i}(1/i)$ contains infinitely many terms of (x_n). Then we choose a subsequence (x_{n_k}) such that $x_{n_k} \in B_k$ for each k. The subsequence (x_{n_k}) is clearly Cauchy and therefore convergent. The rest follows by Proposition 1.22. ▮

Theorem 1.11. *The Baire-Category Theorem.* In a complete metric space X, $\bigcap_{n=1}^{\infty} O_n$, where each O_n is open and dense, is dense in X. Hence a complete metric space is of the second category. ▮

Proof. Suffice it to show that $O \cap (\bigcap_{n=1}^{\infty} O_n)$ is nonempty for any nonempty open O in X. Since O_1 is dense and open, there exist $x_1 \in O \cap O_1$ and $\varepsilon_1 > O$ such that $\overline{S_{x_1}(\varepsilon_1)} \subset O \cap O_1$. Similarly, we can find $x_2 \in S_{x_1}(\varepsilon_1) \cap O_2$ and $0 < \varepsilon_2 < \frac{1}{2}$ such that $\overline{S_{x_2}(\varepsilon_2)} \subset S_{x_1}(\varepsilon_1) \cap O_2$. By induction, we can find sequences $(x_n) \in X$ and real numbers (ε_n) such that $0 < \varepsilon_n < 1/2^{n-1}$ and $\overline{S_{x_{n+1}}(\varepsilon_{n+1})} \subset S_{x_n}(\varepsilon_n) \cap O_n$. Since $\varepsilon_n \to 0$, the sequence (x_n) is Cauchy and has a limit $x \in X$. Then $x \in \bigcap_{n=1}^{\infty} \overline{S_{x_n}(\varepsilon_n)}$, for otherwise there exists some positive integer N such that $x \notin \overline{S_{x_n}(\varepsilon_n)}$ for all $n > N$, which means that $d(x, x_n) > \varepsilon_N$ for all $n > N$, which is a contradiction. Clearly $x \in O \cap (\bigcap_{n=1}^{\infty} O_n)$. ▮

Remarks

1.85. If a complete metric space is the union of a countable number of closed sets, then at least one of them is *not* nowhere dense and therefore must contain a nonempty open set.

1.86. *A Principle of Uniform Boundedness.* Let $\mathscr{F}$ be a family of real-valued continuous functions on a complete metric space X. Suppose that for each $x \in X$ there is a real number $k(x)$ such that $|f(x)| \leq k(x)$ for all $f \in \mathscr{F}$. Then there is a nonempty open set O and a constant K such that $|f(x)| \leq K$ for all $f \in \mathscr{F}$ and $x \in O$.

To see the validity of the principle, we observe that $X = \bigcup_{m=1}^{\infty} E_m$, where $E_m = \{x\colon |f(x)| \leq m \text{ for all } f \in \mathscr{F}\}$. Since E_m is closed for each m, the principle follows by Remark 1.85 above.

In the context of metric spaces, we now introduce one last important topological concept, that of metrizability. A topological space $(X, \mathscr{C})$ is called *metrizable* if there exists a metric d for X that induces $\mathscr{C}$. A natural question is: When is a topological space metrizable? The metrization problem was solved by P. Urysohn in 1924 for second countable topological spaces. He proved the following theorem.

Theorem 1.12. *The Urysohn Metrization Theorem.* Every second countable T_3 space is metrizable. ▮

For the proofs of this and the next two theorems, we refer the reader to [29].

R. H. Bing (1951), J. Nagata (1950), and Y. M. Smirnov (1951) all independently solved the general metrization problem. To state their results, we introduce the following concept: A family $\mathscr{A}$ of subsets of a topological space X is called *discrete* (or *locally finite*, respectively) if every point of X has an open set containing it that intersects *at most one* member (or only finitely many members, respectively) of $\mathscr{A}$. A family $\mathscr{A}$ is called σ *discrete* (or σ *locally finite*, respectively) if it is a countable union of discrete (locally finite) subfamilies.

Theorem 1.13. *The Nagata–Smirnov Metrization Theorem.* A topological space is metrizable if and only if it is a T_3 space with a σ locally finite base. ▮

Theorem 1.14. *The Bing Metrization Theorem.* A topological space is metrizable if and only if it is a T_3 space with a σ discrete base. ▮

Now that we are familiar with the basic theorems in metric spaces, we can study some of the fundamental and interesting fixed point theorems in such spaces. First, we need a definition. In what follows, d will always stand for the metric in any metric space.

Definition 1.40. Let X and Y be metric spaces and T be a mapping from X into Y. Then T is said to be *Lipschitz* if there exists a real number M such that for all x, y in X, we have

$$d(Tx, Ty) \leq M d(x, y).$$

T is said to be *nonexpansive* if $M = 1$ and a *contraction* if $M < 1$. We call T *contractive* if for all x, y, in X and $x \neq y$, we have

$$d(Tx, Ty) < d(x, y).$$ ▮

Observe that contraction ⇒ contractive ⇒ nonexpansive ⇒ Lipschitz, and a mapping satisfying any of these conditions is continuous.

Examples

1.31. *A Contractive T That Is Not a Contraction.* Let $X = [1, \infty)$ and $Tx = x + 1/x$. Then for x, y in X and $x \neq y$, we have

$$d(Tx, Ty) = |x - y|(xy - 1)/xy < d(x, y).$$

Since $\lim_{y \to \infty}(xy - 1)/xy = 1$, T is not a contraction. It may be noted here that T does not have a fixed point.

1.32. *A Contraction.* A simple example is $Tx = x/2$ for x in $(-\infty, \infty)$. A more interesting example is the following: Let $X = C[0, 1]$ with the usual "sup" metric. Then the mapping T from X into X defined by

$$T[f](t) = k \int_0^t f(x)\, dx, \qquad 0 \leq t \leq 1,$$

is a contraction for $0 < k < 1$.

1.33. *A Mapping That Is a Contraction Only after Iteration.* Consider the mapping T from $C[a, b]$ (with the "sup" metric and $-\infty < a < b < \infty$) into itself defined by

$$T[f](t) = \int_a^t f(x)\, dx.$$

Then we have

$$T^m[f](t) = \frac{1}{(m-1)!} \int_a^t (t - x)^{m-1} f(x)\, dx.$$

Now it is clear that for sufficiently large values of m the mapping T^m is a contraction, whereas T itself need not be a contraction if $b - a > 1$.

A very frequently used theorem in proving the existence and uniqueness of solutions of an equation is the following result of S. Banach proven in 1922.

Theorem 1.15. *The Banach Fixed Point Theorem.* Let T be a contraction from a complete metric space X into itself. Then there exists a unique x_0 in X satisfying (i) $Tx_0 = x_0$ and (ii) for any x in X, $\lim_{n\to\infty} T^n x = x_0$. ▮

The reader is encouraged to prove Theorem 1.15. This theorem can also be deduced as an easy corollary to the next theorem.

Theorem 1.16. Suppose T is a continuous mapping from a complete metric space X into itself such that there exists x_0 in X and a real number $k < 1$ satisfying the inequality

$$d(T^{n+1}x_0, T^{n+2}x_0) \leqq k \cdot d(T^n x_0, T^{n+1}x_0)$$

for all non-negative integers n. Then the sequence $T^n x_0$ converges to a fixed point u of T as n tends to infinity. ▮

Proof. First we show that the sequence $T^n x_0$ is a Cauchy sequence. For any positive integer m,

$$\begin{aligned} d(T^n x_0, T^{n+m}x_0) &\leqq \sum_{i=0}^{m-1} d(T^{n+i}x_0, T^{n+i+1}x_0) \\ &\leqq d(x_0, Tx_0) \sum_{i=0}^{m-1} k^{n+i} \\ &\leqq d(x_0, Tx_0) \frac{k^n}{(1-k)}, \end{aligned}$$

which goes to zero as $n \to \infty$. Thus the sequence $T^n x_0$ is a Cauchy sequence and has a limit u in X. Since T is continuous,

$$u = \lim_{n\to\infty} T^{n+1}x_0 = T(\lim_{n\to\infty} T^n x_0) = Tu.$$ ▮

The next theorem slightly improves on Theorem 1.15.

Theorem 1.17. If one of the iterates of a mapping T from a complete metric space into itself is a contraction, then T has a unique fixed point. ▮

Proof. If T^n is a contraction, then by Theorem 1.15 there is a unique x_0 such that $T^n x_0 = x_0$. Since $T^n(Tx_0) = T(T^n x_0) = Tx_0$, the uniqueness of the fixed point of T^n implies that $Tx_0 = x_0$. ▮

Now we present an interesting fixed point theorem of Sehgal [50] for a class of mappings that includes the contractive mappings. Note that Example

1.31 shows that additional conditions are necessary on such mappings to obtain fixed point theorems. Sehgal's theorem was proven for contractive mappings by M. Edelstein in 1962.

Theorem 1.18. Let T a continuous mapping from a metric space X into itself such that for all x, y in X with $x \neq y$,

$$d(Tx, Ty) < \max\{d(x, Tx), d(y, Ty), d(x, y)\}. \tag{1.1}$$

Suppose that, for some z in X, the sequence $T^n z$ has a cluster point u. Then the sequence $T^n z$ converges to u and u is the unique fixed point of T. ▮

Proof. If $T^n z = T^{n+1} z$ for some nonnegative integer n, then it is clear that $\lim_{n\to\infty} T^n z = u$, and by condition (1.1) u is the unique fixed point of T. Therefore, we may assume that for all nonnegative integers n, $d(T^n z, T^{n+1} z) > 0$. Now, let $V(y) = d(y, Ty)$. Then V is a continuous function on X and by condition (1.1), for all positive integers n, $V(T^n z) < V(T^{n-1} z) < V(z)$. Let $a = \lim_{n\to\infty} V(T^n z)$. Let (n_i) be a sequence of positive integers such that $T^{n_i} z$ converges to u. Then by the continuity of V and T, $V(u) = V(Tu) = a$. It follows by condition (1.1) that $u = Tu$ and $a = 0$. We now show that the sequence $T^n z$ converges to u. Given $\varepsilon > 0$, there exists a positive integer k such that

$$\max\{V(T^k z), d(T^k z, u)\} < \varepsilon.$$

Using condition (1.1), we have for all positive integers $n \geq k$,

$$\begin{aligned} d(T^n z, u) &= d(T^n z, T^n u) \\ &\leq \max\{V(T^{n-1} z), d(T^{n-1} z, u)\} \\ &\leq \max\{V(T^{n-2} z), d(T^{n-2} z, u)\} \\ &\leq \max\{V(T^k z), d(T^k z, u)\} \\ &< \varepsilon. \end{aligned}$$

Hence $\lim_{n\to\infty} T^n z = u$ and the theorem follows. ▮

Clearly, a contractive T satisfies condition (1.1). But the converse is not true. For example, let T: $[0, 5] \to [0, 5]$ be defined by $Tx = x/2$ if $x \in [0, 4]$ and $= -2x + 10$ if $x \in [4, 5]$. Since $T4 - T5 = 2$, T is not contractive. However, T satisfies condition (1.1).

We now remark that the fixed point u of a contractive mapping, even if it exists, may not be approximated by the successive iterates of the mapping at an arbitrary point in the space. For example, consider $X = \{0, \frac{3}{2}, \frac{4}{3}, \frac{5}{4},$

$\ldots\}$ with the absolute-value metric, and the mapping T with $T0 = 0$, $T(n+1)/n = (n+2)/(n+1)$ for $n = 2, 3, \ldots$ Our next theorem, due to Sam B. Nadler, Jr., shows that such a situation cannot arise in a locally compact connected metric space.

Theorem 1.19. Let T be a contractive mapping on a locally compact connected metric space X. Suppose that $Tu = u$ for some u in X. Then for each x in X, the sequence $T^n x$ converges to u as n tends to infinity. ∎

Proof. Let $C = \{x: \lim_{n\to\infty} T^n x = u\}$. Then $u \in C$. It suffices to show that the set C is both closed and open.

First, let $x_i \in C$ and $\lim_{i\to\infty} x_i = x_0$ Let $\varepsilon > 0$. Choose integers j and N such that $d(x_j, x_0) < \varepsilon/2$ and $d(T^n x_j, u) < \varepsilon/2$ for all $n \geq N$. Then for $n \geq N$,

$$d(T^n x_0, u) \leq d(T^n x_0, T^n x_j) + \varepsilon/2 \leq d(x_0, x_j) + \varepsilon/2 < \varepsilon.$$

Hence x_0 is in C and C is closed. Now to prove that C is open, let $x \in C$. Choose $r > 0$ such that $K = \{y: d(y, u) \leq r\}$ is compact. Choose a positive integer N and any point y such that $d(x, y) < r/2$ and $d(T^n x, u) < r/2$ for all $n \geq N$. Then since T is contractive,

$$d(T^n y, u) \leq d(T^n y, T^n x) + d(T^n x, u) < r/2 + r/2 = r.$$

Hence $T^n y$ belongs to K for all $n \geq N$. Since K is compact, the sequence $T^n y$ has a cluster point z. By Theorem 1.18, $z = u$ and $\lim_{n\to\infty} T^n y = u$. Hence C is open. ∎

The next theorem is a fixed point theorem due to J. Caristi under a different type of condition. It is often useful in various contexts in fixed point theory and analysis. Here we present J. P. Penot's proof of this theorem.

Theorem 1.20. Let T be a map (not necessarily continuous) from a complete metric space X into itself and g be a nonnegative lower semicontinuous function on X such that for each $x \in X$,

$$d(x, Tx) \leq g(x) - g(Tx).$$

Then T has a fixed point. ∎

Proof. We define a relation $\leq$ in X by setting $x \geq y$ if and only if

$d(x, y) \leq g(y) - g(x)$. Then $\leq$ is a partial order on X. Write

$$X(x) = \{y \in X: y \geq x\}.$$

We define inductively an increasing sequence (x_n) in X as follows: choose x_0 arbitrarily and after $x_0, x_1, \ldots, x_n$ are chosen, choose $x_{n+1} \in X(x_n)$ such that

$$g(x_{n+1}) < \inf g\big(X(x_n)\big) + (1/n).$$

Notice that for each $x \in X(x_{n+1})$, which is contained in $X(x_n)$,

$$g(x) \geq \inf g\big(X(x_n)\big) > g(x_{n+1}) - (1/n),$$

and therefore,

$$d(x, x_{n+1}) \leq g(x_{n+1}) - g(x) < 1/n,$$

so that the diameter of $X(x_{n+1})$ is less than $2/n$. Using the lower semicontinuity of g, it is easily verified that each $X(x_n)$ is closed. Since X is complete, it is clear that there is a unique element x_∞ in X such that $\bigcap_{n=0}^{\infty} X(x_n) = \{x_\infty\}$. Since by hypothesis $Tx_\infty \geq x_\infty$, $Tx_\infty \geq x_n$ for each n so that $Tx_\infty \in \bigcap_{n=0}^{\infty} X(x_n)$. Hence, $Tx_\infty = x_\infty$. ∎

Recently there have been numerous attempts to weaken the hypothesis of the Banach Fixed Point Theorem, but at the same time retaining the convergence property of the successive iterates to the unique fixed point of the mapping. We present two of these generalizations. The first one is due to Sehgal and later improved upon by Guseman and the second one is due to Boyd and Wong. The motivation of the Sehgal–Guseman result comes from the fact that there are mappings T that have at each point an iterate that is a contraction, and yet none of the iterates of T is a contraction. The following example will illustrate this.

Example 1.34. Let $X = [0, 1]$ and for every positive integer n, $I_n = [1/2^n, 1/2^{n-1}]$. Then $X = \{0\} \cup (\bigcup_{n=1}^{\infty} I_n)$. We define $T: X \to X$ as follows. Let

$$Tx = \begin{cases} \dfrac{n+2}{n+3}\left(x - \dfrac{1}{2^{n-1}}\right) + \dfrac{1}{2^n}, & \text{if } x \in \left[\dfrac{3n+5}{2^{n+1}(n+2)}, \dfrac{1}{2^{n-1}}\right], \\ \dfrac{1}{2^{n+1}}, & \text{if } x \in \left[\dfrac{1}{2^n}, \dfrac{3n+5}{2^{n+1}(n+2)}\right], \end{cases}$$

and $T0 = 0$. Then the following properties of T follow easily:

(i) T is a nondecreasing continuous function on $[0, 1]$.

(ii) For x in I_n and any y in X,

$$|Tx - Ty| \leqq \frac{n+3}{n+4} |x - y|.$$

By taking the $(n + 3)$th iterate of T, we have for x in I_n and any y in X,

$$|T^{n+3}x - T^{n+3}y| \leqq \tfrac{1}{2} |x - y|.$$

Also $|T^2 0 - T^2 y| \leqq \frac{1}{2} |0 - y|$. This means that at every point of X, T has an iterate that is a contraction. Now we show that none of the iterates of T is a contraction. Let $0 \leqq k \leqq 1$ and N be a given positive integer. Let $n > (Nk/(1 - k)) - 2$. By the uniform continuity of the iterates of T, there is a $\delta > 0$ such that for $|x - y| < \delta$ and $1 \leqq i \leqq N$, we have

$$|T^i x - T^i y| < \frac{n + N + 3}{2^{n+N+1}(n + N + 2)}.$$

Setting $x = 1/2^{n-1}$ and y any member of I_n such that $0 < |x - y| < \delta$, it can be verified that $T^i x$ and $T^i y$ are both members of

$$\left[\frac{3(n + i) + 5}{2^{n+i+1}(n + i + 2)}, \frac{1}{2^{n+i-1}}\right]$$

for $i = 1, 2, \ldots, N$. Thus we have

$$|Tx - Ty| = \frac{n+2}{n+3} |x - y|,$$

$$|T^2 x - T^2 y| = \frac{n+2}{n+4} |x - y|,$$

$$\cdots$$

$$|T^N x - T^N y| = \frac{n+2}{n+2+N} |x - y| > k |x - y|.$$

This shows that none of the iterates of T can be a contraction.

We now present the theorem due to Sehgal and Guseman for mappings having a contractive iterate at each point.

Theorem 1.21. (*Sehgal–Guseman*) Let T be a mapping from a complete metric space X into itself. Suppose there exists $B \subset X$ such that the following hold.

(i) $TB \subset B$.

(ii) There is k, $0 < k < 1$ such that for each $x \in B$, there is a positive integer $n(x)$ with

$$d(T^{n(x)}x, T^{n(x)}y) \leq kd(x, y) \qquad \text{for all } y \in B. \tag{1.2}$$

(iii) For some $z \in B$, $\overline{\{T^n z: n \geq 1\}} \subset B$.

Then there is a unique $u \in B$ such that $Tu = u$ and

$$T^n x \to u \text{ as } n \to \infty \text{ for each } x \in B. \qquad \blacksquare$$

Proof. First, we claim that for $x \in B$

$$r(x) = \sup_n \{d(T^n x, x)\} < \infty.$$

To see this, let

$$t(x) = \sup \{d(T^n x, x): 1 \leqq n \leqq n(x)\}.$$

If n is any positive integer, there is an integer $s \geq 0$ such that $s \cdot n(x) < n \leq (s + 1) \cdot n(x)$. Then

$$\begin{aligned} d(T^n x, x) &\leq d\big(T^{n(x)}(T^{n-n(x)}x), T^{n(x)}x\big) + d(T^{n(x)}x, x) \\ &\leq kd(T^{n-n(x)}x, x) + t(x) \\ &\leq t(x) + kt(x) + \cdots + k^s t(x) \\ &< t(x)/(1 - k). \end{aligned}$$

This proves our claim.

Next, let $z_1 = T^{n(z)}z$ and $z_{i+1} = T^{n(z_i)}z_i$. Then it follows by routine calculation that

$$d(z_{i+1}, z_i) \leq k^i d(T^{n(z_i)}z, z) \leq k^i r(z)$$

and

$$d(z_j, z_i) \leq \sum_{l=i}^{j-1} d(z_{l+1}, z_l) \leq \frac{k^i}{1-k} r(z) \qquad \text{for } j > i.$$

Then (z_i) is Cauchy. By (iii) and completeness of X, $z_i \to u \in B$ as $i \to \infty$. Clearly, $T^{n(u)}z_i \to T^{n(u)}u$ as $i \to \infty$.

But

$$\begin{aligned} d(T^{n(u)}z_i, z_i) &= d(T^{n(z_{i-1})}(T^{n(u)}z_{i-1}), T^{n(z_{i-1})}z_{i-1}) \\ &\leq kd(T^{n(u)}z_{i-1}, z_{i-1}) \\ &\cdots \\ &\leqq k^i\, d(T^{n(u)}z, z) \to 0 \qquad \text{as } i \to \infty. \end{aligned}$$

Hence

$$d(T^{n(u)}u, u) = \lim_{i\to\infty} d(T^{n(u)}z_i, z_i) = 0.$$

It is clear that u is the unique fixed point for $T^{n(u)}$ in B. Therefore since $Tu = T(T^{n(u)}u) = T^{n(u)}(Tu)$, $Tu = u$.

The theorem will be proved if we show $T^n x \to u$ as $n \to \infty$ for each $x \in B$. To show this, let

$$A(x) = \sup \{d(T^m x, u):\ 1 \leq m \leq n(u) - 1\}.$$

Now, if $n = an(u) + s$ with integers a and s such that $a > 0, 0 \leq s < n(u)$, then

$$\begin{aligned} d(T^n x, u) &= d(T^{an(u)+s}x, T^{n(u)}u) \\ &\leq kd(T^{(a-1)n(u)+s}x, u) \\ &\cdots \\ &\leqq k^a A(x). \end{aligned}$$

It follows that $T^n x \to u$ as $n \to \infty$. ▮

The next theorem is a slightly simplified version of a theorem of Boyd and Wong.

Theorem 1.22. Let T be a mapping from a complete metric space X into itself. Suppose there exists a right continuous mapping f from R^+, the nonnegative reals, into itself such that for all x, y in X

$$d(Tx, Ty) \leqq f(d(x, y)).$$

If $f(t) < t$ for each $t > 0$, then T has a unique fixed point u in X and for each x in X, $\lim_{n\to\infty} T^n x = u$. ▮

Proof. Let $x \in X$, $x_n = T^n x$ and $d_n \doteq d(x_n, x_{n+1})$. We may and do assume that $d_n > 0$ for $n \geq 0$. Then for $n > 1$,

$$d_n = d(Tx_{n-1}, Tx_n) \leqq f(d_{n-1}) < d_{n-1},$$

so that the sequence d_n is decreasing. Let $d = \lim_{n\to\infty} d_n$. Then $d = 0$, since the preceding inequality implies that $d \leqq f(d)$, which is less than d unless

$d = 0$. Now we show that the sequence x_n is Cauchy. If not, then there exists $\varepsilon > 0$ and for each positive integer k there exist positive integers n_k and m_k with $k \leqq m_k < n_k$ such that $d(x_{m_k}, x_{n_k}) \geqq \varepsilon$. With no loss of generality, we can also assume that n_k is the smallest integer greater than m_k satisfying the above inequality. Let $t_k = d(x_{m_k}, x_{n_k})$. Then

$$\varepsilon \leqq t_k \leqq d(x_{m_k}, x_{n_k-1}) + d(x_{n_k-1}, x_{n_k}) \leqq \varepsilon + d_{n_k-1},$$

and therefore $\varepsilon \leqq t_k$ and $\lim_{k\to\infty} t_k = \varepsilon$. We have also

$$\begin{aligned} \varepsilon \leqq t_k &\leqq d(x_{m_k}, x_{m_k+1}) + d(x_{m_k+1}, x_{n_k+1}) + d(x_{n_k+1}, x_{n_k}) \\ &\leqq d_{m_k} + f(t_k) + d_{n_k} \to f(\varepsilon) \qquad \text{as } k \to \infty. \end{aligned}$$

This is a contradiction since $f(\varepsilon) < \varepsilon$. Therefore, the sequence x_n is Cauchy with some limit u in X. Clearly by the continuity of T, $Tu = u$. The uniqueness of the fixed point follows immediately from the definition of f. ▮

In Part B, we shall present a common fixed point theorem (due to Kakutani) for commuting linear mappings. But for two commuting mappings on [0, 1], a simple theorem of this type can be very easily stated and proved.

Theorem 1.23. Let f and g both map [0, 1] into itself such that

(i) $f(g(x)) = g(f(x))$ for each x in [0, 1];
(ii) f is nonexpansive;
(iii) g is continuous.

Then there is a common fixed point for both f and g. ▮

Proof. By Theorem 1.8 the set A of fixed points for f is nonempty. By condition (i), $g(A) \subset A$. Since f is continuous, A is closed. It is sufficient to show that A is an interval since then g has a fixed point in A by Theorem 1.8.

Let $m = \inf A$ and $M = \sup A$. Let $x \in [m, M]$. Then

$$f(x) - m \leqq | f(x) - f(m) | \leqq x - m,$$

which means that $f(x) \leqq x$. Similarly, $f(x) \geqq x$. This means that x is a fixed point of f and therefore, $x \in A$. Hence $A = [m, M]$. The theorem now follows. ▮

Remark 1.87. Theorem 1.23 was proved by R. DeMarr[†] who also proved the following theorem.

[†] R. DeMarr, *Amer. Math. Monthly*, May (1963).

Theorem 1.24. (*DeMarr*). Let f and g be two commuting mappings from $[0, 1]$ into itself such that for all $x, y \in [0, 1]$

(i) $|f(x) - f(y)| \leq \alpha |x - y|$,

(ii) $|g(x) - g(y)| \leq \beta |x - y|$, where $\beta < (\alpha + 1)/(\alpha - 1)$.

Then there exists a common fixed point for both f and g. ∎

Our last fixed point theorem is another interesting extension of the Banach Fixed Point Theorem. It involves mappings that are contractions in a different sense. To clarify this, let us consider the following variants of the contraction condition. (In what follows, T is a mapping from a complete metric space X into itself.)

Condition (K). There is a constant $c < \frac{1}{2}$ such that for all x, y in X

$$d(Tx, Ty) \leqq c \cdot [d(x, Tx) + d(y, Ty)].$$

Condition (K_1). There is a constant $c < \frac{1}{2}$ such that for all x, y in X

$$d(Tx, Ty) \leqq c \cdot [d(x, Ty) + d(y, Tx)].$$

Condition (K_2). There exist nonnegative constants a, b, and c such that $2a + 2b + c < 1$ and for all x, y in X

$$d(Tx, Ty) \leqq a \cdot [d(x, Tx) + d(y, Ty)] + b \cdot [d(x, Ty) + d(y, Tx)] + cd(x, y).$$

Condition (K) was first studied by R. Kannan in 1968. Since then, the other two conditions have been studied independently by various mathematicians. It is clear that the first two conditions imply the third. But none of these conditions, though seemingly contractions of some kind, are contractions. They need not even imply continuity. The following examples clarify this and their relationships to one another.

Examples.† Mappings T satisfying the conditions (K), (K_1), *or* (K_2).

1.35. An example of a discontinuous T satisfying condition (K), but not condition (K_1), is the following: Define $Tx = -x/2$ for x in $(-1, 1)$ and $T1 = T(-1) = 0$.

1.36. An example of a discontinuous T satisfying condition (K_1), but not condition (K) is the following: Define $Tx = x/2$ if $x \in [0, 1)$ and $T1 = 0$. Then T: $[0, 1] \to [0, 1]$.

† Some easy computations are needed to verify these examples. These are left to the reader.

1.37. The mapping $Tx = -x/2$, $x \in [-1, 1]$ is the example of a contraction that does not satisfy condition (K_1).

1.38. The mapping $Tx = x/2$, $x \in [0, 1]$ is a contraction that does not satisfy condition (K).

1.39. An example of a continuous T satisfying condition (K) but neither (K_1) nor the contraction condition is the following: $Tx = -x/2$ if $x \in [-4, 4]$, $= 2x - 10$ if $x \in [4, 5]$, $= 2x + 10$ if $x \in [-5, -4]$. One can verify that T satisfies (K) with c in $(\frac{2}{5}, \frac{1}{2})$.

1.40. An example of a discontinuous T satisfying condition (K_2) but neither condition (K) nor condition (K_1) is the following: $Tx = x/2$ if $x \in [0, 1)$, $= -\frac{1}{8}$ if $x = 1$, $= -x/2$ if $x \in [-2, 0)$.

Now we present our last fixed point theorem in this section.

Theorem 1.25. Let T be a mapping from a complete metric space X into itself satisfying condition (K_2). Then for any x in X, $\lim T^n x$ exists as n tends to infinity and this limit is the unique fixed point of T. ∎

Proof. For all x, y in X, T satisfies the inequality

$$d(Tx, Ty) \leqq a[d(x, Tx) + d(y, Ty)] + b[d(x, Ty) + d(y, Tx)] + cd(x, y), \quad (1.3)$$

where $2a + 2b + c < 1$.

For each positive integer n and x in X, we replace x by $T^{n-1}x$ and y by $T^n x$ in (1.3). Then using the triangle inequality, we have

$$\begin{aligned} d(T^n x, T^{n+1}x) &\leqq a[d(T^{n-1}x, T^n x) + d(T^n x, T^{n+1}x)] + b[d(T^{n-1}x, T^{n+1}x) \\ &\quad + d(T^n x, T^n x)] + cd(T^{n-1}x, T^n x) \\ &\leqq a[d(T^{n-1}x, T^n x) + d(T^n x, T^{n+1}x)] + b[d(T^{n-1}x, T^n x) \\ &\quad + d(T^n x, T^{n+1}x)] + cd(T^{n-1}x, T^n x). \end{aligned}$$

Simplifying, we have

$$d(T^n x, T^{n+1}x) \leqq kd(T^{n-1}x, T^n x),$$

where $k = (a + b + c)/(1 - a - b)$, which is less than 1. Repeating the above process n times, we have

$$d(T^n x, T^{n+1}x) \leqq k^n d(x, Tx)$$

and therefore, for each positive integer m we can write

$$d(T^n x, T^{n+m} x) \leqq \sum_{i=0}^{m-1} d(T^{n+i} x, T^{n+i+1} x)$$
$$\leqq \sum_{i=0}^{m-1} k^{n+i} d(x, Tx).$$

Thus the sequence $(T^n x)$ is a Cauchy sequence, and since X is complete, the sequence has a limit ξ.

To prove that $T\xi = \xi$, we use condition (1.3) and find that

$$\begin{aligned} d(\xi, T\xi) &\leqq d(\xi, T^n x) + d(T^n x, T\xi) \\ &\leqq d(\xi, T^n x) + a[d(T^{n-1} x, T^n x) + d(\xi, T\xi)] \\ &\quad + b[d(T^{n-1} x, T\xi) + d(\xi, T^n x)] + cd(T^{n-1} x, \xi). \end{aligned} \tag{1.4}$$

Noting that $d(T^{n-1}x, T\xi) \leqq d(T^{n-1}x, \xi) + d(\xi, T\xi)$, we see from condition (1.4) that $(1 - a - b)d(\xi, T\xi)$ can be made as small as we please. Hence $\xi = T\xi$. To prove the uniqueness of the fixed point, let $T\xi = \xi$ and $T\eta = \eta$. Then from condition (1.3), writing $x = \xi$ and $y = \eta$, we have

$$d(\xi, \eta) = d(T\xi, T\eta) \leqq (2b + c)d(\xi, \eta).$$

Since $2b + c$ is less than 1, $\xi = \eta$. ∎

Theorem 1.25 can be further extended, as the reader can verify in Problem 1.5.45. Though condition (K_2) is quite different from the contraction condition, as Examples 1.35–1.40 demonstrated, it can be shown that in a compact metric space, a continuous mapping T satisfying this condition is a contraction with respect to an equivalent metric, i.e., a metric generating the same topology as the original metric. The reader can verify this in Problem 1.5.43.

Before we close this section, we present two applications of the Banach Fixed Point Theorem. First, we prove one form of the classical implicit function theorem.

Theorem 1.26. Let f be a continuous real-valued function defined on $|x| \leq a$, $|y| \leq b$ such that

(i) $f(0, 0) = 0$,

(ii) $|f(x_1, y_1) - f(x_2, y_2)| \leq k(|x_1 - x_2| + |y_1 - y_2|)$,

where k is a fixed number in (0, 1). Then the equation $y = cx + f(x, y)$

has a unique solution $y = h(x)$ with $h(0) = 0$ and h defined in

$$|x| \leq s < \min\left\{a, \frac{1-k}{|c|+k} \cdot b\right\}.$$

Furthermore,

$$|h(x_2) - h(x_1)| \leq \frac{|c|+k}{1-k} |x_2 - x_1|. \qquad \blacksquare$$

Proof. Let X be the family of all continuous functions g defined on $[-s, s]$ (s defined as in the theorem) such that $g(0) = 0$ and $|g(x)| \leqq b$. With the usual "sup" metric d, X is a complete metric space.

We define $T: X \to X$ by

$$T(g)(x) = cx + f(x, g(x)).$$

Then T is well defined since

(i) $T(g)(0) = f(0, 0) = 0$,

(ii) $T(g)$ is continuous,

(iii)
$$\begin{aligned} |T(g)(x)| &\leqq |cx| + |f(x, g(x))| \\ &\leq |cs| + k(s+b) \\ &< b, \qquad \text{by the definition of } s. \end{aligned}$$

Moreover,

$$\begin{aligned} |T(g_1)(x) - T(g_2)(x)| &= |f(x, g_1(x)) - f(x, g_2(x))| \\ &\leq k\,|g_1(x) - g_2(x)|. \end{aligned}$$

This means that $d(T(g_1), T(g_2)) \leq kd(g_1, g_2)$. By the Banach Fixed Point Theorem, T has a unique fixed point $h(x) \in X$. The rest of the proof is now clear. $\blacksquare$

Our next application of the Banach Fixed Point Theorem deals with the question of existence and uniqueness of solutions to ordinary differential equations. Let us consider the initial-value problem

$$x'(t) = f(t, x(t)), \qquad x(0) = x_0, \tag{1.5}$$

where $x'(t)$ denotes the derivative of $x(t)$. We are interested in the existence of a function $x(t)$ satisfying equation (1.5) on some closed interval $[-\sigma, \sigma]$, $\sigma > 0$.

Theorem 1.27. *The Picard Existence Theorem.* Let $f(t, s)$ be a continuous function from $D = [-a, a] \times [-b, b]$ into R. Suppose $|f(t, s)| \leq M$ in D and

$$|f(t, s_1) - f(t, s_2)| \leq K |s_1 - s_2|$$

for some positive number K, $t \in [-a, a]$ and $s_1, s_2 \in [-b, b]$. Then equation (1.5) with $x_0 = 0$ has a unique solution defined for $|t| \leq r < \min(a, b/M, 1/K)$. ∎

Proof. Let $X = \{g \in C[-r, r]: g(0) = 0 \text{ and } |g(t)| \leq b \text{ if } -r \leq t \leq r\}$. Then X is a complete metric space with the usual "sup" metric d. Clearly, the integral

$$T(g)(s) = \int_0^s f(t, g(t))\, dt$$

is well defined for $g \in X$ and is a continuous function of s for $s \in [-r, r]$. Also, by the choice of r, $|T(g)(s)| \leq M|s| \leq Mr < b$, and therefore $T(g) \in X$. Now

$$\begin{aligned} |T(g_1)(s) - T(g_2)(s)| &\leq \int_0^s |f(t, g_1(t)) - f(t, g_2(t))|\, dt \\ &\leq K \int_0^s |g_1(t) - g_2(t)|\, dt, \end{aligned}$$

so that

$$d(T(g_1), T(g_2)) \leq Krd(g_1, g_2).$$

Since $Kr < 1$, T is a contraction, and therefore it has a unique fixed point $x(t) \in X$ so that

$$x(s) = \int_0^s f(t, x(t))\, dt.$$

Clearly $x(0) = 0$, and by the continuity of the integrand, $x'(s)$ exists and

$$x'(s) = f(s, x(s)), \qquad s \in (-r, r).$$ ∎

Problems

1.5.1. (a) Show that the sum of two real-valued functions on the reals need not have the intermediate-value property, though each one has this property. [Hint: Consider $[f'(x)]^2 + [g'(x)]^2$, where $f(x) = x^2 \sin(1/x)$ for

$x \neq 0$ and $f(x) = 0$ for $x = 0$, and $g(x) = x^2 \cos(1/x)$ for $x \neq 0$, $g(x) = 0$ for $x = 0$.] Is this true if one of the functions is also continuous?

(b) *The Order Topology*. Let $(X, <)$ be a totally ordered set. Then the class of all sets of the form $\{x: x < a\}$ or $\{x: a < x\}$ for some $a \in X$ forms a subbase for a topology, called *the order topology* of X. Consider X with this topology and prove the following assertions:

(i) Given $a < b$ in X, there are open sets U and V such that $a \in U$, $b \in V$, and $x < y$ whenever $x \in U$ and $y \in V$;

(ii) X is connected if and only if every continuous real function on X has the intermediate value property if and only if X has the following two properties: (1) X is order complete (2) given $a < b$, there is x in X such that $a < x < b$.

1.5.2. (*Rottmann*). Suppose $d(x, y) = 0$ for $x = y$ and $= \max\{|x|, |y|\}$ for $x \neq y$. Show that d defines a metric for the real R and that (R, d) is a disconnected but complete metric space. Show also that the mapping $x \to -x$ is continuous, whereas the mapping $(x, y) \to x + y$ from $R \times R$ into R is not continuous.

1.5.3. Let X be a topological space. Suppose every point of X has a path-connected open set containing it. Show that every path-connected component (i.e., maximal path-connected set) is then open and therefore closed. Use this to show that any open set in R^n is an at most countable disjoint union of open connected sets.

1.5.4. Use (1.5.3) to show that X is path connected if and only if X is connected and every point of X has a path-connected open set containing it. Use this to show that an open set in R^2 is connected if and only if it is path connected.

1.5.5. Let X be a completely regular, Hausdorff, and connected topological space consisting of more than one point. Then show that every nonempty open subset of X is uncountable.

1.5.6. Show that the product of two locally connected topological spaces is locally connected.

1.5.7. Prove the following result (due to S. S. Mitra): If there is a continuous open map from $I = [0, 1]$ (with usual topology) *onto* a Hausdorff space X containing at least two elements, then X and I are homeomorphic. [Hint: Let a be the smallest number $0 < a \leq 1$ such that $f([0, a]) = X$; show that $f|_{[0,a]}$ is one-to-one.]

1.5.8. Let f be a real-valued, strictly increasing function defined on a set D of real numbers. Show that f^{-1} is a continuous, strictly increasing function if D is an open, closed, or connected set. Show also that any real-valued continuous, one-to-one function defined on $[a, b]$ is strictly monotonic.

1.5.9. Let $f: R \to R$ be such that its graph $\{(x, f(x)): x \in R\}$ is a closed and connected subset of R^2. Prove that f is continuous.

★ **1.5.10.** Prove that a metric space is connected if and only if every nonempty proper subset has a nonempty boundary.

★ **1.5.11.** Prove that a metric space X is complete if there is a positive number ε such that for each $x \in X$, the set $\{y: d(x, y) \leq \varepsilon\}$ is compact.

1.5.12. Let D be the set of all nondecreasing, left-continuous functions f defined on R with $0 \leq f \leq 1$. Suppose for $f, g \in D$, $d(f, g) = \inf\{\delta: f(x - \delta) - \delta \leq g(x) \leq f(x + \delta) + \delta$ for every $x\}$. Show that

(i) d is a metric on D;

(ii) $d(f_n, f) \to 0$ as $n \to \infty$ if and only if $f_n(x) \to f(x)$ as $n \to \infty$ whenever x is a point of continuity of f.

[This metric (useful in probability theory) is called the Lèvy metric, named after the great French probabilist Paul Lèvy.]

1.5.13. (a) Let (X, d) be a metric space. If $d_1(x, y) = d(x, y)/[1 + d(x, y)]$, then show that (X, d_1) is a bounded metric space with the same topology as induced by d. Use this to prove the last assertion in Remark 1.77.

(b) Prove that R^∞ is not locally compact, whereas the subset $Q^\infty = \{(x_1, x_2, \ldots): |x_i| \leq 1/i$ for each $i\}$ is compact as a subspace of R^∞.

(c) Prove that any separable metric space (X, d) can be mapped homeomorphically onto a subset of Q^∞. [Hint: Assume $d(x, y) \leq 1$ for every x, y in X. Let (x_i) be a countable dense subset of X. Let $F: X \to Q^\infty$ be defined by $(F(x))_i = (1/i)d(x, x_i)$.]

1.5.14. Suppose X and Y are any two metric spaces and the metric in Y is bounded. Let $\mathscr{F}$ be the set of all maps from X into Y. Define for $f, g \in \mathscr{F}$, $d(f, g) = \sup_{x \in X} d(f(x), g(x))$. Show that this d defines a metric in $\mathscr{F}$ that is complete if Y is complete.

1.5.15. (a) *Lebesgue's Covering Lemma.* Let X be a compact metric space and (U_t) be an open covering for X. Then there exists a positive number β, called *the Lebesgue number* of the covering, such that

$$d(x, y) < \beta \Rightarrow \{x, y\} \subset U_t \text{ for some } t.$$

[Hint: For $x \in X$, there exists $\beta(x) > 0$ such that $d(x, y) < 2\beta(x)$ implies that $\{x, y\} \subset U_t$ for some t. There exist $x_1, \ldots, x_n$ such that $X = \bigcup_{i=1}^n \{x: d(x, x_i) < \beta(x_i)\}$. Let $\beta = \min\{\beta(x_i): 1 \leq i \leq n\}$.]

(b) Use (a) to prove that every continuous map f of a compact metric space into another metric space is *uniformly continuous*, i.e., given $\varepsilon > 0$, there exists $\beta > 0$ such that $d(x, y) < \beta \Rightarrow d(f(x), f(y)) < \varepsilon$.

1.5.16. Let X and Y be metric spaces and $f: X \to Y$ be a homeomorphism. Show that if f is uniformly continuous (see Problem 1.5.15) and Y is complete, then X is complete; but completeness of Y need not be implied by that of X and the uniform continuity of f.

1.5.17. Let E be a dense subset of a metric space X and f be a uniformly continuous mapping of X into a complete metric space Y. Show that there exists a unique uniformly continuous mapping $g: X \to Y$ such that for $x \in E$, $g(x) = f(x)$.

× **1.5.18.** *Dini's Theorem.* Let X be a compact metric space and $f_n \in C(X)$ such that $f_n(x) \leq f_{n+1}(x)$ and $f(x) = \lim_{n\to\infty} f_n(x)$, for each $x \in X$. Then $f_n(x) \to f(x)$ uniformly as $n \to \infty$, if $f \in C(X)$.

★ **1.5.19.** *Limit Points of Sequences and Permutations.* Let (X, d) be a compact metric space and (x_n), (y_n) two sequences. If $d(x_n, y_n) \to 0$, then these sequences have the same set of limit points. Prove the converse (due to Von Neumann): If the sequences have the same set of limit points, then there exists a permutation π of the natural numbers such that $d(x_n, y_{\pi(n)}) \to 0$ as $n \to \infty$. [The following hint is due to P. R. Halmos: Let $\varepsilon_n = 1/n + d(x_n, C)$, where C is the common set of limit points. Then $S_{x_n}(\varepsilon_n)$ contains infinitely many y_m. Let $\sigma(1) > 1$ be the smallest positive integer with $y_{\sigma(1)} \in S_{x_1}(\varepsilon_1)$. Inductively, $\sigma(n+1)$ is the smallest positive integer $> \sigma(n)$ and $y_{\sigma(n+1)} \in S_{x_{n+1}}(\varepsilon_{n+1})$. Then $d(x_n, y_{\sigma(n)}) \to 0$ as $n \to \infty$. Similarly, there is a one-to-one map τ of the naturals into itself such that $d(x_{\tau(n)}, y_n) \to 0$ as $n \to \infty$. Now by the Schroeder–Bernstein theorem (see [5], pp. 88, 89; for a simple proof, see the hint for Problem 1.1.10), there is a permutation π of the naturals such that for each n, $\pi(n) = \sigma(n)$ or $\tau^{-1}(n)$. Then $d(x_n, y_{\pi(n)}) \to 0$ as $n \to \infty$.]

× **1.5.20.** *The Cantor Set and the Cantor Function.* (a) The *Cantor set* C is a subset of $[0, 1]$ obtained by first removing the middle third $(1/3, 2/3)$ from $[0, 1]$, then removing the middle thirds $(1/9, 2/9)$ and $(7/9, 8/9)$ of the remaining intervals, and continuing indefinitely. Prove the following assertions:

(i) C is compact, nowhere dense, and has no isolated points.

(ii) $C = \{\sum_{n=1}^{\infty} x_n/3^n : x_n = 0 \text{ or } 2 \text{ for each } n\}$ and card $C = c$.

(b) Define a function f on $[0, 1]$ as follows: $f(1/3, 2/3) = 1/2$, $f(1/9, 2/9) = 1/4$, $f(7/9, 8/9) = 3/4$, $f(1/27, 2/27) = 1/8$, $f(7/27, 8/27) = 3/8$, $f(19/27, 20/27) = 5/8$, $f(25/27, 26/27) = 7/8$, and so on; and for $x \in C$, let $f(x) = \sup\{f(y): y \in [0, 1] - C, y < x\}$. Show that

(i) f is a continuous nondecreasing function from $[0, 1]$ *onto* $[0, 1]$;

(ii) for $x \in C$ and $x = \sum_{n=1}^{\infty} 2x_n/3^n$, where each $x_n = 0$ or 1, $f(x) = \sum_{n=1}^{\infty} x_n/2^n$.

This f is called the *Cantor function.*

(c) Obtain a Cantor set, as in (a), by using the factor β, $0 < \beta < 1$, instead of the factor $1/3$. Verify its different topological properties. Show that if C is the middle-two-thirds Cantor set, then $x \in C$ and $y \in C$ imply that $\frac{1}{2}(x + y) \notin C$. Finally show that all the Cantor sets obtained in this manner are homeomorphic to each other by showing that each is homeomorphic to $\times_{i=1}^{\infty} X_i$, where each $X_i = X$ is a two-point discrete space.

1.5.21. *Convex Sets.* A subset $A \subset R^2$ is called *convex* if for *all* t in $(0, 1)$, A has the condition (c): $x, y \in A \Rightarrow tx + (1 - t)y \in A$. Prove that if A is closed and if the condition (c) holds for only a single fixed t in $(0, 1)$, then A is convex. What can you say if the set A is open? (These results also hold in much more general topological spaces with a vector space structure.)

1.5.22. *Convex Functions.* A real-valued function f defined on (a, b) is said to be *convex* if for all $t \in (0, 1)$, f satisfies the condition (c'): $f\big(tx + (1 - t)y\big) \leq tf(x) + (1 - t)f(y)$ for all $x, y \in (a, b)$. Prove the following assertions.

(i) If f is convex then f is continuous.

(ii) Suppose f is continuous and the condition (c′) holds for only a single fixed t in $(0, 1)$, then f is convex.

(iii) Suppose f is convex on (a, b). Then f has a left-hand derivative $D^-f(x)$ as well as a right-hand derivative $D^+f(x)$ at each point x of (a, b) and $D^-f(x) \leq D^+f(x)$. Furthermore, on any $[c, d] \subset (a, b)$, f satisfies the Lipschitz condition. [Hint: For $x < y < z$, note that

$$\frac{f(y) - f(x)}{y - x} \leq \frac{f(z) - f(x)}{z - x} \leq \frac{f(z) - f(y)}{z - y};$$

to see this, write

$$y = \frac{z - y}{z - x}\, x + \frac{y - x}{z - x}\, z.$$

Then for $c < x < y < d$, we easily have

$$\frac{f(x) - f(c)}{x - c} \leq \frac{f(y) - f(x)}{y - x} \leq \frac{f(d) - f(y)}{d - y}.$$

Now show that $|\,[f(y) - f(x)]/(y - x)\,| \leq \max\{|\,D^-f(c)\,|, |\,D^+f(d)\,|\}$.]

(iv) If f has a nonnegative second derivative in (a, b), then f is convex. The converse is also true if f is twice differentiable.

1.5.23. Find a complete metric for $X = \{0\} \cup [1, \infty)$ such that $[1, \infty)$ does not contain a closest point to 0. [Hint: Define for $x, y \in [1, \infty)$, $d(x, y) = |x - y|/[1 + |x - y|]$ and $d(0, x) = [2 + |x|]/[1 + |x|]$. Then d is the desired metric.]

1.5.24. Let X be a compact metric space and $T: X \to X$ such that $d(x, y) = d(Tx, Ty)$ for all $x, y \in X$. Show that T is *onto*.

1.5.25. *A Fixed-Set Theorem.* Let X be a compact Hausdorff space and $T: X \to X$ be a continuous map. Show that there exists a nonempty closed set $A \subset X$ such that $T(A) = A$.

1.5.26. *The Topology of Pointwise Convergence.* Let $\mathscr{F}(X, Y)$ be the family of all maps from a topological space X into a topological space Y. Then $\mathscr{F}(X, Y) = \times_{x \in X} Y_x$, where $Y_x = Y$ for each x. The product topology from this product is called the topology of pointwise convergence. Show that $f_n \to f$ as $n \to \infty$ in $\mathscr{F}(X, Y)$ with this topology if and only if for each x in X, $f_n(x) \to f(x)$ as $n \to \infty$ in Y. Also show that $\mathscr{F}(X, Y)$ with the topology of pointwise convergence is compact, T_2, T_3, or connected if and only if Y has the same property.

1.5.27. *The Quotient Topology.* The largest topology for a set Y that would make a given mapping f from a topological space X onto Y continuous is called the *quotient topology* for Y (relative to f). Show the following.

(i) If f is a continuous open map from a topological space X onto a topological space Y, then the topology for Y is its quotient topology (relative to f).

(ii) If X is compact, separable, connected, or locally connected, then so is Y with the quotient topology.

1.5.28. Suppose there is an equivalence relation r on a topological space X, X/r is the set of all equivalence classes, and $\psi(x) = [x]$, the equivalence class containing x. X/r with the quotient topology (relative to ψ) is called the *quotient space*. Show the following.

(i) X/r is T_1 if and only if the equivalence classes in X are closed.

(ii) A set A in X/r is open if and only if the union of all the equivalence classes in A is open in X.

1.5.29. *A Pseudometric Space.* A mapping $d: X \times X \to R^+$ (the nonnegative reals) is called a *pseudometric* for X if for all x, y and $z \in X$, d satisfies (i) $d(x, x) = 0$, (ii) $d(x, y) = d(y, x)$, and (iii) $d(x, y) + d(y, z) \geq d(x, z)$. A pseudometric d induces a natural topology for X, by considering as a base the family of all sets of the form $\{y: d(y, x) < k\}$; and X, with this topology, is called a *pseudometric space*. Show that in a pseudometric space (X, d)

(i) the relation xry if and only if $d(x, y) = 0$ is an equivalence relation; and

(ii) the quotient space X/r is metricizable with metric d_0 well defined by $d_0([x], [y]) = d(x, y)$. (See 1.5.28.) [It is relevant to mention here that *pseudometric spaces are paracompact*. For a proof of this nontrivial fact, the reader should consult [29], p. 160.]

1.5.30. Prove that a contractive mapping T of a complete metric space X into a *compact* subset of itself has a fixed point x_0 such that for every $x \in X$, $\lim_{n\to\infty} T^n x = x_0$.

1.5.31. Prove the following result due to V. M. Sehgal and J. W. Thomas.

A Common Fixed Point Theorem. Let M be a closed subset of a complete metric space X and let $\mathscr{F}$ be a commutative semigroup of mappings of M into M such that for each $x \in M$ there is $f_x \in \mathscr{F}$ with

$$d\big(f_x(y), f_x(x)\big) \leq \psi\big(d(y, x)\big)$$

for all $y \in M$, where ψ is a nondecreasing, right-continuous function such that $\psi(r) < r$ for all $r > 0$. Suppose for some $x_0 \in M$, $\sup\{d\big(f(x_0), x_0\big)$: $f \in \mathscr{F}\} = d_0 < \infty$. Then there exists a unique $\xi \in M$ such that $f(\xi) = \xi$ for each $f \in \mathscr{F}$. Moreover, there is a sequence $g_n \in \mathscr{F}$ such that for each $x \in M$, $\lim_{n\to\infty} g_n(x) = \xi$. [Hint: First show that $\lim_{n\to\infty} \psi^n(d_0) = 0$. Let $f_0 = f_{x_0}$, $f_n = f_{x_n}$ where $x_{n+1} = f_n(x_n)$; show that (x_n) is Cauchy and $\lim_{n\to\infty} x_n = \xi$ with $f_\xi(\xi) = \xi$. Then ξ is the unique fixed point for f_ξ and hence for each $f \in \mathscr{F}$.]

1.5.32. Let X be a topological space with relative topology and $Y \subset X$. Then Y is called a *retract* of X if there exists a continuous map f: $X \to Y$ such that $f(y) = y$, $y \in Y$. Prove the following.

(i) If Y is a retract of X and X is T_2, then Y is closed in X.

(ii) A necessary and sufficient condition that Y be a retract of X is that every continuous map from Y to any topological space Z can be extended to X.

(iii) If Y is a retract of X and X has the fixed point property (i.e., any continuous map from X into X has a fixed point), then Y also has this property.

1.5.33. Prove the following result due to M. Edelstein. Let T: $X \to X$, a complete metric space. Suppose there exists $\varepsilon > 0$ such that $0 < d(x, y) < \varepsilon$ implies $d(Tx, Ty) < d(x, y)$. If ξ is a limit point of the sequence $T^n x$, then ξ is a fixed point for some T^k.

★ **1.5.34.** [*Definition.* Let T: $X \to X$, a complete metric space. Let $O(x)$ denote the orbit of $x \in X$, that is the set $\{T^n x$: $n \geq 0$, $T^0 x = x\}$ and $\varrho(x) = \sup\{d(z, w)$: $z, w \in O(x)\}$. Then the mapping T is said to be of *di-*

minishing orbital diameter if $\varrho(x) < \infty$ for each $x \in X$ and $\varrho(x) > 0 \Rightarrow \lim_{n\to\infty}\varrho(T^n x) < \varrho(x)$.] Prove the following assertions.

(i) Let $X = \{(x, y): x \geq 0, y \geq 0\}$ with the usual "distance" metric and let $T: X \to X$ be defined by $T(x, y) = (x, x^2)$. Show that T has diminishing orbital diameter and $\varrho(x)$ is continuous on X.

(ii) *An Extension of a Result of W. A. Kirk by V. M. Sehgal.* Suppose T is continuous and has diminishing orbital diameter. If $\varrho(x)$ is continuous on X, then each limit point $\xi \in X$ of the sequence $(T^n x)$, $x \in X$, is a fixed point of T and $\xi = \lim_{n\to\infty} T^n x$.

1.5.35. *A Limited Contraction Fixed Point Theorem* (S. Weingram). Prove that if $f: R^n \to R^n$ is a continuous map such that for some compact set K there is a constant $k(0 < k < 1)$ satisfying

$$| f(x) - f(y) | \leq k \, | x - y |$$

for any two points x, y *not* in K, then f has a fixed point.

1.5.36. Let T be a mapping from a complete metric space (X, d) into itself such that

(i) the map $x \to d(x, Tx)$ is lower semicontinuous;
(ii) there exists $(x_n) \in X$ such that $d(x_n, Tx_n) \to 0$ as $n \to \infty$;
(iii) there exist $a > 0$, $b > 0$ and $0 < c < 1$ such that

$$d(Tx, Ty) \leq ad(x, Tx) + bd(y, Ty) + cd(x, y).$$

Show that T has a unique fixed point and none of the conditions (i), (ii), or (iii) can be omitted.

1.5.37. *A Common Fixed Point Theorem* (J. Cano). Suppose f and g are continuous commuting mappings from a metric space (X, d) into itself. Suppose that $\{x: g(x) = x\}$ is a nonempty compact set and whenever $f(y) \neq y$, $d(f^2(y), f(y)) < d(f(y), y)$. Then f and g have a common fixed point.

1.5.38. *Necessary Condition for a Set of Reals to Have the Fixed Point Property* (Dotson). If every continuous function from $S(\subset R)$ into S has a fixed point, then S is either a singleton or a closed bounded interval.

1.5.39. *A Fixed Point Theorem in R^n* (Reich). For $x = (x_1, x_2, \ldots, x_n) \in R^n$, let $\| x \| = \sup_{1\leq i\leq n} | x_i |$. Suppose f is a continuous map from K^n into R^n, where $K^n = \{x \in R^n: \| x \| \leq 1\}$ such that for each y in $S^{n-1} = \{x \in R^n: \| x \| = 1\}$, there is *no* $m > 1$ with $f(y) = my$. Then f has a fixed point. [Hint: Suppose f has no fixed point; then the mapping g defined by $g(x) = [f(x) - x]/\| f(x) - x \|$ is continuous. Use Brouwer's Fixed Point Theorem to get a contradiction.]

1.5.40. *Cesaro Means and Fixed Points* (Dotson). Suppose T is a function from R into R such that for some x, the sequence $(1/n)\sum_{k=0}^{n-1}T^k x$ is bounded. Then T has a fixed point. Here the fixed point need not be the limit of the sequence $(1/n)\sum_{k=0}^{n-1}T^k y$ for any y different from the fixed point.

1.5.41. Prove the following result due to J. Cano. Let f be a continuous function from a compact set $K(\subset R)$ into itself. Suppose that there is some $x \in K$ such that every cluster point of $[f^n(x)]$ is a fixed point of f. Then the sequence $[f^n(x)]$ is convergent. This result fails in R^2.

★ **1.5.42.** *A Converse of the Banach Fixed Point Theorem.* It follows from Theorem 1.15 that in a bounded complete metric space (X, d), $\bigcap_{n=1}^{\infty} T^n X = \{a\}$ for every contraction T with unique fixed point a. Show that the following converse holds: Suppose (X, d) is a compact metric space. T is a continuous map from X into itself such that $\bigcap_{n=1}^{\infty} T^n X = \{a\}$, a singleton. Then there is a metric d_1, generating the given topology of X, such that T is a contraction with respect to d_1. [Hint: Define $\bar{d}(x, y) = \sup\{d(T^n x, T^n y): 0 \leqq n < \infty\}$. Then $(X, \bar{d})$ is a metric space. T is nonexpansive with respect to $\bar{d}$, which induces the same topology as d. Now let $n(y) = \sup\{n: y \in T^n X\}$ and $n(x, y) = \inf\{n(x), n(y)\}$. Given k in $(0, 1)$, define $p(x, y) = k^{n(x,y)} \cdot \bar{d}(x, y)$ and let $d_1(x, y) = \inf \sum_{n=1}^{m} p(x_i, x_{i+1})$, where the infimum is taken over all finite subsets $x_1, x_2, \ldots, x_{m+1}$ in X such that $x = x_1$ and $y = x_{m+1}$. Then d_1 is the desired metric.] This result is due to L. Janos. For a more general result, the reader can consult the 1967 paper of P. R. Meyers.† He proved the following: Let T be a continuous mapping from a complete metric space into itself with a unique fixed point u such that $\lim_{n\to\infty} T^n x = u$ for any x. Then if u has an open neighborhood with compact closure, T is a contraction with respect to an equivalent metric. *This result is false if u does not have such a neighborhood.* Can the reader find an example? [It is relevant to point out here that C. Bessaga [*Colloq. Math.* 7, 41–43 (1959)] proved the following converse: Let T be a map of a set X into itself such that each iterate T^n, $n \geq 1$, has a unique fixed point. Then given any number k, $0 < k < 1$, there exists a complete metric for X such that T is a contraction with the constant k.]

1.5.43. Let (X, d) be a compact metric space and T be a continuous map from X into itself. Suppose that T satisfies condition (K_2) as in Theorem 1.25. Show that there is a metric d_1 on X, generating the same topology as d, such that T is a contraction with respect to d_1. (Hint: Use Problem 1.5.42.)

1.5.44. *Another Extension of the Banach Fixed Point Theorem.* Let T be a map from a complete metric space X into itself satisfying the following

† P. R. Meyers, *J. Res. Nat. Bur. Standards Sect. B*, **71** (1967).

condition. Given $\varepsilon > 0$, there exists $\delta > 0$ such that $\varepsilon \leqq d(x, y) < \varepsilon + \delta$ implies $d(Tx, Ty) < \varepsilon$. Then for any x in X, $\lim T^n x$ exists and this limit is the unique fixed point of T. [Hint: Notice that T is contractive and $c_n = d(T^n x, T^{n+1} x)$ is a decreasing sequence with limit 0. If the sequence $T^n x$ is not Cauchy, then for infinitely many n and m, $d(T^n x, T^m x)$ is greater than 2ε (> 0). For this ε, let δ be as in the hypothesis above. Let N be such that for $n > N$, $c_n < \delta'/3$, where $\delta' = \min\{\delta, \varepsilon\}$. Let n ($> N$) and k be such that $d(T^n x, T^{n+k} x) > 2\varepsilon$. Since for all j with $1 \leqq j \leqq k$, $| d(T^n x, T^{n+j} x) - d(T^n x, T^{n+j+1} x) | \leqq c_{n+j} < \delta'/3$, it follows that for some j, $1 \leqq j \leqq k$, we must have $\varepsilon + 2\delta'/3 < d(T^n x, T^{n+j} x) < \varepsilon + \delta'$. But for this j, $d(T^n x, T^{n+j} x) \leqq d(T^n x, T^{n+1} x) + d(T^{n+1} x, T^{n+j+1} x) + d(T^{n+j} x, T^{n+j+1} x) < \delta'/3 + \varepsilon + \delta'/3$, which is a contradiction.] This result is due to A. Meir and E. Keeler.

1.5.45. *An Extension of Theorem 1.25.* Suppose T maps a complete metric space X into itself such that for some constant $c < 1$ and for all x, y in X, $d(Tx, Ty) \leqq c \max\{d(x, Tx), d(y, Ty), d(x, Ty), d(y, Tx), d(x, y)\}$. Then for any x in X, $\lim T^n x$ exists and this limit is the unique fixed point of T. This result is due to L. B. Ćirić.

1.5.46. *A Common Fixed Point Theorem.* Let (T_n) be a sequence of mappings of a complete metric space X into itself. Suppose that for each pair T_i, T_j, there are nonnegative constants a, b, and c (depending upon i and j) such that for all x, y in X,

$$d(T_i x, T_j y) \leqq a[d(x, T_i x) + d(y, T_j y)] + b[d(x, T_j y) + d(y, T_i x)] + c d(x, y),$$

where $2a + 2b + c < 1$. Then the sequence T_n has a unique common fixed point. [Hint: Let $x_0 \in X$ and $x_n = T_n x_{n-1}$, $n = 1, 2, \ldots$. Then the sequence (x_n) is Cauchy and its limit is the unique common fixed point of the T_n's.] This is a result of B. N. Roy extended by K. Iseki. This result can be extended to an arbitrary family of mappings.

1.5.47. Prove the following result due to T. K. Hu. The following assertions are equivalent in a metric space X:

(i) X is complete.

(ii) If S is a nonempty closed subset of X and $f\colon S \to S$ a contraction, then f has a fixed point.

Use the above result to prove the following converse of Janos' result (Problem 1.5.42) due to S. Kasahara. Let S be any nonempty closed subset of a metric space X. Suppose that every continuous f from S into S such that $\bigcap_{n=1}^{\infty} f^n(S)$ is a singleton, is a contraction with respect to an equivalent metric on S. Then X is compact.

1.6. The Stone–Weierstrass Theorem and the Ascoli Theorem

The two theorems in the title are extremely important tools for the analyst for the proof of many general results on continuous functions. The Ascoli theorem provides the basis for most proofs of compactness in function spaces, whereas the Stone–Weierstrass theorem helps us to extend certain results holding for functions of a special type to all continuous functions by an approximation argument.

Around 1883, G. Ascoli introduced the notion of equicontinuity for a set of continuous functions and then used this concept to find a sufficient condition for the compactness of a set of continuous functions with "sup" metric. Later on in 1889, C. Arzela proved the necessity of this condition.

In regard to the other main theorem in this section, it was K. Weierstrass who, around 1900, first showed that the continuous functions on [0, 1] could be approximated uniformly by the polynomials. This beautiful result was then extended in many different directions by various mathematicians. The most important of them all was achieved by M. H. Stone, first in 1937 and then, more completely, in 1948.

We will begin with two definitions and then proceed to prove Stone's extension of Weierstrass' theorem and several important consequences.

Definitions 1.41. (a) A linear subspace S of $C_b(X)$, the space of real-valued bounded continuous functions on a topological space X, is called an *algebra* if the product of any two elements in S is again in S.

(b) A family $\mathscr{F}$ of real-valued functions on X is said *to separate points* of X if, given x, y in X with $x \neq y$, there is $f \in \mathscr{F}$ with $f(x) \neq f(y)$. ∎

In what follows, $C_b(X)$ is always considered a complete metric space with the metric $d(f, g) = \sup\{|f(x) - g(x)|: x \in X\}$.

Theorem 1.28. *The Stone–Weierstrass Theorem.* Suppose X is compact and $\mathscr{A}$ is an algebra contained in $C(X)$ that separates points of X and contains the constant functions; then $\bar{\mathscr{A}} = C(X)$. [$C(X)$ denotes the continuous functions on X, which coincides with $C_b(X)$ when X is compact.] ∎

Proof. First, we notice that given $\varepsilon > 0$, there is a polynomial $P(x)$ such that $|P(x) - |x|| < \varepsilon$ for all $x \in [-1, 1]$. The reason is that considering the binomial series† for $(1 - x)^{1/2}$, we can find a positive integer N

† Notice that for any $\delta > 0$, the binomial series for $(1 + \delta - x)^{1/2}$ has radius of convergence $1 + \delta$ and therefore this series converges uniformly on $[-1, 1]$.

and constants C_n such that for all $x \in [-1, 1]$,

$$\left| (1 - x)^{1/2} - \sum_{n=0}^{N} C_n x^n \right| < \varepsilon;$$

and then replacing $(1 - x)$ with x^2, we have

$$\left| | x | - \sum_{n=0}^{N} C_n (1 - x^2)^n \right| < \varepsilon.$$

Because of the above, given $g \in \mathscr{A}$ and $\varepsilon > 0$, we can find a polynomial $P(g)$ in g such that $\| g(x) | - P(g)(x) | < \varepsilon$ for all $x \in X$. In other words, $g \in \mathscr{A} \Rightarrow | g | \in \bar{\mathscr{A}}$. Hence for $f, g \in \mathscr{A}$, the functions

$$f \vee g = \tfrac{1}{2}[f + g + | f - g |]$$

and

$$f \wedge g = \tfrac{1}{2}[f + g - | f - g |]$$

are also in $\bar{\mathscr{A}}$.

Next we note that given $f \in C(X)$ and any $x, y \in X$, we can find $g_{x,y} \in \mathscr{A}$ such that $g_{x,y}(x) = f(x)$ and $g_{x,y}(y) = f(y)$. Here one simply needs to take

$$g_{x,y} = \frac{f(x) - f(y)}{h(x) - h(y)} \cdot h + \frac{f(x)h(y) - f(y)h(x)}{h(y) - h(x)},$$

where $h \in \mathscr{A}$ such that $h(x) \neq h(y)$.

Now let $f \in C(X)$ and $\varepsilon > 0$. Let $y \in X$. For each $x \subset X$, there is an open set V_x containing x such that $g_{x,y}(t) > f(t) - \varepsilon$ for $t \in V_x$. Let $X = \bigcup_{i=1}^{p} V_{x_i}$ and $\bigvee_{i=1}^{p} g_{x_i,y} = g_y$. Then $g_y(t) > f(t) - \varepsilon$ for all $t \in X$, $g_y \in \bar{\mathscr{A}}$, and $g_y(y) = f(y)$. Again, for each $y \in X$, there is an open set U_y containing y such that $g_y(t) < f(t) + \varepsilon$, $t \in U_y$. Let $X = \bigcup_{i=1}^{q} U_{y_i}$ and $\bigwedge_{i=1}^{q} g_{y_i} = g$. Then $g \in \bar{\mathscr{A}}$ and $| g(t) - f(t) | < \varepsilon$ for all t $\in X$. The proof is complete. ▮

Corollary 1.1. Any real-valued continuous function on a compact subset E of R^n is the uniform limit of a sequence of polynomials (in the n coordinates) on E. ▮

Proof. Corollary 1.1 follows easily from Theorem 1.26, since the polynomials clearly form an algebra, and also, they separate points, since for two distinct points on E at least one of the coordinates will be different for both. ▮

We note that Theorem 1.28 does *not* hold for complex-valued functions. The reason is that any sequence of polynomials in the complex variable z converging uniformly on $\{z: |z| \leq 1\}$ must have its limit differentiable in the open unit disk. However, we present the following weaker result.

Theorem 1.29. Let $\mathscr{A}$ be an algebra $\subset C_1(X)$, the complex-valued continuous functions on a compact space X, such that

(i) $\mathscr{A}$ contains the constant functions,
(ii) $\mathscr{A}$ separates points of X,
(iii) $f \in \mathscr{A} \Rightarrow$ the conjugate $\bar{f}$ is in $\mathscr{A}$.

Then $\bar{\mathscr{A}} = C_1(X)$. ∎

Proof. For $f \in \mathscr{A}$, $R(f) = \frac{1}{2}(f + \bar{f})$ and $I(f) = (1/2i)(f - \bar{f})$ are both real valued and in $\mathscr{A}$. Let $\mathscr{A}_0$ consist of the real-valued functions in $\mathscr{A}$. Then $\mathscr{A}_0$ is a subalgebra of $C(X)$. Since $f(x) \neq f(y)$ implies either $R(f)(x) \neq R(f)(y)$ or $I(f)(x) \neq I(f)(y)$, $\mathscr{A}_0$ separates points of X. By Theorem 1.28, $\bar{\mathscr{A}}_0 = C(X)$. Clearly $\bar{\mathscr{A}} = C_1(X)$. ∎

Corollary 1.2. Let f be a continuous periodic real-valued function on R with period 2π, i.e., $f(x + 2\pi) = f(x)$ for all $x \in R$. Then given $\varepsilon > 0$, there exist constants a_n, b_n such that

$$\left| f(x) - \sum_{n=0}^{N} (a_n \cos nx + b_n \sin nx) \right| < \varepsilon$$

for all $x \in R$. ∎

Proof. For $x \in R$, let $e^{ix} = z$ and let $\Gamma = \{z: |z| = 1\}$ be the unit circle on the complex plane. Then the mapping $\Phi: x \to e^{ix}$ is a homeomorphism of $(0, 2\pi)$ onto $\Gamma - e^{2\pi i}$. Define the continuous function $g: \Gamma \to R$ by $g(\Phi(x)) = f(x)$ and $g(e^{2\pi i}) = f(2\pi) = f(0)$. Now the algebra generated by $\{e^{ix}, e^{-ix}, 1\}$ satisfies the conditions of Theorem 1.29. Hence there exist complex numbers (α_n) such that

$$\left| g(e^{ix}) - \sum_{k=-N}^{N} \alpha_k e^{ikx} \right| < \varepsilon$$

for all $x \in [0, 2\pi]$ and therefore for all $x \in R$ (by periodicity). The corollary follows by considering the real parts inside the "absolute value" expression. ∎

Remarks

1.88. By Weierstrass' theorem, all real linear combinations of the functions $1, x, x^2, \ldots, x^n, \ldots$ are dense in $C[0, 1]$. This theorem remains true if instead of considering all positive powers of x, we only consider the infinite set $1, x^{n_1}, x^{n_2}, \ldots, x^{n_k}, \ldots$, where (n_k) is a strictly increasing sequence of positive integers and $\sum_{k=1}^{\infty}(1/n_k)$ diverges. This remarkable result is due to Ch. H. Muntz, obtained in 1914. A form of this theorem can be found in a paper by Clarkson and Erdös.†

1.89. $C(X)$ is separable if X is a compact metric space. To see this, let (V_n) be a countable basis for the topology of X and let $f_n(x) = d(x, X - V_n)$. Clearly, if $x \neq y$ and $x \in V_n$, $y \notin V_n$, then $f_n(x) \neq 0$ and $f_n(y) = 0$. Now to apply Theorem 1.28, consider the algebra of all finite linear combinations of $f_1^{k_1} f_2^{k_2} \cdots f_n^{k_n}$ (where k_i's are nonnegative integers) which separates points of X. This algebra, by Theorem 1.28, is dense in $C(X)$. Clearly, the above finite linear combinations with rational coefficients form a countable dense set for the algebra and hence for $C(X)$.

1.90. Let $\mathscr{A}$ be an algebra $\subset C(X)$, X a compact Hausdorff space. Suppose $\mathscr{A}$ separates points of X and given $x \in X$, there exists $f \in \mathscr{A}$ such that $f(x) \neq 0$. Then $\bar{\mathscr{A}} = C(X)$. [This will follow exactly as in the proof of Theorem 1.28, noting that $|x|$ on $[-1, 1]$ can be uniformly approximated by polynomials in x having no constant term.]

1.91. Let $\mathscr{A}$ be as in Remark 1.90, separating points of X but all $f \in \mathscr{A}$ vanishing at some $x_0 \in X$. Then $\bar{\mathscr{A}} = \{f \in C(X): f(x_0) = 0\}$. The reason is that $\mathscr{A}_0 = \{f + c: f \in \mathscr{A}$ and c is a constant$\}$ is an algebra satisfying the conditions in Remark 1.90 and therefore $\bar{\mathscr{A}}_0 = C(X)$. So given $\varepsilon > 0$ and $g \in C(X)$ with $g(x_0) = 0$, there is $f \in \mathscr{A}$ and some constant c such that $|f(x) + c - g(x)| < \varepsilon/2$ for all $x \in X$. Since $f(x_0) = g(x_0) = 0$, $|c| < \varepsilon/2$; and therefore $|f(x) - g(x)| < \varepsilon$ for all $x \in X$.

1.92. Let $C_0(X)$ be the real-valued continuous functions f on a *locally compact Hausdorff* space X that vanish at infinity (i.e., for each $\varepsilon > 0$, there is compact $K \subset X$ such that $|f(x)| < \varepsilon$ if $x \in X - K$). If $X = R$, $1 \notin C_0(R)$ but $1/(1 + x^2) \in C_0(R)$. The reader can easily verify the following facts:

(a) $C_0(X)$ is a closed subalgebra of $C(X)$.

(b) If $X_\infty = X \cup \{\infty\}$ is the one-point compactification of X, then $C_0(X) = \{f|_X: f \in C(X_\infty)$ and $f(\infty) = 0\}$.

(c) If $\mathscr{A}$ is an algebra $[\subset C_0(X)]$ separating points of X and contains for each x a function that does not vanish at x, then $\bar{\mathscr{A}} = C_0(X)$.

† J. A. Clarkson and P. Erdös, *Duke Math. J.* **10**, 5 (1943).

Next we present a version of the Ascoli theorem. For a more general version of the theorem, the reader should consult [29].

First we need a definition.

Definition 1.42. Let X be a topological space and Y a metric space. A family $\mathscr{F}$ of continuous maps from X into Y is called *equicontinuous* at a point $x_0 \in X$ if given $\varepsilon > 0$, there exists an open set $V(x_0)$ such that $x_0 \in V(x_0)$ and $d(f(x), f(x_0)) < \varepsilon$ whenever $x \in V(x_0)$ and $f \in \mathscr{F}$. If $\mathscr{F}$ is equicontinuous at every point on X, then $\mathscr{F}$ is called equicontinuous on X. ▮

In what follows, X is a compact Hausdorff space and Y is a complete metric space. Then $C(X, Y)$, the set of all continuous maps from X into Y, is a complete metric space with the metric $d(f, g) = \sup_{x \in X} d(f(x), g(x))$.

Theorem 1.30. *The Arzela–Ascoli Theorem.* A subset $\mathscr{F}$ of $C(X, Y)$ has compact closure if and only if $\mathscr{F}$ is equicontinuous on X and for each $x \in X$ the set $\{f(x): f \in \mathscr{F}\}$ has compact closure in Y. ▮

Proof. The "only if" part is left to the reader. For the "if" part, it is sufficient to prove that $\mathscr{F}$ is totally bounded.

Let $\varepsilon > 0$. By compactness of X and equicontinuity of $\mathscr{F}$, there exist open sets $V(x_i)$, $1 \leq i \leq n$, whose union is X such that $x_i \in V(x_i)$ and $d(f(y), f(x_i)) < \varepsilon$ for any $y \in V(x_i)$ and all $f \in \mathscr{F}$. Now the set $B = \{f(x_i): 1 \leq i \leq n \text{ and } f \in \mathscr{F}\}$ is totally bounded and therefore there exist $y_j \in Y$, $1 \leq j \leq m$, such that $B \subset \bigcup_{j=1}^{m} S_{y_j}(\varepsilon)$. For any mapping $\pi: \{1, 2, \ldots, n\} \to \{1, 2, \ldots, m\}$, let $\mathscr{F}_\pi = \{f \in \mathscr{F}: f(x_i) \in S_{y_{\pi(i)}}(\varepsilon), 1 \leq i \leq n\}$. Clearly for $f_1, f_2 \in \mathscr{F}_\pi$, if $x \in V(x_i)$, then

$$d(f_1(x), f_2(x)) \leq d(f_1(x), f_1(x_i)) + d(f_1(x_i), y_{\pi(i)}) + d(y_{\pi(i)}, f_2(x_i)) + d(f_2(x), f_2(x_i)) < 4\varepsilon;$$

this means that diameter $(\mathscr{F}_\pi) \leq 4\varepsilon$. Since $\mathscr{F}$ is the union of finitely many such $\mathscr{F}_\pi$'s, the proof is complete. ▮

Problems

1.6.1. Prove the following result due to J. L. Walsh: Let $\Phi \in C[0, 1]$. Then a necessary and sufficient condition that an arbitrary $f \in C[0, 1]$ can be uniformly approximated by a polynomial in Φ is that Φ be *strictly monotonic* on [0, 1].

1.6.2. Let X be a locally compact Hausdorff space and $C_c(X)$ be the set of all real-valued continuous functions on X that vanish outside compact sets. Suppose $\mathscr{A}$ is an algebra $\subset C_c(X)$, separating points of X, and is such that for every compact set K there is $f \in \mathscr{A}$ such that $f(x) = 1$ for $x \in K$. Show that $\mathscr{A}$ is dense in $C_c(X)$ with the "sup" metric.

1.6.3. Let X and Y be locally compact Hausdorff spaces and let $f \in C_c(X \times Y)$. Show that given $\varepsilon > 0$ there exist $g_i \in C_c(X)$ and $h_i \in C_c(Y)$, $1 \leq i \leq n$, such that

$$\left| f(x, y) - \sum_{i=1}^{n} g_i(x) h_i(y) \right| < \varepsilon$$

for all $(x, y) \in X \times Y$. (See Problem 1.6.2.)

1.6.4. Prove (with details) Remark 1.92.

★ **1.6.5.** Let $f \in C[0, 1]$ and $f(0) = 0$. Show that f can be uniformly approximated by a sequence of polynomials $P_n(x) = \sum_{m=0}^{k_n} a_{nm} x^m$ such that for each m, $\lim_{n \to \infty} a_{nm} = 0$.

★ **1.6.6.** Prove the following extension of Problem 1.6.1—also due to Walsh. Let Φ be a real-valued function on [0, 1]. Then an arbitrary $f \in C[0, 1]$ can be uniformly approximated by a polynomial in Φ if and only if Φ is one-to-one and bounded and Φ^{-1} can be extended to a continuous function on the closure of the range of Φ.

1.6.7. Let f be a real-valued continuous function on R, and let h be any real-valued function that has positive infimum on each compact subset of R. Show that there exists an infinitely differentiable function g on R such that $| f(x) - g(x) | < h(x)$ for all $x \in R$. [Hint: Let p_n be a polynomial such that $| p_n(x) - f(x) | \leq \frac{1}{3} h(x)$ for all x with $| x | \leq n + \frac{1}{2}$. Find infinitely differentiable g_n such that $0 \leq g_n \leq 1$, $g_n(x) = 1$ if $n - 1 \leq | x | \leq n$ and $= 0$ if $| x | \leq n - \frac{3}{2}$ or $| x | \geq n + \frac{1}{2}$. Take $g = \sum_{n=1}^{\infty} (1 - g_{n-1}) g_n p_n$.] This result is due to T. Carleman. There is an extension of this result to R^n due to A. Boghossian.

1.6.8. Prove the "only if" part of Theorem 1.30.

1.6.9. Let $\mathscr{F}_k = \{f\colon [0, 1] \to [0, 1]$ and $| f(x) - f(y) | \leq k | x - y |$ for $0 \leq x, y \leq 1\}$, where $k > 0$. Show that $\mathscr{F}_k$ is compact as a subset of $C([0, 1], [0, 1])$.

1.6.10. Show that the family of functions $1/[1 + (x - n)^2]$ of R into [0, 1] is equicontinuous but not compact.

1.6.11. Let X, Y be compact metric spaces and $f\colon X \times Y \to Z$ (a metric space). For $x \in X$, let $f_x(y) = f(x, y)$. Show that if f is continuous, then the mapping $x \to f_x$ of X into $C(Y, Z)$ is continuous and the family $\{f_x: x \in X\}$ is equicontinuous.

★ **1.6.12.** *Stone–Weierstrass Theorem for Lattices* (Kakutani–Krein): Let X be a compact Hausdorff space. Then a subset $S \subset C(X)$ is called a *lattice* if $f \vee g$ and $f \wedge g$ are in S for all $f, g \in S$. Prove that (i) any closed subalgebra of $C(X)$ containing 1 is a lattice, and (ii) any lattice $S \subset C(X)$ that is a closed subspace, contains 1, and separates points of X, is all of $C(X)$.

★ **1.6.13.** Let f be a continuous real-valued function on [0, 1]. Show that there is a monotonic increasing sequence (P_n) of polynomials converging uniformly to f on [0, 1].

1.6.14. Prove that if $f \in C(R)$ can be approximated uniformly throughout R by polynomials, then f itself is a polynomial.

2

Measure

The concept of measure is an extension of the concept of length: The measure of an interval in R is usually its length; the measure of a polygon in R^2 is usually its area; and so on. The problem of extending the classes of sets for which the notions of length, area, etc. are defined to larger classes of sets for which these notions are still defined gave rise to the general theory of measure.

The first definitions of the measure of an arbitrary set in R^n seem to have been given by G. Cantor (in 1883), O. Stolz (in 1884), and A. Harnack (in 1885). These definitions were substantially improved later by G. Peano (in 1887) and C. Jordan (in 1892). By means of his concept of measure, Peano determined a necessary and sufficient condition for a bounded nonnegative function on a closed and bounded interval in R to be Riemann integrable. In 1898, in his book *Leçons sur la Théorie des Fonctions*, E. Borel formulated several postulates (which outline essential properties of length) for defining measures of sets; these led to the following:

(1) A measure is always nonnegative.
(2) The measure of the union of a countable number of nonoverlapping sets equals the sum of their measures.
(3) The measure of the difference of a set and a subset is equal to the difference of their measures.
(4) Every set whose measure is not zero is uncountable.

H. Lebesgue[†] presented a mathematically rigorous description of the class of sets for which he defined a measure satisfying Borel's postulates. This

[†] H. Lebesgue, Intégrale, longueur, aire, *Ann. Mat. Pura Appl.* **7**(3), 231–359 (1902).

measure is now well known as Lebesgue measure on R^n and is perhaps the most important and useful concept of a measure that can be found in R^n to date. A theory of measure similar to that of Lebesgue was also independently developed by W. H. Young.[†]

It is now well known that measure theory and its techniques are indispensible tools for many parts of modern analysis. Measure theory provides the proper framework for the study of discontinuous and nondifferentiable functions. For instance, thc concept of measure provides us with a precise idea about the set of nondifferentiable points of a monotonic function or the set of points of discontinuity of a Riemann-integrable function. Different concepts of integrals are either based on or inseparably connected with the concept of a measure; and these, in turn, play an important part in the application of mathematical analysis to present day science, including theoretical physics and probability theory.

This chapter presents the theory roughly as follows.

First, a measure is defined on an algebra of subsets of an arbitrary abstract nonempty set and different properties of the measure based on its definition are derived. Then, as concrete examples, we construct Lebesgue measure on intervals in R. Next, with a view to obtaining a measure on a σ-algebra, we introduce outer measures and measurable sets. We show how to construct outer measures (and therefore measures in general) and extend a measure on an algebra to a σ-algebra. In order to obtain a large class of measurable sets, we introduce the concept of a metric outer measure on a metric space. These ideas of constructing a measure from an outer measure, considering a metric outer measure to obtain a large class of measurable sets as the domain space of a measure, etc., were noted first by C. Carathéodory.[‡] Finally, to take a concrete example, we discuss Lebesgue measure in R and its properties at length.

Throughout this chapter, X is a nonempty abstract set. When X is a topological space, we will occasionally speak about Borel subsets of X. In this chapter as well as in Chapters 3 and 4, the Borel sets will always mean the sets in the smallest σ-algebra containing the open subsets of X. In Chapter 5, however, we will speak about many different σ-algebras based on the topology of X; there the sets in the σ-algebra generated by the open subsets will be the weakly Borel subsets. The reader may note, however, that when X is σ-compact (for example, when X is R^n), both the definitions coincide.

[†] W. H. Young, Open sets and the theory of content, *Proc. London Math. Soc.* **2**(2), 16–51 (1904).

[‡] C. Carathéodory, *Nachr. Ges. Wiss. Göttingen*, 404–426 (1914).

2.1. Measure on an Algebra

In this section, we will define a measure on an algebra and establish several properties of a measure based upon its definition.

Definition 2.1. Let $\mathscr{A}$ be an algebra† of subsets of X and μ be an extended real-valued function on $\mathscr{A}$. Then μ is called a measure on $\mathscr{A}$ if

(a) $\mu(\varnothing) = 0$,
(b) $\mu(A) \geq 0 \quad$ if $A \in \mathscr{A}$,
(c) $\mu(\bigcup_{n=1}^{\infty} A_n) = \sum_{n=1}^{\infty} \mu(A_n)$, for every sequence A_n of pairwise disjoint sets of $\mathscr{A}$ with $\bigcup_{n=1}^{\infty} A_n \in \mathscr{A}$.

The property (c) is known as the *countable additivity* property of μ. μ is called a *finitely additive measure* if (c) is replaced by

(c′) $\mu(\bigcup_{n=1}^{k} A_n) = \sum_{n=1}^{k} \mu(A_n)$, for every finite sequence (A_n) of pairwise disjoint sets in $\mathscr{A}$. ∎

Example 2.1. Let $\mathscr{A} = 2^X$. Let μ be defined on $\mathscr{A}$ by

$$\mu(A) = \begin{cases} \text{number of points in } A, & \text{if } A \text{ is finite;} \\ \infty, & \text{if } A \text{ is infinite.} \end{cases}$$

Then μ is a measure. (Prove this assertion.) This measure is known as the *counting measure* in X.

Example 2.2. Let X be uncountable and $\mathscr{A}$ be the class defined by

$$\mathscr{A} = \{A \subset X\colon \text{either } A \text{ or } X - A \text{ is countable}\}.$$

Then $\mathscr{A}$ is an algebra (in fact, a σ-algebra). Let μ be defined on $\mathscr{A}$ by

$$\mu(A) = \begin{cases} 0, & \text{if } A \text{ is countable;} \\ 1, & \text{if } X - A \text{ is countable.} \end{cases}$$

Then μ is a measure on $\mathscr{A}$. (Prove this assertion.)

† If $\mathscr{A}$ is a ring, then this definition defines a measure on a ring.

Example 2.3. Let $X = Z^+$, $\mathscr{A} = 2^X$, and $\sum_{n=1}^{\infty} a_n$ be a convergent series of positive real numbers. Let μ be defined on $\mathscr{A}$ by

$$\mu(A) = \begin{cases} \sum_{n \in A} a_n, & \text{if } A \text{ is nonempty and finite,} \\ \infty, & \text{if } A \text{ is infinite,} \\ 0, & \text{if } A \text{ is empty.} \end{cases}$$

Then μ is not a measure, since

$$X = \bigcup_{n=1}^{\infty} \{n\}, \qquad \mu(X) = \infty,$$

but

$$\sum_{n=1}^{\infty} \mu\{n\} = \sum_{n=1}^{\infty} a_n < \infty.$$

However, μ is clearly a finitely additive measure.

Several important properties of a measure that follow as consequences of its definition are given in the next two propositions.

Proposition 2.1. Let μ be a measure on an algebra $\mathscr{A}$. Then

(a) $A \subset B$, $A \in \mathscr{A}$, $B \in \mathscr{A}$ imply $\mu(A) \leq \mu(B)$.

(b) $\mu(\bigcup_{n=1}^{\infty} A_n) \leq \sum_{n=1}^{\infty} \mu(A_n)$ (the *countable subadditive property*), if $A_n \in \mathscr{A}$, $1 \leq n < \infty$ and $\bigcup_{n=1}^{\infty} A_n \in \mathscr{A}$. ∎

Proof. (a) $B = A \cup (B - A)$; so, if $A_1 = A$, $A_2 = B - A$, and $A_n = \varnothing$ for $n > 2$, then with μ being countably additive, we have $\mu(B) = \mu(A) + \mu(B - A)$. Hence $\mu(B) \geq \mu(A)$.

(b) Let $B_1 = A_1$ and $B_n = A_n - \bigcup_{i=1}^{n-1} A_i$, for $n > 1$. Then $B_n \in \mathscr{A}$ for each n, $\bigcup_{n=1}^{\infty} B_n = \bigcup_{n=1}^{\infty} A_n \in \mathscr{A}$, $B_n \subset A_n$, and the B_n's are pairwise disjoint. Hence

$$\mu\left(\bigcup_{n=1}^{\infty} A_n\right) = \mu\left(\bigcup_{n=1}^{\infty} B_n\right) = \sum_{n=1}^{\infty} \mu(B_n) \leqq \sum_{n=1}^{\infty} \mu(A_n).$$ ∎

Proposition 2.2. Let μ be a measure on an algebra $\mathscr{A}$. Then

(a) If $A_n \subset A_{n+1}$, $A_n \in \mathscr{A}$ for $1 \leqq n < \infty$ and $\bigcup_{n=1}^{\infty} A_n \in \mathscr{A}$, then

$$\mu\left(\bigcup_{n=1}^{\infty} A_n\right) = \lim_{n \to \infty} \mu(A_n).$$

(b) If $A_n \supset A_{n+1}$, $A_n \in \mathscr{A}$ for $1 \leqq n < \infty$, $\mu(A_1) < \infty$ and $\bigcap_{n=1}^{\infty} A_n \in \mathscr{A}$, then

$$\mu\left(\bigcap_{n=1}^{\infty} A_n\right) = \lim_{n\to\infty} \mu(A_n).$$

Furthermore, for a finitely additive μ, countable additivity is implied by (a). ∎

Proof. First, we prove the last assertion. So let μ be a finitely additive measure with property (a). Let $(A_n)_{n=1}^{\infty}$ be a sequence of pairwise disjoint sets in $\mathscr{A}$ with $\bigcup_{n=1}^{\infty} A_n \in \mathscr{A}$. Let $B_k = \bigcup_{n=1}^{k} A_n$. Then $B_k \subset B_{k+1}$, $B_k \in \mathscr{A}$, and $\bigcup_{k=1}^{\infty} B_k = \bigcup_{n=1}^{\infty} A_n \in \mathscr{A}$. Hence by the assumption of (a),

$$\mu\left(\bigcup_{n=1}^{\infty} A_n\right) = \mu\left(\bigcup_{k=1}^{\infty} B_k\right) = \lim_{k\to\infty} \mu(B_k) = \lim_{k\to\infty} \sum_{n=1}^{k} \mu(A_n) = \sum_{n=1}^{\infty} \mu(A_n).$$

Next, we establish properties (a) and (b) of a measure. To prove (a), let $A_n \subset A_{n+1}$ and $\bigcup_{n=1}^{\infty} A_n \in \mathscr{A}$. Let $B_1 = A_1$, $B_n = A_n - A_{n-1}$ for $n > 1$. Then $B_n \cap B_m = \varnothing$ $(n \neq m)$ and $\bigcup_{n=1}^{\infty} B_n = \bigcup_{n=1}^{\infty} A_n \in \mathscr{A}$. Then

$$\mu\left(\bigcup_{n=1}^{\infty} A_n\right) = \mu\left(\bigcup_{n=1}^{\infty} B_n\right) = \sum_{n=1}^{\infty} \mu(B_n)$$

$$= \lim_{k\to\infty} \sum_{n=1}^{k} \mu(B_n) = \lim_{k\to\infty} \mu(A_k), \text{ since } A_k = \bigcup_{n=1}^{k} B_n.$$

To prove (b), let $A_n \supset A_{n+1}$, $\mu(A_1) < \infty$, $\bigcap_{n=1}^{\infty} A_n \in \mathscr{A}$. Then

$$A_1 - \bigcap_{n=1}^{\infty} A_n = \bigcup_{n=1}^{\infty} (A_1 - A_n), \qquad A_1 - A_n \subset A_1 - A_{n+1}.$$

Therefore by (a),

$$\mu\left(A_1 - \bigcap_{n=1}^{\infty} A_n\right) = \lim_{n\to\infty} \mu(A_1 - A_n).$$

Since $\mu(A_1) < \infty$,

$$\mu\left(A_1 - \bigcap_{n=1}^{\infty} A_n\right) = \mu(A_1) - \mu\left(\bigcap_{n=1}^{\infty} A_n\right)$$

and

$$\mu(A_1 - A_n) = \mu(A_1) - \mu(A_n).$$

Now part (b) follows. ∎

A measure μ on an algebra $\mathscr{A}$ of subsets of X is called *finite* if $\mu(X) < \infty$. It is called *σ-finite* if there is a sequence (X_n) of sets in $\mathscr{A}$ with $\mu(X_n) < \infty$ and $X = \bigcup_{n=1}^{\infty} X_n$. As was the case in the proof of Proposition

2.1(b), it is always possible to take $(X_n)_{n=1}^{\infty}$ as pairwise disjoint. This finiteness condition on a measure is often, though not always, needed for many parts of the general theory. The most important example of a σ-finite measure on an algebra of subsets of R is that of Lebesgue measure on intervals (which is discussed in the next section). Of course not every measure is σ-finite. For instance, the counting measure on 2^X (X uncountable) is not σ-finite.

Another useful concept is that of *semifiniteness*—a notion weaker than that of σ-finiteness. A measure μ on an algebra $\mathscr{A}$ of subsets of X is called *semifinite* if for each $A \in \mathscr{A}$, with $\mu(A) = \infty$, there is $B \subset A$, with $B \in \mathscr{A}$ and $0 < \mu(B) < \infty$. Thus every σ-finite measure is semifinite (Problem 2.1.7). The counting measure is always semifinite. But the 0–∞ measure (which is zero on countable subsets of X and infinity on uncountable subsets of X) is not semifinite. Measures that are not semifinite are very wild when restricted to certain sets. Every measure is, in a sense, semifinite, once its 0–∞ part (the wild part) is taken away. The next proposition demonstrates this.

Proposition 2.3. Let μ be a measure on a σ-algebra $\mathscr{A}$ of subsets of X.

(a) If μ semifinite, then every set in $\mathscr{A}$ with infinite measure contains sets in $\mathscr{A}$ with arbitrarily large finite measure.

(b) Each measure μ on $\mathscr{A}$ is the sum $\mu_1 + \mu_2$ of a semifinite measure μ_1 and a measure μ_2 that assumes only the values 0 and ∞. The measure μ_1 need not be unique, but there is a smallest μ_2 in the sense that if $\mu = \lambda_1 + \lambda_2$ is another such decomposition, then $\mu_2(E) \leq \lambda_2(E)$ for each $E \in \mathscr{A}$. ∎

Proof. (a) If the assertion is false, then for some $A \in \mathscr{A}$ with $\mu(A) = \infty$ we have

$$0 < \sup\{\mu(B)\colon B \subset A,\ B \in \mathscr{A},\ \mu(B) < \infty\} = a < \infty. \tag{2.1}$$

Now since $a < \infty$, we can find $B_n \subset A$, $B_n \in \mathscr{A}$, and $B_n \subset B_{n+1}$ such that $a \geqq \mu(B_n) > a - 1/n$. Let $B = \bigcup_{n=1}^{\infty} B_n$. Then $B \subset A$ and $B \in \mathscr{A}$. By Proposition 2.2, $\mu(B) = \lim_{n\to\infty} \mu(B_n)$. Therefore $\mu(B) = a$. Since $\mu(A) = \infty$, $\mu(A - B) = \infty$. Since μ is semifinite, there is $C \subset A - B$ such that $0 < \mu(C) < \infty$. Then $B \cup C \subset A$ and $\mu(B \cup C) = \mu(B) + \mu(C) > a$, in contradiction to equation (2.1). The assertion now follows.

(b) We define

$$\mu_1(E) = \sup\{\mu(B)\colon B \subset E,\ B \in \mathscr{A},\ \mu(B) < \infty\}$$

and $\mu_2(E) = \sup\{\mu(B) - \mu_1(B): B \subset E, \ B \in \mathscr{A}, \mu_1(B) < \infty\}$. These μ_1 and μ_2 will meet the requirements of the theorem. We will only show that μ_1 is a measure and leave the rest of the proof to the reader (Problem 2.1.10).

Notice that we can always find a sequence (B_n) of elements of $\mathscr{A}$ such that $B_n \subset B_{n+1} \subset E$, $\mu(B_n) < \infty$ and $\lim_{n\to\infty}\mu(B_n) = \mu_1(E)$; hence if $B = \bigcup_{n=1}^{\infty} B_n$, then $B \in \mathscr{A}$, $B \subset E$, and $\mu(B) = \mu_1(E)$. Also if $\mu(E) < \infty$, then $\mu(E) = \mu_1(E)$. Let $(E_i)_{i=1}^{\infty}$ be a sequence of pairwise disjoint sets in $\mathscr{A}$. Then if $\mu_1(\bigcup_{i=1}^{\infty} E_i) < \infty$, we can find $B_i \subset E_i$, $B_i \in \mathscr{A}$, such that $\mu(B_i) = \mu_1(E_i) < \infty$; therefore,

$$\mu_1\left(\bigcup_{i=1}^{\infty} E_i\right) \geqq \mu_1\left(\bigcup_{i=1}^{n} E_i\right) \geqq \mu\left(\bigcup_{i=1}^{n} B_i\right) = \sum_{i=1}^{n} \mu(B_i) = \sum_{i=1}^{n} \mu_1(E_i),$$

for each n. This means that $\mu_1(\bigcup_{i=1}^{\infty} E_i) \geqq \sum_{i=1}^{\infty} \mu_1(E_i)$. To prove the converse inequality, we notice that there exist $B_n \subset \bigcup_{i=1}^{\infty} E_i$, $B_n \in \mathscr{A}$, $\mu(B_n) < \infty$, such that $\mu_1(\bigcup_{i=1}^{\infty} E_i) = \lim_{n\to\infty}\mu(B_n)$. Since $B_n = \bigcup_{i=1}^{\infty}(B_n \cap E_i)$ and μ is a measure, $\mu(B_n) = \sum_{i=1}^{\infty} \mu_1(B_n \cap E_i) \leqq \sum_{i=1}^{\infty} \mu_1(E_i)$. Hence $\mu_1(\bigcup_{i=1}^{\infty} E_i) \leqq \sum_{i=1}^{\infty} \mu_1(E_i)$ and μ_1 is a measure. ∎

Problems

× **2.1.1.** Let μ be a finitely additive measure on an algebra $\mathscr{A}$. For A, $B \in \mathscr{A}$, show that

(a) $\mu(A) \leqq \mu(B)$, if $A \subset B$,

(b) $\mu(A \cup B) + \mu(A \cap B) = \mu(A) + \mu(B)$,

(c) $\mu(B - A) = \mu(B) - \mu(A)$, if $A \subset B$ and $\mu(A) < \infty$.

2.1.2. Show that in Proposition 2.2, (b) is false unless $\mu(A_n) < \infty$ for some n. [Hint: Take $X = R$, $A_n = (n, \infty)$, $\mathscr{A} = 2^X$, $\mu =$ counting measure.]

× **2.1.3.** Let $X = Z^+$, $\mathscr{A} = \{A \subset X$: either A or $X - A$ is finite$\}$. Let $\mu(A) = 0$, if A is finite, $= \infty$, if A is infinite. Then show that (*a*) $\mathscr{A}$ is an algebra and (b) μ is a finitely additive measure, which is not a measure.

2.1.4. Show that in Proposition 2.2, for a finitely additive μ, (b) does not imply countable additivity. [Hint: Use Problem 2.1.3.]

× **2.1.5.** Let $(\mu_n)_{n=1}^{\infty}$ be a sequence of measures on an algebra $\mathscr{A}$ such that for $A \in \mathscr{A}$, $\mu_n(A) \leq \mu_{n+1}(A)$. Show that μ defined on $\mathscr{A}$ by $\mu(A) = \lim_{n\to\infty}\mu_n(A)$ is a measure on $\mathscr{A}$.

2.1.6. Let μ be a measure on a σ-algebra $\mathscr{A}$. Let $E_i \in \mathscr{A}$, $1 \leq i < \infty$, and $\sum_{i=1}^{\infty} \mu(E_i) < \infty$. Show that $\mu(\overline{\lim}_{i\to\infty} E_i) = 0$. (This is known as the Borel–Cantelli Lemma and is widely used in probability theory.)

× **2.1.7.** Show that every σ-finite measure is semifinite.

× **2.1.8.** Let μ be a measure on an algebra $\mathscr{A}$. Suppose $A \in \mathscr{A}$ and for every $E \in \mathscr{A}$, $\mu_A(E) = \mu(A \cap E)$. Show that μ_A is a measure on the algebra $\{E \in \mathscr{A}: E \subset A\}$. ($\mu_A$ is called the restriction of μ to A.)

2.1.9. Suppose μ_1 and μ_2 are two measures on a σ-algebra $\mathscr{A}$ such that for each $E \in \mathscr{A}$, $\mu_1(E) \geq \mu_2(E)$. Show that there is a measure μ_3 on $\mathscr{A}$ such that $\mu_1(E) = \mu_2(E) + \mu_3(E)$ for $E \in \mathscr{A}$. This μ_3 is unique if μ_2 is σ-finite. [Hint: Take $\mu_3(E) = \sup\{\mu_1(B) - \mu_2(B): B \subset E, \mu_2(B) < \infty\}$.]

2.1.10. Complete the proof of Proposition 2.3(b).

2.1.11. For $E \subset Z^+$, define $\beta(E) = \lim_{n\to\infty}(1/n) \cdot \text{card}(E \cap [0, n])$, whenever the limit exists. Let $\mathscr{A}$ be the class of all such subsets of Z^+ for which β exists. Prove the following:

(a) if A, B and $A \cup B$ are all in $\mathscr{A}$, then $\beta(A \cup B) = \beta(A) + \beta(B)$, whenever $A \cap B = \varnothing$;

(b) $\mathscr{A}$ is not an algebra;

(c) for each real number u in $[0, 1]$, there is an $E \in \mathscr{A}$ such that $\beta(E) = u$.

[Hint for (b): Consider the sets $E_1 = \{\text{all odd positive integers}\}$, $E_2 = \bigcup_{n=0}^{\infty} \{\text{all odd integers in } [2^{2n}, 2^{2n+1}] \text{ and all even integers in } [2^{2n+1}, 2^{2n+2}]\}$, and $E_1 \cap E_2$.]

2.1.12. Let μ be a finite measure on an algebra $\mathscr{A}$. Let $A_1, A_2, \ldots, A_n$ be in $\mathscr{A}$. Show that

(a) $$\mu\left(\bigcup_{k=1}^{n} A_k\right) = \sum_{k=1}^{n} \mu(A_k) - \sum_{i_1<i_2} \mu(A_{i_1} \cap A_{i_2}) + \cdots,$$

(b) $$\mu\left(\bigcup_{k=1}^{n} A_k\right) \geq \sum_{k=1}^{n} \mu(A_k) - \sum_{i_1<i_2} \mu(A_{i_1} \cap A_{i_2}).$$

2.1.13. Let μ be a σ-finite measure on a σ-algebra $\mathscr{A}$ of subsets of X. Find a finite measure β on $\mathscr{A}$ such that for $E \in \mathscr{A}$, $\beta(E) = 0$ if and only if $\mu(E) = 0$.

2.1.14. Let $\mu(X) = 1$ in 2.1.13. Suppose that $A_n \in \mathscr{A}$, $B_n \in \mathscr{A}$, and $\mu(\lim_n \sup A_n) = 1$ and $\mu(\lim_n \inf B_n) = 1$. Show that $\mu(\lim_n \sup(A_n \cap B_n)) = 1$. What if $\mu(\lim_n \sup B_n) = 1$ instead of $\mu(\lim_n \inf B_n) = 1$?

2.1.15. *Atoms in* $(X, \mathscr{A}, \mu)$. Let μ be a measure on a σ-algebra of subsets of X. By an *atom*, we mean a set in $\mathscr{A}$ such that (i) $\mu(A) > 0$ and (ii) $A \supset B \in \mathscr{A} \Rightarrow$ either $\mu(B) = 0$ or $\mu(A - B) = 0$. Prove the following assertions:

(a) If μ is σ-finite, then $\mathscr{A}$ can have at most countably many atoms.

(b) Suppose $\mathscr{A}$ has no atoms. Let $A \in \mathscr{A}$ with $0 < \mu(A) < \infty$. Then for each u, $0 < u < \mu(A)$, there exists $B \subset A$, $B \in \mathscr{A}$ such that $\mu(B) = u$.

[Hint for (b): Since A is not an atom, given $\varepsilon > 0$, there exists $B_\varepsilon \subset A$ such that $0 < \mu(B_\varepsilon) < \varepsilon$. Let $\mathscr{F} = \{B \subset A: B \in \mathscr{A}$ and $\mu(B) \leq u\}$. Then $\mathscr{F}$ is partially ordered under inclusion. Identifying sets of equal measure, each linearly ordered subclass of $\mathscr{F}$ can contain only an at most countable number of distinct sets in $\mathscr{A}$. Use Zorn's lemma to obtain a maximal element B_u in $\mathscr{F}$. Then $\mu(B_u) = u$.]

2.1.16. Let μ_1 and μ_2 be two nonatomic measures on a σ-algebra $\mathscr{A}$ of subsets of X such that $\mu_1(X) = \mu_2(X) = 1$. Show that given $0 < u < 1$, there exists $A \in \mathscr{A}$ such that $\mu_1(A) \geq u$ and $\mu_2(X - A) \geq 1 - u$. Find a similar result for n such measures.

2.1.17. Let μ be a finite measure on a σ-algebra $\mathscr{A}$ of subsets of X. Define for any A, B in $\mathscr{A}$, $d(A, B) = \mu(A \triangle B)$. Prove the following:

(a) $(\mathscr{A}, d)$ is a complete pseudometric space.

[Hint: Let (A_n) be Cauchy in $\mathscr{A}$. Then there is a subsequence (A_{n_i}) such that $\mu(A_{n_i} \triangle \bigcup_{j=i}^\infty A_{n_j}) < 1/2^i$. If $A = \bigcup_{n=1}^\infty \bigcap_{m=n}^\infty A_m$, then $\mu(A_n \triangle A) \to 0$ so that $\lim_{n\to\infty} A_n = \lim_n \inf A_n$ (see p. 8). Notice also that since $(X - A_n)$ is also a Cauchy sequence, it follows similarly that $\lim_{n\to\infty}(X - A_n) = \lim_n \inf(X - A_n)$. Thus, $\lim_n \inf A_n = \lim_n \sup A_n$, when (A_n) converges.]

(b) If $\mathscr{A}$ is separable (see Problem 1.3.15), then $(\mathscr{A}, d)$ is a separable pseudometric space.

(c) Suppose that $\mathscr{A}$ is nonatomic. Then $(\mathscr{A}, d)$ is a convex pseudometric space; in other words, given $0 < r < 1$ and $A, B \in \mathscr{A}$, there exists $C_r \in \mathscr{A}$ such that $d(A, C_r) = r \cdot d(A, B)$ and $d(C_r, B) = (1 - r) \cdot d(A, B)$.

(d) Let $F(B) = d(A, B)$, where $A \in \mathscr{A}$. Then F is a continuous function from $(\mathscr{A}, d)$ into the nonnegative reals.

2.2. Lebesgue Measure on Intervals

The most important example of a measure is, perhaps, the Lebesgue measure on R^n $(n \geq 1)$, named after the celebrated French mathematician Henri Lebesgue (1875–1941), one of the original founders of the theory of measure and integration. Lebesgue measure is the subject of this section. Though usually the domain of Lebesgue measure is a σ-algebra containing all open sets and numerous other complicated sets, in this section we restrict ourselves to the simpler problem of defining Lebesgue measure on an algebra of half-open intervals.

Let $X = R$ and $\mathscr{A}$ be the class of all finite disjoint unions of right-closed, left-open intervals including intervals of the form $(-\infty, a]$, (a, ∞) and $(-\infty,$

$\infty)$, $-\infty < a < \infty$, and the empty set. These types of intervals are considered because, unlike open intervals, they form an algebra.

Lemma 2.1. $\mathscr{A}$ is an algebra. (See Problem 2.2.1.) ▌

To define a measure m_0 on $\mathscr{A}$, we need the following observation. If $(I_n)_{n=1}^k$ and $(J_m)_{m=1}^s$ are two finite sequences of pairwise disjoint intervals in $\mathscr{A}$ such that $\bigcup_{n=1}^k I_n = \bigcup_{m=1}^s J_m = A$, then $\sum_{n=1}^k l(I_n) = \sum_{m=1}^s l(J_m)$, where l denotes the length, $l((a, b]) = b - a$ and the length of an infinite interval is infinite. This follows from the equalities

$$l(I_n) = \sum_{m=1}^{s} l(I_n \cap J_m), \qquad l(J_m) = \sum_{n=1}^{k} l(I_n \cap J_m).$$

Thus for $A \in \mathscr{A}$, we can define

$$m_0(A) = \sum_{n=1}^{k} l(I_n), \qquad \text{if } A = \bigcup_{n=1}^{k} I_n.$$

We also note that if $A = \bigcup_{n=1}^k I_n$, $B = \bigcup_{j=1}^s J_m$, and $A \subset B$ (the intervals I_n and J_m being in $\mathscr{A}$), then we have for A and B in $\mathscr{A}$,

$$\text{(i)} \quad I_n = \bigcup_{m=1}^{s} (I_n \cap J_m) \qquad \text{and} \qquad J_m \supset \bigcup_{n=1}^{k} (I_n \cap J_m),$$

$$\text{(ii)} \quad l(I_n) = \sum_{m=1}^{s} l(I_n \cap J_m) \qquad \text{and} \qquad l(J_m) \geq \sum_{n=1}^{k} l(I_n \cap J_m),$$

and therefore it follows easily that $m_0(A) \leq m_0(B)$. Thus we have the following lemma.

Lemma 2.2. For $A \in \mathscr{A}$, let $m_0(A) = \sum_{n=1}^k l(I_n)$ if $A = \bigcup_{n=1}^k I_n$. Then m_0 is well defined, and for $B, C \in \mathscr{A}$ and $B \subset C$, $m_0(B) \leq m_0(C)$. ▌

Now we can prove the following theorem.

Theorem 2.1. m_0 is a measure on $\mathscr{A}$. (m_0 is called the Lebesgue measure on intervals.) ▌

Proof. Let $(A_i)_{i=1}^\infty$ be a sequence of pairwise disjoint sets in $\mathscr{A}$ such that $\bigcup_{i=1}^\infty A_i \in \mathscr{A}$. Then for each i, $1 \leq i < \infty$, there is a finite sequence of pairwise disjoint intervals $I_{ij} \in \mathscr{A}$ such that $A_i = \bigcup_{j=1}^{m_i} I_{ij}$. Also since $\bigcup_{i=1}^\infty A_i \in \mathscr{A}$, there are pairwise disjoint intervals $I_k \in \mathscr{A}$ such that $\bigcup_{i=1}^\infty A_i = \bigcup_{k=1}^n I_k$.

Now for each k,

$$\bigcup_{i=1}^{\infty} A_i \supset \bigcup_{i=1}^{k} A_i = \bigcup_{i=1}^{k} \bigcup_{j=1}^{m_i} I_{ij}.$$

So by Lemma 2.2.

$$m_0\left(\bigcup_{i=1}^{\infty} A_i\right) \geq m_0\left(\bigcup_{i=1}^{k} A_i\right) = \sum_{i=1}^{k} \sum_{j=1}^{m_i} l(I_{ij}) = \sum_{i=1}^{k} m_0(A_i).$$

Hence $m_0(\bigcup_{i=1}^{\infty} A_i) \geq \sum_{i=1}^{\infty} m_0(A_i)$. To prove the converse inequality, we observe that $I_k = \bigcup_{i=1}^{\infty} \bigcup_{j=1}^{m_i} (I_k \cap I_{ij})$. If $l(I_k \cap I_{ij}) = \infty$ for some i, j, then $m_0(A_i) = \infty$ and consequently $m_0(\bigcup_{i=1}^{\infty} A_i) = \sum_{i=1}^{\infty} m_0(A_i)$. So then we assume that $l(I_k \cap I_{ij}) < \infty$ for all i, j, and k.

Let $\varepsilon > 0$. We write $I_k \cap I_{i_j} = (a_{ij}^k, b_{ij}^k]$. Let $[a_k, b_k] \subset I_k$. Then

$$[a_k, b_k] \subset \bigcup_{i=1}^{\infty} \bigcup_{j=1}^{m_i} \left(a_{ij}^k, b_{ij}^k + \frac{\varepsilon}{2^i m_i}\right).$$

By compactness of $[a_k, b_k]$, there is a positive integer p such that

$$[a_k, b_k] \subset \bigcup_{i=1}^{p} \bigcup_{j=1}^{m_i} \left(a_{ij}^k, b_{ij}^k + \frac{\varepsilon}{2^i m_i}\right).$$

Therefore,

$$b_k - a_k \leq \sum_{i=1}^{p} \sum_{j=1}^{m_i} \left(b_{ij}^k - a_{ij}^k + \frac{\varepsilon}{2^i m_i}\right) \leq \sum_{i=1}^{\infty} \sum_{j=1}^{m_i} l(I_k \cap I_{ij}) + \varepsilon.$$

Since $\varepsilon (> 0)$ is arbitrary and $[a_k, b_k]$ is an arbitrary closed subinterval of I_k,

$$l(I_k) \leq \sum_{i=1}^{\infty} \sum_{j=1}^{m_i} l(I_k \cap I_{ij}).$$

Hence

$$\begin{aligned} m_0\left(\bigcup_{i=1}^{\infty} A_i\right) &= \sum_{k=1}^{n} l(I_k) \\ &\leq \sum_{k=1}^{n} \sum_{i=1}^{\infty} \sum_{j=1}^{m_i} l(I_k \cap I_{ij}) = \sum_{i=1}^{\infty} \sum_{j=1}^{m_i} \sum_{k=1}^{n} l(I_k \cap I_{ij}) \\ &= \sum_{i=1}^{\infty} \sum_{j=1}^{m_i} l(I_{ij}), \qquad I_{ij} = \bigcup_{k=1}^{n} I_k \cap I_{ij} \\ &= \sum_{i=1}^{\infty} m_0(A_i). \end{aligned}$$

∎

Remarks

2.1. In R^n $(n > 1)$, the class of all finite disjoint unions of sets of the form

$$I = \{(x_1, x_2, \ldots, x_n) \colon x_i \in I_i, \text{ an interval in } \mathscr{A} \text{ above}, 1 \le i \le n\}$$

form an algebra, as before. I is generally called a generalized interval and the interval I_j in its definiton is called its jth side. By defining $m_0(I) = l(I_1) \cdots l(I_n)$, we can again obtain (as before) a measure (known as n-dimensional Lebesgue measure on the generalized intervals). (Serious readers are encouraged to construct proofs for these facts. The proofs should be similar to those in the case of $n = 1$.)

2.2. For any given monotonic increasing function F on R that is continuous on the right, it is now possible to construct a measure m_F on the algebra $\mathscr{A}$ such that $m_F(a, b] = F(b) - F(a)$. This measure m_F is usually called the Lebesgue–Stieltjes measure on $\mathscr{A}$ induced by F. (See Problem 2.2.5.)

Problems

(In the following problems, $\mathscr{A}$ is always the algebra as defined in the beginning of this section.)

× **2.2.1.** Prove Lemma 2.1.

× **2.2.2.** Prove that the Lebesgue measure m_0 is translation invariant on the intervals in $\mathscr{A}$ [that is, $m_0(I) = m_0(I + x)$ for each interval I in $\mathscr{A}$ and any point $x \in R$].

× **2.2.3.** Let λ be a measure on $\mathscr{A}$ such that $\lambda(I + d) = \lambda(I)$ for each interval I in $\mathscr{A}$ and for each d in a dense subset D of R. Let $\lambda(I) < \infty$ for each finite interval in $\mathscr{A}$. Show that there is a constant k such that $\lambda(A) = k \cdot m_0(A)$ for each $A \in \mathscr{A}$, m_0 being the Lebesgue measure. [Hint: Show that λ is translation invariant (Problem 2.2.3) by using Proposition 2.2. If $\lambda(0, 1] = k$, then $\lambda(0, p] = kp$ for every positive integer p. Thus, if $l(I)$ is a rational number, then $\lambda(I) = k \cdot l(I)$.]

2.2.4. Let λ be a measure on $\mathscr{A}$ such that $\lambda(I) < \infty$ for every finite interval I. Let $y \in R$ and F_y be the function on R given by

$$F_y(x) = \begin{cases} -\lambda(x, y], & \text{if } x < y, \\ 0, & \text{if } x = y, \\ \lambda(y, x] & \text{if } x > y. \end{cases}$$

Show that F_y is a monotonic increasing function, which is continuous on the right. (F_y is usually called a distribution function induced by λ.)

× **2.2.5.** Let F be a monotonic increasing function on R that is continuous on the right. Then show that

(a) If $(a, b] \subset \bigcup_{i=1}^{\infty} (a_i, b_i)$, then

$$F(b) - F(a) \leq \sum_{i=1}^{\infty} F(b_i) - F(a_i).$$

(b) Let

$$m_F(a, b] = F(b) - F(a),$$

$$m_F(a, \infty) = \lim_{x\to\infty} F(x) - F(a),$$

$$m_F(-\infty, a] = F(a) - \lim_{x\to-\infty} F(x).$$

Show that m_F defines a measure on $\mathscr{A}$ in an obvious manner. [This measure is usually called the *Lebesgue–Stieltjes measure* on $\mathscr{A}$ induced by F. Note that $F(x) = x$ yields the Lebesgue measure on $\mathscr{A}$.]

2.2.6. Let $Z_1 = \{1/n: n \in Z^+\}$. For $B \in \mathscr{A}$, let $\mu_1(B) = \lambda(B \cap Z^+)$ and $\mu_2(B) = \lambda(B \cap Z_1)$, where λ is the counting measure. Show that μ_1 is a Lebesgue–Stieltjes measure m_F for some right continuous nondecreasing function F on R [see Problem 2.2.5(b)], whereas μ_2 is not.

2.2.7. *The Setwise Limit of a Sequence of Finite Measures on an Algebra Need Not Be a Measure.* Consider the following example to show this. For each positive integer n, define $\mu_n(E) = n \cdot m_0(E \cap (0, 1/n])$. Then each μ_n is a measure and for each $E \in \mathscr{A}$, $\lim_{n\to\infty}\mu_n(E) = \mu(E)$ exists, where μ is a set function having only the values 0 or 1. This μ is not countably additive. (Compare Problem 2.1.5. If the algebra is replaced by a σ-algebra, then the limit must be a measure; see Problem 4.3.17.)

2.3. Construction of Measures: Outer Measures and Measurable Sets

In the previous two sections, we defined the concept of a measure on an algebra and then constructed Lebesgue measure on the algebra generated by the left-open, right-closed intervals in R. To develop a suitable theory of integration, whether on R or on any abstract set, we need a measure to be defined on a σ-algebra. This is usually done through the introduction of outer measures, a concept somewhat different from that of measure and due to C. Carathéodory, a distinguished Greek mathematician.

Our main concerns in this section are (1) how to construct an outer measure on 2^X and then to derive from it a measure on a σ-algebra, and (2) to extend a given measure on an algebra to one on a σ-algebra containing it and, in particular, to extend the domain of Lebesgue measure from the algebra generated by the half-open intervals to a much larger class of sets, a σ-algebra of Lebesgue-measurable sets.

Definition 2.2. By an outer measure μ^* we mean an extended real-valued set function defined on 2^X, having the following properties:

(a) $\mu^*(\varnothing) = 0$,

(b) $\mu^*(A) \leq \mu^*(B)$, for $A \subset B$,

(c) $\mu^*(\bigcup_{n=1}^{\infty} E_n) \leq \sum_{n=1}^{\infty} \mu^*(E_n)$. ∎

Note that an outer measure is always nonnegative.

Examples

2.4. Suppose

$$\mu^*(E) = \begin{cases} 0, & E = \varnothing; \\ 1, & E \neq \varnothing. \end{cases}$$

Then μ^* is an outer measure, which is not a measure on 2^X, if X has at least two points.

2.5. Let X be uncountable. Suppose

$$\mu^*(E) = \begin{cases} 0, & E \text{ countable}, \\ 1, & E \text{ uncountable}. \end{cases}$$

Then μ^* is an outer measure.

More examples of outer measures will follow the next proposition, which outlines a method of constructing an outer measure on 2^X.

Proposition 2.4. Let $\mathscr{F}$ be a class of subsets of X containing the empty set such that for every $A \subset X$ there exists a sequence $(B_n)_{n=1}^{\infty}$ from $\mathscr{F}$ such that $A \subset \bigcup_{n=1}^{\infty} B_n$. Let τ be an extended real-valued function on $\mathscr{F}$ such that $\tau(\varnothing) = 0$ and $\tau(A) \geq 0$ for $A \in \mathscr{F}$. Then μ^* defined on 2^X by

$$\mu^*(A) = \inf\left\{ \sum_{n=1}^{\infty} \tau(B_n) : B_n \in \mathscr{F},\ A \subset \bigcup_{n=1}^{\infty} B_n \right\}$$

is an outer measure.

Proof. First, clearly $\mu^*(\varnothing) = 0$. Next, if $A_1 \subset A_2$ and $A_2 \subset \bigcup_{n=1}^{\infty} B_n$, then $A_1 \subset \bigcup_{n=1}^{\infty} B_n$. This means that $\mu^*(A_1) \leq \mu^*(A_2)$.

Finally, let $E_n \subset X$ for each natural number n. Then if $\mu^*(E_n) = \infty$ for some n, $\mu^*(\bigcup_{n=1}^{\infty} E_n) \leq \sum_{n=1}^{\infty} \mu^*(E_n)$. Suppose that $\mu^*(E_n) < \infty$ for each n. Then given $\varepsilon > 0$, there exists $(B_{nm})_{m=1}^{\infty}$ from $\mathscr{F}$ such that

$$E_n \subset \bigcup_{m=1}^{\infty} B_{nm}$$

and

$$\sum_{m=1}^{\infty} \tau(B_{nm}) < \mu^*(E_n) + \varepsilon/2^n.$$

Now

$$\bigcup_{n=1}^{\infty} E_n \subset \bigcup_{n=1}^{\infty} \bigcup_{m=1}^{\infty} B_{nm}$$

and, therefore,

$$\mu^*\left(\bigcup_{n=1}^{\infty} E_n \right) \leq \sum_{n=1}^{\infty} \sum_{m=1}^{\infty} \tau(B_{nm}) < \sum_{n=1}^{\infty} \mu^*(E_n) + \varepsilon.$$

Since $\varepsilon > 0$ is arbitrary, the proposition follows. ∎

In the next four examples, $\mathscr{F}$, τ, and μ^* are defined as in Proposition 2.4. These examples will illustrate this proposition.

Examples

2.6. Let $X = Z_+$, $\mathscr{F} = \{\{x\}: x \in X\} \cup \{\varnothing\}$.
Suppose

$$\tau(E) = \begin{cases} 0, & E = \varnothing; \\ 1, & E \neq \varnothing. \end{cases}$$

Then

$$\mu^*(A) = \begin{cases} \infty, & \text{if } A \text{ is infinite;} \\ \text{the number of points in } A, & \text{if } A \text{ is finite.} \end{cases}$$

2.7. In the previous example, let $\mathscr{F} = \{X, \varnothing\}$ and $\tau(X) = 1$, $\tau(\varnothing) = 0$. Then

$$\mu^*(A) = \begin{cases} 1, & \text{if } A \neq \varnothing; \\ 0, & \text{if } A = \varnothing. \end{cases}$$

2.8. Let $X = R$ and $\mathscr{F} = \{A: A \neq X, A \subset X\}$.
Suppose

$$\tau(A) = \begin{cases} 0, & \text{if } A = \varnothing; \\ 1, & \text{if } A \neq \varnothing. \end{cases}$$

Then

$$\mu^*(B) = \begin{cases} 2, & \text{if } B = X, \\ 1, & \text{if } B \neq \emptyset, \quad B \neq X, \\ 0, & \text{if } B = \emptyset. \end{cases}$$

2.9. *Lebesgue Outer Measure.* Let $X = R$ and $\mathscr{F}$ be the class of all left-open, right-closed intervals {finite or infinite of the form $(-\infty, a]$, $(a, b]$, or (a, ∞)}, including the empty set.

Let $\tau(\emptyset) = 0$ and $\tau(I) = l(I)$ for every interval $I \in \mathscr{F}$. Then the outer measure μ^*, induced by τ, on 2^R is called the Lebesgue outer measure on R and

$$\mu^*(A) = \inf\left\{ \sum_{n=1}^{\infty} l(I_n): A \subset \bigcup_{n=1}^{\infty} I_n,\, I_n \in \mathscr{F} \right\}.$$

Later in this section, it will follow that $\mu^*(I) = l(I)$ for every interval $I \in \mathscr{F}$ and the restriction of μ^* to a special class of subsets of R will yield a measure known as the Lebesgue measure on R, which will be an extension of the Lebesgue measure on intervals discussed in Section 2.2.

The purpose of introducing outer measures is to construct a measure on a σ-algebra. Outer measures are, in general, not measures on 2^X, as can be seen in Examples 2.4 and 2.5. But it so happens that an outer measure when restricted to a suitable class of subsets (usually called measurable sets) becomes a measure on a σ-algebra. The next theorem will demonstrate this fact. First, we need the following definition. (In what follows μ^* is always an outer measure.)

Definition 2.3. $E \subset X$ is called μ^*-measurable if for every $A \subset X$

$$\mu^*(A) = \mu^*(A \cap E) + \mu^*(A \cap E^c).$$

[This is equivalent to requiring only $\mu^*(A) \geq \mu^*(A \cap E) + \mu^*(A \cap E^c)$, since the converse inequality is obvious from the subadditive property of μ^*.] ∎

Notice that a μ^*-measurable set E splits no set A in such a way that μ^* fails to be additive on $\{A \cap E, A \cap E^c\}$. Since we are looking for a class of sets on which μ^* can act at least additively, the preceding definition seems to be meaningful to achieve that end. The next theorem justifies its meaningfulness.

Theorem 2.2. The class $\mathscr{B}$ of μ^*-measurable sets is a σ-algebra. Also $\bar{\mu}$, the restriction of μ^* to $\mathscr{B}$ is a measure. ∎

Proof. First, $\varnothing \in \mathscr{B}$; also $E \in \mathscr{B}$ implies $E^c \in \mathscr{B}$, by the symmetry of Definition 2.3.

Next, we wish to show that $\mathscr{B}$ is closed under finite unions. Let $E \in \mathscr{B}$, $F \in \mathscr{B}$, and $A \subset X$. Since $F \in \mathscr{B}$,

$$\mu^*(A \cap E^c) = \mu^*(A \cap E^c \cap F) + \mu^*(A \cap E^c \cap F^c).$$

Also,

$$A \cap (E \cup F) = (A \cap E) \cup (A \cap E^c \cap F)$$

and therefore

$$\mu^*\big(A \cap (E \cup F)\big) \leq \mu^*(A \cap E) + \mu^*(A \cap E^c \cap F),$$

so that

$$\begin{aligned}\mu^*&\big(A \cap (E \cup F) + \mu^*\big(A \cap (E \cup F)^c\big)\\ &\leq \mu^*(A \cap E) + \mu^*(A \cap E^c \cap F) + \mu^*(A \cap E^c \cap F^c)\\ &= \mu^*(A \cap E) + \mu^*(A \cap E^c), \qquad \text{using the first equality,}\\ &= \mu^*(A), \qquad \text{since } E \in \mathscr{B}.\end{aligned}$$

This proves that $E \cup F \in \mathscr{B}$ and $\mathscr{B}$ is now an algebra.

Now we show that μ^* is a measure on the algebra $\mathscr{B}$. Let $(E_n)_{n=1}^{\infty}$ be a sequence of pairwise disjoint sets from $\mathscr{B}$. Then if $A \subset X$,

$$\mu^*\big(A \cap (E_1 \cup E_2)\big) = \mu^*(A \cap E_1) + \mu^*(A \cap E_2)$$

(using the fact that $E_2 \in \mathscr{B}$). Hence by induction (verify this), for each n

$$\mu^*\left[A \cap \left(\bigcup_{i=1}^{n} E_i\right)\right] = \sum_{i=1}^{n} \mu^*(A \cap E_i). \tag{2.2}$$

Now if $A = X$, for each n

$$\mu^*\left(\bigcup_{i=1}^{\infty} E_i\right) \geq \mu^*\left(\bigcup_{i=1}^{n} E_i\right) = \sum_{i=1}^{n} \mu^*(E_i).$$

This means that $\mu^*(\bigcup_{i=1}^{\infty} E_i) \geq \sum_{i=1}^{\infty} \mu^*(E_i)$ or μ^* is countably additive on $\mathscr{B}$ and hence a measure on $\mathscr{B}$.

Finally, we show that $\mathscr{B}$ is a σ-algebra. Let $(F_n)_{n=1}^{\infty}$ be a sequence of sets in $\mathscr{B}$. Write $G_1 = F_1$, $G_n = F_n - \bigcup_{i=1}^{n-1} F_i$ for $n > 1$. Then G_n's are pairwise disjoint sets in $\mathscr{B}$ and $\bigcup_{n=1}^{\infty} G_n = \bigcup_{n=1}^{\infty} F_n$. (Recall that $\mathscr{B}$ is already an algebra.) Let $A \subset X$. Then since $\bigcup_{n=1}^{m} G_n \in \mathscr{B}$, we have [using equation (2.2)]

$$\mu^*(A) = \mu^*\left[A \cap \left(\bigcup_{n=1}^{m} G_n\right)\right] + \mu^*\left[A \cap \left(\bigcup_{n=1}^{m} G_n\right)^c\right]$$

$$\geq \sum_{n=1}^{m} \mu^*(A \cap G_n) + \mu^*\left[A \cap \left(\bigcup_{n=1}^{\infty} G_n\right)^c\right]$$

for every positive integer m. This means that

$$\mu^*(A) \geq \sum_{n=1}^{\infty} \mu^*(A \cap G_n) + \mu^*\left[A \cap \left(\bigcup_{n=1}^{\infty} G_n\right)^c\right]$$

$$\geq \mu^*\left[A \cap \left(\bigcup_{n=1}^{\infty} G_n\right)\right] + \mu^*\left[A \cap \left(\bigcup_{n=1}^{\infty} G_n\right)^c\right],$$

and so

$$\bigcup_{n=1}^{\infty} G_n = \bigcup_{n=1}^{\infty} F_n \in \mathscr{B}.$$

∎

Remark 2.3. Note that $\mu^*(A) = 0$ implies A is μ^* measurable; and therefore the measure $\bar{\mu}$ (the restriction of μ^* on $\mathscr{B}$) has the following property: $E \in \mathscr{B}$, $\bar{\mu}(E) = 0$, and $F \subset E$ imply $F \in \mathscr{B}$. A measure having this property is called a *complete measure*. Completeness is a useful property for many technical considerations. (This will be seen when we discuss measurable functions in the next chapter.) Not every measure on a σ-algebra is complete (see Problems 2.4.3 and 2.4.4.) However, every measure on a σ-algebra can be completed in the following sense.

Proposition 2.5. Let μ be a measure on a σ-algebra $\mathscr{A}$. Suppose

$$\mathscr{A}_0 = \{A \cup B \colon A \in \mathscr{A}, B \subset C, C \in \mathscr{A}, \text{ and } \mu(C) = 0\}$$

and $\mu_0(A \cup B) = \mu(A)$ for $A \cup B \in \mathscr{A}_0$. Then $\mathscr{A}_0$ is a σ-algebra and μ_0 is a complete measure on $\mathscr{A}_0$, which is an extension of μ. ∎

The proof of this proposition is left to the reader.

The following examples will now illustrate Theorem 2.2.

Examples

2.10. Consider the outer measure μ^* in Example 2.7. Then if E is a nonempty proper subset of X and $A = X$,

$$\mu^*(A) = 1 \neq 2 = \mu^*(A \cap E) + \mu^*(A \cap E^c).$$

Hence $\{\emptyset, X\}$ is the class of all μ^*measurable subsets of X.

2.11. If μ^* is the counting measure on 2^X, then the class of all μ^*-measurable sets is 2^X.

If the class of μ^*-measurable sets turns out to be trivial or small, the measure $\bar{\mu}$ (induced by μ^*) does not seem to be very useful for many practical purposes. So it is important to know when sufficiently many sets will be μ^*-measurable. The next theorem provides an interesting result in this direction. We need first the following definition illustrated by an example.

Definition 2.4. Let (X, d) be a metric space and μ^* be an outer measure on 2^X such that $\mu^*(A \cup B) = \mu^*(A) + \mu^*(B)$ whenever $d(A, B)^\dagger > 0$. Then μ^* is called a *metric outer measure.* ∎

Example 2.12. The *Lebesgue outer measure* is a metric outer measure. To show this, let $X = R$, $A \subset R$, $B \subset R$, $d(A, B) > 0$ (d being the usual real-line distance) and μ^*, Lebesgue outer measure defined in Example 2.9.

It suffices to show that $\mu^*(A \cup B) \geq \mu^*(A) + \mu^*(B)$. If $\mu^*(A \cup B) = \infty$, we are done. Suppose $\mu^*(A \cup B) < \infty$. Then given $\varepsilon > 0$, there exists a sequence of intervals $((a_n, b_n])_{n=1}^{\infty}$ such that $A \cup B \subset \bigcup_{n=1}^{\infty}(a_n, b_n]$ and $\sum_{n=1}^{\infty}(b_n - a_n) \leq \mu^*(A \cup B) + \varepsilon$. Now since $d(A, B) > 0$, there is a positive integer n_0 such that $d(A, B) > 1/n_0$. Now for each n, we can write $(a_n, b_n] = \bigcup_{i=1}^{k_n} I_{ni}$; where $(I_{ni})_{i=1}^{k_n}$ are pairwise disjoint left-open, right-closed intervals and $l(I_{ni}) \leq 1/n_0$. Then $A \cup B \subset \bigcup_{n=1}^{\infty}\bigcup_{i=1}^{k_n} I_{ni}$. Since $d(A, B) > 1/n_0$, each I_{ni} can only intersect one of the two sets A and B. Hence, some of the I_{ni}'s will cover A while others will cover B. Therefore,

$$\mu^*(A) + \mu^*(B) \leq \sum_{n=1}^{\infty} \sum_{i=1}^{k_n} l(I_{ni})$$

$$= \sum_{n=1}^{\infty} (b_n - a_n) \leq \mu^*(A \cup B) + \varepsilon.$$

Since $\varepsilon > 0$ is arbitrary, the claim is proven.

Theorem 2.3. Every Borel set in a metric space (X, d) is μ^* measurable with respect to an outer measure μ^* on 2^X if and only if μ^* is a metric outer measure. (Note that the class of Borel sets is the smallest σ-algebra containing the open sets of X. See Problem 1.3.13 in Chapter 1.) ∎

Proof. For the "only if" part, suppose every open set in X is μ^*-measurable. Let $d(A, B) = \delta > 0$. Then $A \subset G = \bigcup_{x \in A}\{y: d(x, y) < \delta\}$, which

† $d(A, B) = \inf\{d(x, y) : x \in A, y \in B\}$.

is an open set and therefore μ^*-measurable. Clearly $G \cap B = \emptyset$. Therefore,

$$\mu^*(A \cup B) = \mu^*\big((A \cup B) \cap G\big) + \mu^*\big((A \cup B) \cap G^c\big) = \mu^*(A) + \mu^*(B).$$

Hence μ^* is a metric outer measure.

For the "if" part, suppose μ^* is a metric outer measure. Since the class of μ^*-measurable sets is a σ-algebra by Theorem 2.2, it is sufficient to show that every closed set is μ^*-measurable. Let B be a closed set and A any set. We must show that

$$\mu^*(A) \geq \mu^*(A \cap B) + \mu^*(A \cap B^c).$$

To do this, we consider $B_n = \{x \in A \cap B^c\colon d(x, B) \geq 1/n\}$. Clearly, $B_n \subset B_{n+1} \subset A \cap B^c$ and $d(B_n, B) \geq 1/n$. Since

$$\mu^*(A) \geq \mu^*\big((A \cap B) \cup B_n\big) = \mu^*(A \cap B) + \mu^*(B_n),$$

it is sufficient to show that

$$\lim_{n\to\infty} \mu^*(B_n) \geq \mu^*(A \cap B^c).$$

To show this, we observe that

$$A \cap B^c = \bigcup_{n=1}^{\infty} B_n = B_n \cup (B_{n+1} - B_n) \cup (B_{n+2} - B_{n+1}) \cup \cdots$$

Hence

$$\mu^*(A \cap B^c) \leq \mu^*(B_n) + \sum_{k=n}^{\infty} \mu^*(B_{k+1} - B_k).$$

If

$$\sum_{k=1}^{\infty} \mu^*(B_{k+1} - B_k) < \infty,$$

then clearly

$$\mu^*(A \cap B^c) \leq \lim_{n\to\infty} \mu^*(B_n),$$

and we are done.

Therefore, we suppose that

$$\sum_{k=1}^{\infty} \mu^*(B_{k+1} - B_k) = \infty.$$

We notice that, for $n \geq 2$,

$$B_1 \cup (B_3 - B_2) \cup \cdots \cup (B_{2n-1} - B_{2n-2}) \subset B_{2n}$$

and

$$(B_2 - B_1) \cup \cdots \cup (B_{2n} - B_{2n-1}) \subset B_{2n}.$$

Also

$$d(B_{2n-1} - B_{2n-2}, B_{2n-3} - B_{2n-4}) > 0$$

as well as

$$d(B_{2n} - B_{2n-1}, B_{2n-2} - B_{2n-3}) > 0.$$

Since μ^* is a metric outer measure, this means that we have, for every positive integer n,

$$\begin{aligned} 2\mu^*(B_{2n}) &\geq \sum_{k=1}^{n-1} \mu^*(B_{2k+1} - B_{2k}) + \sum_{k=1}^{n} \mu^*(B_{2k} - B_{2k-1}) \\ &= \sum_{k=1}^{2n-1} \mu^*(B_{k+1} - B_k), \end{aligned}$$

and hence $\lim_{n\to\infty}\mu^*(B_{2n}) = \infty$. Therefore, $\lim_{n\to\infty}\mu^*(B_n) = \infty$ and

$$\mu^*(A \cap B^c) \leq \lim_{n\to\infty} \mu^*(B_n),$$

which was to be shown. ∎

Now we recall the definition of *Lebesgue outer measure* from Example 2.9. Since it is a very special (and perhaps the most important outer measure on R) we will denote it, from now on, by m^*. Hence for $A \subset R$,

$$m^*(A) = \inf\left\{ \sum_{n=1}^{\infty} l(I_n) : A \subset \bigcup_{n=1}^{\infty} I_n, \text{ where the } I_n \text{ are intervals of the form } (-\infty, a], (a, b] \text{ or } (a, \infty)\right\}.$$

Let $\mathscr{A}$ be the algebra of finite disjoint union of these intervals and m_0 be the Lebesgue measure on $\mathscr{A}$, as discussed in Section 2.2. Then it is clear that

$$m^*(A) = \inf\left\{ \sum_{n=1}^{\infty} m_0(B_n) : A \subset \bigcup_{n=1}^{\infty} B_n, B_n \in \mathscr{A}\right\}.$$

In other words, the outer measure m^* is induced on 2^R by the measure m_0 on $\mathscr{A}$. By Theorem 2.2, m^* (restricted to the m^*-measurable sets, called the

Lebesgue-measurable sets and denoted by $\mathscr{M}$) is a measure. This is the well-known *Lebesgue measure* on R. We denote it by m. Now by Example 2.12, m^* is a metric outer measure, and by Theorem 2.3, $\mathscr{M}$ contains all Borel sets on R. (Caution: Not every set in $\mathscr{M}$ is a Borel set and not every set in 2^R is in $\mathscr{M}$. See Problem 2.3.3, Theorem 2.5, and Problem 2.4.3.) Since m^* is induced by m_0 and $\mathscr{M} \supset \mathscr{A}$, a natural question is whether m is an extension of m_0. That indeed this is so even in the general situation is demonstrated by the following important theorem.

Theorem 2.4. *Carathéodory Extension Theorem.* Let μ be a measure on an algebra[†] $\mathscr{A} \subset 2^X$. Suppose for $E \subset X$

$$\mu^*(E) = \inf\left\{ \sum_{i=1}^{\infty} \mu(E_i) \colon E \subset \bigcup_{i=1}^{\infty} E_i,\ E_i \in \mathscr{A} \right\}.$$

Then the following properties hold:

(a) μ^* is an outer measure.
(b) $E \in \mathscr{A}$ implies $\mu(E) = \mu^*(E)$.
(c) $E \in \mathscr{A}$ implies E is μ^*-measurable.
(d) The restriction $\bar{\mu}$ of μ^* to the μ^*-measurable sets is an extension of μ to a measure on a σ-algebra containing $\mathscr{A}$.
(e) If μ is σ-finite,[‡] then $\bar{\mu}$ is the only measure (on the smallest σ-algebra containing $\mathscr{A}$) that is an extension of μ. ∎

Proof. (a) This follows from Proposition 2.4.

(b) Let $E \in \mathscr{A}$. Clearly $\mu^*(E) \leq \mu(E)$. Conversely, given $\varepsilon > 0$, there exists $E_i \in \mathscr{A}$, $1 \leq i < \infty$ such that

$$E \subset \bigcup_{i=1}^{\infty} E_i \qquad \text{and} \qquad \sum_{i=1}^{\infty} \mu(E_i) \leq \mu^*(E) + \varepsilon.$$

But

$$E = \bigcup_{i=1}^{\infty} (E \cap E_i).$$

Therefore,

$$\mu(E) \leq \sum_{i=1}^{\infty} \mu(E \cap E_i).$$

This means that $\mu(E) \leq \mu^*(E) + \varepsilon$. Since $\varepsilon > 0$ is arbitrary, (b) follows.

[†] It is sufficient to let μ be a nonnegative countably additive extended real-valued set function and $\mathscr{A}$ be a ring of sets satisfying Proposition 2.4.

[‡] Problem 2.3.20 describes what happens when μ is semifinite.

(c) Let $E \in \mathscr{A}$. To prove that E is μ^*measurable, it suffices to show the following:

$$\mu^*(A) \geq \mu^*(A \cap E) + \mu^*(A \cap E^c) \qquad \text{for } A \subset X. \tag{2.3}$$

Given $\varepsilon > 0$, there exists $A_i \in \mathscr{A}$, $1 \leq i < \infty$ such that

$$\sum_{i=1}^{\infty} \mu(A_i) \leq \mu^*(A) + \varepsilon, \qquad A \subset \bigcup_{i=1}^{\infty} A_i. \tag{2.4}$$

Now

$$A \cap E \subset \bigcup_{i=1}^{\infty} (A_i \cap E) \qquad \text{and} \qquad A \cap E^c \subset \bigcup_{i=1}^{\infty} (A_i \cap E^c).$$

Therefore,

$$\mu^*(A \cap E) \leq \sum_{i=1}^{\infty} \mu(A_i \cap E) \tag{2.5}$$

and

$$\mu^*(A \cap E^c) \leq \sum_{i=1}^{\infty} \mu(A_i \cap E^c). \tag{2.6}$$

From inequalities (2.4)–(2.6), the inequality (2.3) follows.

(d) This assertion follows from above and Theorem 2.2.

(e) Let $\mathscr{B}$ be the smallest σ-algebra containing $\mathscr{A}$ and μ_1 be another measure on $\mathscr{B}$ such that $\mu_1(E) = \mu(E)$ for $E \in \mathscr{A}$. We need to show the following:

$$\mu_1(A) = \bar{\mu}(A) \qquad \text{for } A \in \mathscr{B}. \tag{2.7}$$

Since μ is σ-finite, we can write $X = \bigcup_{i=1}^{\infty} E_i$, $E_i \in \mathscr{A}$, $E_i \cap E_j = \varnothing$ $(i \neq j)$ and $\mu(E_i) < \infty$, $1 \leq i < \infty$. For $A \in \mathscr{B}$,

$$\bar{\mu}(A) = \sum_{i=1}^{\infty} \bar{\mu}(A \cap E_i) \qquad \text{and} \qquad \mu_1(A) = \sum_{i=1}^{\infty} \mu_1(A \cap E_i).$$

So to prove equation (2.7), it is sufficient to show the following:

$$\mu_1(A) = \bar{\mu}(A) \qquad \text{for } A \in \mathscr{B}, \text{ whenever } \bar{\mu}(A) < \infty. \tag{2.8}$$

Let $A \in \mathscr{B}$ with $\bar{\mu}(A) < \infty$. Given $\varepsilon > 0$, there are $E_i \in \mathscr{A}$, $1 \leq i < \infty$, $A \subset \bigcup_{i=1}^{\infty} E_i$, and

$$\bar{\mu}\left(\bigcup_{i=1}^{\infty} E_i\right) \leq \sum_{i=1}^{\infty} \mu(E_i) < \bar{\mu}(A) + \varepsilon. \tag{2.9}$$

Since

$$\mu_1(A) \le \mu_1\left(\bigcup_{i=1}^{\infty} E_i\right) \le \sum_{i=1}^{\infty} \mu_1(E_i) = \sum_{i=1}^{\infty} \mu(E_i),$$

it follows from inequality (2.9) that

$$\mu_1(A) \le \bar{\mu}(A). \tag{2.10}$$

Now considering the sets E_i from inequality (2.9), $F = \bigcup_{i=1}^{\infty} E_i \in \mathscr{B}$ and so F is μ^*-measurable. Since $A \subset F$, $\bar{\mu}(F) = \bar{\mu}(A) + \bar{\mu}(F - A)$ or $\bar{\mu}(F - A) = \bar{\mu}(F) - \bar{\mu}(A) < \varepsilon$ [from inequality (2.9)]. Since $\mu_1(E) = \bar{\mu}(E)$ for each $E \in \mathscr{A}$, $\mu_1(F) = \bar{\mu}(F)$. (Why?) Then

$$\bar{\mu}(A) \le \bar{\mu}(F) = \mu_1(F) = \mu_1(A) + \mu_1(F - A) \le \mu_1(A) + \bar{\mu}(F - A) \tag{2.11}$$

[by inequality (2.10), since inequality (2.10) is true if A is replaced by any set in $\mathscr{B}$ with finite $\bar{\mu}$-measure]. Then from inequality (2.11) it follows that $\bar{\mu}(A) \le \mu_1(A)$. This inequality along with inequality (2.10) completes the proof. ∎

We are now in a position to summarize the main properties of m, the Lebesgue measure on R:

(1) m is a measure on the σ-algebra of Lebesgue-measurable sets, which includes all Borel sets properly. (An example of a Lebesgue-measurable set that is not a Borel set is given in the next section. See also Problem 2.3.13.) Also not every set on R is Lebesgue measurable, which is the subject matter of the next section.

(2) $m(I) = l(I)$ for every interval I (open, half-open, or closed), since m is an extension of m_0 (Theorem 2.4) and $m(\{x\}) = 0$ for every singleton $\{x\}$ (Problem 2.3.4).

(3) m is translation invariant (Problems 2.3.5 and 2.3.6), i.e., for each Lebesgue-measurable set $A \subset R$ and $x \in R$, $A + x$ is Lebesgue measurable and $m(A + x) = m(A)$.

(4) The Lebesgue-measurable sets can be approximated in Lebesgue measure by open sets containing them and by closed sets contained in them. (See Problem 2.3.9.)

While we do not present the proofs of these properties here, the proofs of similar properties of m_n (the n-dimensional Lebesgue measure) will be sketched later in this section. The student must, however, work out the details of the problems that outline the properties of m.

If we now recall Problem 2.2.5, we see that the Lebesgue measure m_0 on intervals is only a special case of the more general (and very useful in probability theory) measure called the Lebesgue–Stieltjes measure m_F on intervals induced by a monotonic increasing function F that is continuous on the right. Moreover

$$m_F(a, b] = F(b) - F(a),$$

$$m_F(a, \infty) = \lim_{x\to\infty} F(x) - F(a),$$

$$m_F(-\infty, a] = F(a) - \lim_{x\to-\infty} F(x).$$

Now this m_F is a measure on the algebra $\mathscr{A}$ of finite disjoint union of the above half-open intervals. Following the Extension Theorem, Theorem 2.4, we can form $m_F{}^*$, the outer measure induced by m_F. Consequently, this outer measure yields a unique measure $\bar{m}_F$ (since m_F is σ-finite on $\mathscr{A}$) on the Borel sets of R, which is an extension of m_F. By Proposition 2.5, the Borel measure $\bar{m}_F$ (that means $\bar{m}_F$ restricted to the Borel sets) can be completed. This completion is called the Lebesgue–Stieltjes measure on R induced by F and is the same as the restriction of $m_F{}^*$ to the $m_F{}^*$-measurable sets. (See Problem 2.3.19.)

Like the Lebesgue measure m in R, there is an n-dimensional Lebesgue measure m_n in R^n, obtained through the method of Theorem 2.4 from the measure m_n defined in Remark 2.1 on the algebra of finite disjoint union of intervals in R^n of the type $I_1 \times I_2 \times \cdots \times I_n$, where each I_j is an one-dimensional interval of the form $(-\infty, a]$, (a, ∞), $(a, b]$ or $(-\infty, \infty)$. (An alternative way of obtaining m_n will be clear in Chapter 3 via Lemmas 3.3 and 3.4.) Thus, for $I = I_1 \times I_2 \times \cdots \times I_n$,

$$m_n(I) = m(I_1)m(I_2) \cdots m(I_n)$$

defines the volume of I. The sets in the smallest σ-algebra containing the intervals in R^n are m_n-measurable and known as n-dimensional Borel subsets of R^n. The measure m_n has the following basic properties:

(1) m_n is σ-finite.

(2) Given $\varepsilon > 0$ and any n-dimensional Lebesgue measurable set A, there is an open set G_ε and a closed set K_ε such that $G_\varepsilon \supset A \supset K_\varepsilon$ and $m_n(G_\varepsilon - K_\varepsilon) < \varepsilon$.

(3) If m_n coincides with a measure μ (defined on the Borel subsets of R^n) on the intervals I described above, then $m_n = \mu$ on the Borel subsets.

(4) m_n is translation invariant.

(5) If μ is a translation-invariant measure on the Borel subsets of R^n and if μ is finite on each bounded Borel set, then there is a positive number k such that $\mu(B) = k \cdot m_n(B)$ for each Borel set B.

(6) If T is a nonsingular linear transformation from R^n into R^n, then for each Borel subset B of R^n, $m_n(T(B)) = |\det T| \cdot m_n(B)$, where $\det T$ represents the determinant of the matrix representing T.

We now present proofs of these properties.

Proof of (1). Let $I_{(k)} = \{(x_1, x_2, \ldots, x_n)\colon\ -k < x_j \leq k\ \forall\, j\}$. Then $m_n(I_{(k)})$ is finite for each k and $R^{(n)} = \bigcup_{k=1}^{\infty} I_{(k)}$. ∎

Proof of (2). Note that it is enough to prove the approximation property when $m_n(A) < \infty$. (Why ?) Let $m_n(A)$ be finite. Then by Theorem 2.4, there is a sequence of intervals (J_i) in R^n such that $A \subset \bigcup_{i=1}^{\infty} J_i$ and $m_n(\bigcup_{i=1}^{\infty} J_i - A) < \varepsilon/4$. Let

$$J_i = \{(x_1, x_2, \ldots, x_n)\colon a_j < x_j \leq b_j\ \forall\, j\}.$$

Then $J_i = \bigcap_{k=1}^{\infty} J_{i,k}$, where $J_{i,k}$ is the open set defined by

$$J_{i,k} = \{(x_1, x_2, \ldots, x_n)\colon a_i < x_i < b_i + (1/k)\ \forall\, i\}.$$

By Proposition 2.2(b), there is a $k(i)$ such that $m_n(J_{i,k(i)} - J_i) < \varepsilon/2^{i+2}$. Let $G_\varepsilon = \bigcup_{i=1}^{\infty} J_{i,k(i)}$. Then it follows that $m_n(G_\varepsilon - A) < \varepsilon/2$. Similarly we can also show that there is an open set $H_\varepsilon \supset A^c$ such that $m_n(H_\varepsilon - A^c) < \varepsilon/2$. Since $H_\varepsilon - A^c = A - H_\varepsilon{}^c \subset A$, the proof is complete. ∎

Proof of (3). This property is immediate from Theorem 2.4(e). ∎

Proof of (4). m_n is translation invariant on the intervals in R^n and therefore, $m_n{}^*$ is translation invariant (because of its definition). ∎

Proof of (5). Let us first establish that any open set in R^n is the union of a countable collection of disjoint intervals of the type

$$L_k = \{(x_1, x_2, \ldots, x_n)\colon r_j/2^k < x_j \leq (r_j + 1)/2^k,\ 1 \leq j \leq n\},$$

where $k, r_1, r_2, \ldots, r_n$ are positive integers. To see this, notice that for $k < p$, either $L_p \subset L_k$ or $L_p \cap L_k = \varnothing$. For fixed k, let $\mathscr{F}_k$ be the class of all such intervals in R^n. Let G be an open subset of R^n. Since for each x in G, there is a k such that x belongs to some L_k in $\mathscr{F}_k$, we can now select inductively members of $\mathscr{F}_k$ that are contained in G, but disjoint from the intervals already chosen from $\mathscr{F}_{k-1}$. The resulting collection is the desired countable collection.

Next, we notice that for each positive integer k, each member of $\mathscr{F}_k$ has the same μ and m_n measure by translation invariance; also, the interval $I = \{(x_1, x_2, \ldots, x_n): 0 < x_j \leq 1 \; \forall \, j\}$ is the disjoint union of 2^{nk} members of $\mathscr{F}_k$. Let $\mu(I) = \alpha$. If J_k is a typical member of $\mathscr{F}_k$, then we have

$$2^{nk} \cdot \mu(J_k) = \mu(I) = \alpha \cdot m_n(I) = \alpha \cdot 2^{nk} m_n(J_k).$$

This means that the measures μ and $\alpha \cdot m_n$ coincide on the open sets in R^n and therefore, also on the half-open intervals in R^n [see the proof of (2) above]. By Theorem 2.4(e), μ and $\alpha \cdot m_n$ must coincide on the Borel sets. ∎

Proof of (6). It is known that a nonsingular linear transformation T from R^n into R^n is of the form $T = T_1 T_2 \ldots T_r$ where each T_j is a nonsingular linear transformation from R^n into R^n and one of the following three types of elementary linear transformations:

(i) T_j is a permutation;

(ii) $T_j(x_1, x_2, \ldots, x_n) = (x_1 + x_2, x_2, \ldots, x_n)$;

(iii) $T_j(x_1, x_2, \ldots, x_n) = (ax_1, x_2, \ldots, x_n)$, $a \in R$.

In this case, $\det T = (\det T_1) \cdots (\det T_r)$. It is now easy to verify that $m_n(T(A)) = |\det T| \cdot m_n(A)$, when A is a half-open interval in R^n and T is an elementary linear transformation of the above three types. This means that the measures μ_1 and μ_2 defined by

$$\mu_1(A) = m_n(T(A)) \text{ and } \mu_2(A) = |\det T| \cdot m_n(A)$$

coincide on the half-open intervals of R^n; therefore, they must coincide on the Borel sets of R^n by Theorem 2.4(e). ∎

We remark that for a singular linear transformation T from R^n into R^n, $m_n(T(A)) = 0$ for any Borel subset A of R^n. To see this, let $T = T_1 T_2 \cdots T_r$, where the T_j's are elementary linear transformations of the types described above. Let j be the first index such that T_j is singular. Then T_j must be of type (iii) above with $a = 0$ so that $m_n(T_j(R^n)) = 0$. Since $T_1 \cdots T_{j-1}$ is nonsingular, by property (6),

$$m_n(T_1 T_2 \ldots T_{j-1}(T_j(R^n))) = 0.$$

It follows that $m_n(T(R^n)) = 0$.

Problems

2.3.1. Let $X = Z^+$. For $A \subset Z^+$, let $a = \sup A$. Suppose

$$\mu^*(A) = \begin{cases} \dfrac{a}{(a+1)}, & \text{if } A \text{ is finite,} \\ 0, & \text{if } A = \varnothing, \\ 1, & \text{if } A \text{ is infinite.} \end{cases}$$

Show that μ^* is an outer measure. Find the μ^*-measurable sets.

× **2.3.2.** (i) Suppose μ^* is an outer measure on 2^X, and for every $A \subset X$ there is a μ^*-measurable set $E \supset A$ such that $\mu^*(A) = \mu^*(E)$. (E is called a *μ^*-measurable cover* of A.) Show that

(a) if $A_n \subset A_{n+1} \subset X$, then $\mu^*(\bigcup_{n=1}^{\infty} A_n) = \lim_{n\to\infty}\mu^*(A_n)$;

(b) if $B_n \subset X$, then $\mu^*(\underline{\lim}_n B_n) \leq \underline{\lim}_n \mu^*(B_n)$;

(c) if $E \subset X$ is *locally measurable* (that is, $B \cap E$ is μ^*-measurable whenever $\mu^*(B) < \infty$ and B is μ^*-measurable), then E must be μ^*-measurable.

(ii) Let X be an uncountable set. Define μ^* as follows: $\mu^*(\varnothing) = 0$, $\mu^*(E) = 1$ if E is nonempty and countable and $\mu^*(E) = \infty$ otherwise. Show that μ^* is an outer measure. Find the class of μ^*-measurable subsets and the class of locally measurable subsets.

2.3.3. Suppose μ^* is an outer measure on 2^X, where X is a topological space. Show that every Borel set is μ^*-measurable if and only if $\mu^*(A \cup B) = \mu^*(A) + \mu^*(B)$ whenever $A \cap \bar{B}$ is empty.

× **2.3.4.** Show that $m^*(A) = 0$ if A is countable. (Recall that m^* is the Lebesgue outer measure.)

× **2.3.5.** Show that m^* is translation invariant, i.e., for $A \subset R$ and $x \in R$, $m^*(A) = m^*(A + x)$.

× **2.3.6.** For each Lebesgue-measurable set $E \subset R$, show that $E + x$ is also Lebesgue measurable for $x \in R$. [Hint: Use Problem 2.3.5.]

× **2.3.7.** (a) Show that the Cantor set has Lebesgue measure zero. [Hint: Compute the measure of its complement in [0, 1].]

(b) Show that for every $1 > \varepsilon > 0$, there exists a nowhere dense perfect† set in [0, 1] that has Lebesgue measure greater than $1 - \varepsilon$. (Hint: Construct the set in the same manner as the usual Cantor set except that each of the intervals removed at the nth step has length $\varepsilon \cdot 3^{-n}$.)

† For the definition and properties of perfect sets, see Problem 2.3.17.

(c) Show that there is a set of Lebesgue measure zero that is of second category in [0, 1].

× **2.3.8.** Show that for $A \subset R$

$$m^*(A) = \inf\left\{\sum_{n=1}^{\infty} l(I_n): A \subset \bigcup_{n=1}^{\infty} I_n,\ I_n\text{'s are open intervals}\right\}.$$

(Recall that in the definition of m^* in Example 2.9, I_n's are left-open right-closed intervals.)

× **2.3.9.** (i) Suppose $E \subset R$ and $m^*(E) < \infty$. Then show that the following conditions are equivalent:

(a) E is Lebesgue measurable.

(b) Given $\varepsilon > 0$, there is an open set $0 \supset E$ with $m^*(0 - E) < \varepsilon$.

(c) Given $\varepsilon > 0$, there is a finite union U of open intervals such that $m^*(U \triangle E) < \varepsilon$.

(ii) Show that for arbitrary $E \subset R$, each of the first two statements in (i) is equivalent to each of the following statements.

(d) Given $\varepsilon > 0$, there is a closed set $F \subset E$ with $m^*(E - F) < \varepsilon$.

(e) There is a G_δ set‡ $G \supset E$ such that $m^*(G - E) = 0$.

(f) There is a F_σ set§ $F \subset E$ such that $m^*(E - F) = 0$.

(iii) Prove that if $g: R \to R$ is continuously differentiable with $g' > 0$, then $g^{-1}(E)$ is a Lebesgue measurable set for each Lebesgue measurable set E. [Use (f) above.]

2.3.10. Show that every set on R has a m^*-measurable cover (Problem 2.3.2) and therefore $m^*(\bigcup_{n=1}^{\infty} A_n) = \lim_{k\to\infty} m^*(\bigcup_{n=1}^{k} A_n)$, $A_n \subset R$.

2.3.11. Show that the σ-finiteness assumption is essential in Theorem 2.4 for the uniqueness of the extension of μ on the smallest σ-algebra containing $\mathscr{A}$. {Hint: Let $X = (0, 1]$, and let $\mathscr{A}$ be the algebra of all finite unions of intervals of the form $(a, b] \subset (0, 1]$ and $\mu(A) = \infty$ if $A \neq \varnothing$, $= 0$ if $A = \varnothing$.}

× **2.3.12.** (a) Suppose μ is a finite measure on the Borel sets of R. Then show that for each Borel set B

$$\begin{aligned}\mu(B) &= \sup\{\mu(K): K \subset B, K \text{ compact}\}\\ &= \inf\{\mu(V): B \subset V, V \text{ open}\}.\end{aligned}$$

(Hint: The class $[B: \mu(B) = \sup\{\mu(K): K \subset B, K \text{ compact}\}]$ is a monotone class containing all half-open intervals. Use Theorem 1.4.) It is relevant to

‡ A G_δ set is a countable intersection of open sets.

§ A F_σ set is a countable union of closed sets.

point out that using the same hint and noting that a closed set in a metric space is a G_δ-set, one can also prove a similar approximation property in any metric space with respect to closed sets from inside and open sets from outside. (For more details, see Chapter 5.)

(b) Let μ be a finitely additive finite measure on the Borel subsets of a metric space X such that for each Borel set $B \subset X$, $\mu(B) = \sup\{\mu(K)$: K is a compact subset of $B\}$. Prove that μ is countably additive.

2.3.13. Show that there exists a Lebesguc-measurable set on R that is not a Borel set. (Hint: Every subset with Lebesgue outer measure zero is Lebesgue measurable. Since c is the cardinality of the Cantor set, there are 2^c Lebesgue-measurable sets, whereas there are c Borel sets. See Problem 1.3.13.)

2.3.14. Let φ be an isometry of R into R (i.e., $|\varphi(x) - \varphi(y)| = |x - y|$, $\forall\, x, y \in R$). Then show that

(a) $\varphi(x) = x + d(\forall\, x \in R)$ or $\varphi(x) = -x + d(\forall\, x \in R)$, for some $d \in R$.

(b) If $A \subset R$ is Lebesgue measurable, then $\varphi(A)$ is Lebesgue measurable and $m(A) = m(\varphi(A))$.

[Hint: If T is an isometry from R^n into R^n, then there exists $a \in R^n$ such that for each $x \in R^n$, $T(x) = a + T_0(x)$ for some linear map T_0 from R^n into R^n. To see this, let $T(0) = 0$. Since T is an isometry, the point $T(\frac{1}{2}(x + y))$ must be the midpoint between $T(x)$ and $T(y)$. Thus, if $T(cx) = cT(x)$ for $x \in R^n$ and $c \in R$, then also $T(x + y) = T(x) + T(y)$ for x, y in R^n. Let us then prove that $T(cx) = cT(x)$ for $c \in R$. It is no loss of generality to assume that $|c| \geq 1$. Since T is an isometry, we have

$$(*)\ |T(cx)| = |T(x)| + |T(cx) - T(x)|.$$

In R^n, we know that $|y + z| = |y| + |z|$ and $y \neq 0 \Rightarrow z = ty$ for some $t \in R$. It follows from $(*)$ that there is a $d \in R$ such that

$$T(cx) - T(x) = dT(x).$$

Using the isometry property, it then follows that $|c| = |d + 1|$. If $d + 1 = c$, then $T(cx) = cT(x)$. If $d + 1 = -c$, then $T(cx) = -cT(x)$; substituting this in $(*)$, we have then $|c| = 1 + |1 + c|$. But this is impossible unless $c < 0$. Thus, for $c > 0$, $T(cx) = cT(x)$. If $c < 0$ and $T(cx) = -cT(x)$, then $T(-cx) = -cT(x) = T(cx)$, which is impossible for $x \neq 0$ since T is one-to-one.]

2.3.15. Construct a real-valued function on [0, 1] whose set of discontinuities has Lebesgue measure zero but has an uncountable intersection with every open subinterval. {Hint: Let K_1 be the Cantor set $\subset [0, 1]$ of Lebesgue measure zero. Let K_2 be the union of similar Cantor sets constructed in each of the intervals of $[0, 1] - K_1$. The sequence (K_n) is constructed inductively, and let $K = \bigcup_{n=1}^{\infty} K_n$. Define $f(x) = 2^{-n}$ for $x \in K_n$, $= 0$ for $x \notin K$.}

2.3.16. Suppose E is a Lebesgue measurable subset of R, and for each x in a dense set of R, $m\big(E \triangle (E + x)\big) = 0$. Prove that $m(E) = 0$ or $m(R - E) = 0$.

2.3.17. *Perfect Sets and Measures on R.* A set $A \subset R$ is called *perfect* if it is closed and every $x \in A$ is a limit point of A. Verify the following assertions:

(i) Every (uncountable) closed set of real numbers is the union of a perfect set and an at most countable set.

(ii) A set of real numbers is perfect if and only if it is the complement of an at most countable number of disjoint open intervals, no two of which have a common endpoint.

(iii) Every nonempty perfect set of real numbers is uncountable.

(iv) The Cantor sets (in Problem 2.3.7) are perfect.

★ (v) Every perfect set of real numbers contains a perfect subset of Lebesgue measure zero.

★ (vi) Every closed set of positive Lebesgue measure contains a perfect subset of Lebesgue measure zero.

(vii) Let μ be a nonzero measure defined on the Borel sets of R such that $\mu([-n, n]) < \infty$ for every positive integer n. Let $S_\mu = \{x \in R : \mu(V_x) > 0$ for every open set V_x containing $x\}$. Then (a) S_μ is closed, (b) $\mu(R - S_\mu) = 0$, and (c) S_μ is perfect, when points have zero measure. [Note that $\mu(R - S_\mu) = \sup\{\mu(K) : K \subset R - S_\mu,\ K \text{ compact}\}$ by Problem 2.3.12.]

2.3.18. *Representation of a Finite Borel Measure in Terms of Its Distribution Function and the Lebesgue Measure* (J. J. Higgins). Let μ be a finite measure on the Borel sets of R. Let f be a real-valued, bounded, nondecreasing function defined on R such that it is continuous from the right and $f(x) \to 0$ as $x \to -\infty$. Such a function is sometimes called a *distribution function.* (Compare Problem 2.2.4.) Show the following:

(i) If f is a distribution function as above, then for any Borel set A, $f(A)$ is a Borel set; also $m\big(f(A)\big) = \lambda(A)$ (m being the Lebesgue measure) defines a finite Borel measure. [Hint: $A \cap B = \varnothing$ means that $f(A) \cap f(B)$ is at most countable.]

(ii) Let μ be a finite Borel measure. Then $f(x) = \mu\big((-\infty, x]\big)$ is a distribution function. If

$$j(x_i) = f(x_i) - f(x_i -),$$

where the (x_i) are the discontinuities of f and if

$$f_c(x) = f(x) - \sum_{x_i \le x} j(x_i),$$

then

$$\mu(A) = m(f_c(A)) + \sum_{x_i \in A} j(x_i)$$

for all Borel sets A.

2.3.19. *Completion of μ and the Outer Measure μ^*.* Let μ be a measure on a σ-algebra $\mathscr{A}$ of subsets of X, and let $\bar{\mu}$ be its completion on the σ-algebra $\bar{\mathscr{A}}$. Let $\mathscr{B}$ denote the μ^*-measurable subsets of X. Then $\mathscr{A} \subset \bar{\mathscr{A}} \subset \mathscr{B}$. Show that $\bar{\mathscr{A}} = \mathscr{B}$ when μ is σ-finite, and that this equality need not be valid otherwise. Note that a similar result holds when μ is a measure on an algebra $\mathscr{E}$ and $\bar{\mu}$ is the completion of μ^* restricted to $\mathscr{A}$, the smallest σ-algebra containing $\mathscr{E}$.

2.3.20. *Semifiniteness and the Extension Theorem.* Let μ be a semifinite measure on an algebra $\mathscr{A}$ of subsets of X. Show that there is always an extension of μ to a semifinite measure on $\sigma(\mathscr{A})$, and the extension of μ to a measure on $\sigma(\mathscr{A})$ is unique if and only if μ^* is semifinite on $\sigma(\mathscr{A})$. Give an example of a semifinite measure on an algebra $\mathscr{A}$ with infinitely many semifinite extensions on $\sigma(\mathscr{A})$. [Hint: Take $X =$ the reals, $Q =$ the rationals, and $\mathscr{A} = \{A \subset X \mid \text{either } A \text{ or } X - A \text{ is finite}\}$. Let $\mu(A) = \text{card}(A \cap Q)$. For any nonnegative real number s, define ν_s on $\sigma(\mathscr{A})$ by

$$\nu_s(A) = \begin{cases} \text{card}(A \cap Q), & \text{if } A \cap (X - Q) \text{ is at most countable,} \\ s + \text{card}(A \cap Q), & \text{if } (X - A) \cap (X - Q) \text{ is at most countable.} \end{cases}$$

Then ν_s is a semifinite extension of μ on $\sigma(\mathscr{A})$.]

2.3.21. If $A_n = \{x \in (0, 1)$: the nth digit in the binary expansion of x is $1\}$, then find $m(\bigcap_{i=1}^k A_{n_i})$ for $n_1 < n_2 < \cdots < n_k$.

2.3.22. Let $I = (0, 1)$ and $A \subset I$. Then A is called a *comb* if there exist numbers a and b with $0 < a \le b < 1$ such that for every nonempty open subinterval J of I,

$$a \le m^*(A \cap J)/m^*(J) \le b.$$

Prove that there does not exist any comb. (For some interesting information on combs, see [52].)

2.3.23. Let μ be a measure on a σ-algebra $\mathscr{A}$ of subsets of X such that $\mu(X) = 1$. Let $\mathscr{A}_{(n)}$ ($\mathscr{A}^{(n)}$) be a decreasing (increasing) sequence of sub-σ-algebras of $\mathscr{A}$. Suppose that $\mathscr{A}^\infty$ is the smallest σ-algebra containing $\bigcup_{n=1}^\infty \mathscr{A}^{(n)}$. Let $\mathscr{A}_\infty = \bigcap_{n=1}^\infty \mathscr{A}_{(n)}$. Show that

(i) $\mathscr{A}_\infty$ is a σ-algebra;

(ii) for each $A \in \mathscr{A}^\infty$ and each $\varepsilon > 0$, there exists $B \in \mathscr{A}^{(m)}$ (where m is a positive integer depending on ε) such that $\mu(A \triangle B) < \varepsilon$. [Hint: Consider $\mathscr{F} = \{A \in \mathscr{A}^\infty$: given $\varepsilon > 0$, there exists $B \in \bigcup_{n=1}^\infty \mathscr{A}^{(n)}$ such that $\mu(A \triangle B) < \varepsilon\}$. Show that $\mathscr{F}$ is a monotone class.]

2.3.24. Let T: $[0, 1) \to [0, 1)$ be defined by $T(x) = 2x$ if $x \in [0, \frac{1}{2})$ and $T(x) = 2x - 1$ if $x \in [\frac{1}{2}, 1)$. Show that T is measure preserving, that is, $m(E) = m(T^{-1}(E))$ for every Borel set $E \subset [0, 1)$. [Hint: Prove it for an interval first.]

2.3.25. *The Recurrence Theorem.* Let μ be a measure on a σ-algebra $\mathscr{A}$ of subsets of X and $\mu(X) = 1$. Let T: $X \to X$ be such that $E \in \mathscr{A} \Rightarrow T^{-1}(E) \in \mathscr{A}$ and $\mu(E) = \mu(T^{-1}(E))$. Prove that if $E \in \mathscr{A}$, then almost every x in E returns to E infinitely often [in other words, there exists $A \subset E$ such that $\mu(A) = \mu(E)$ and $x \in A \Rightarrow T^n(x) \in E$ for infinitely many positive integral values of n]. [Hint: First prove that for almost every point x in E, there is a positive integer n such that $T^n(x) \in E$. To this end, consider $B = E \cap (\bigcap_{n=1}^\infty T^{-n}(X - E))$ and show that the sets B, $T^{-1}(B)$, $T^{-2}(B)$, ... are pairwise disjoint.]

2.3.26. *An Application of the Baire Category Theorem* (splitting of a family of sets of positive measure). Let μ be a measure on a σ-algebra $\mathscr{A}$ of subsets of X and $\mathscr{F}$ be a family of sets in $\mathscr{A}$ with positive measure. Then $A \in \mathscr{A}$ is said to split $\mathscr{F}$ if $0 < \mu(A \cap B) < \mu(B)$ for every $B \in \mathscr{F}$. Prove the following result due to R. B. Kirk: Suppose that μ is nonatomic and countable. (See 2.1.15.) Then there exists $A \in \mathscr{A}$ which splits $\mathscr{F}$. In particular, there exists a Lebesgue-measurable set $A \subset R$ which splits all open sets with respect to Lebesgue measure. [Hint: With no loss of generality, assume that $\mu(X) < \infty$. Let $d(A, B) = \mu(A \triangle B)$. Then by Problem 2.1.17, $(\mathscr{A}, d)$ is a complete pseudometric space. If $\mathscr{F} = (B_n)$, let $\mathscr{F}_n = \{A \in \mathscr{A}$: $\mu(A \cap B_n) = 0$ or $\mu(A \cap B_n) = \mu(B_n)\}$. Then $\mathscr{F}_n$ is closed. Also since μ is nonatomic, by Problem 2.1.15(b), $\mathscr{F}_n$ is nowhere dense. By the Baire Category Theorem, there exists $A \in \bigcap_{n=1}^\infty (\mathscr{A} - \mathscr{F}_n)$. This A splits $\mathscr{F}$.]

2.4. Non-Lebesgue-Measurable Sets and Inner Measure

In this section, we show the existence of a *non-Lebesgue-measurable set* on R by invoking the Axiom of Choice. Such sets, indeed, rarely arise naturally in practical situations. R. Solovay† has recently shown that the

† R. Solovay, *Ann. of Math.* (2) 92, 1–56 (1970).

proof of the existence of such a set on R must depend on the Axiom of Choice. Often nonmeasurable sets are used to construct different counterexamples to understand different aspects of the theory well. For instance, using the existence of a nonmeasurable set one can construct a Lebesgue-measurable set on R that is not a Borel set (Problem 2.4.3).

We will also introduce in this section the concept of inner measure. This concept has some historical significance since the concept of measurability was originally characterized in terms of both inner and outer measure. Aside from this it is also useful for the purpose of extending a measure on an algebra to an algebra containing the given algebra and *any* given set. Using this concept, we will also present a proper simple extension of the Lebesgue measure that is complete and translation invariant.

First we give an example of a non-Lebesgue-measurable set.

Theorem 2.5. Let A be a Lebesgue-measurable set with $m(A) > 0$. Then there exists $E \subset A$ such that E is not Lebesgue measurable. ∎

Proof. Since $A = \bigcup_{n=1}^{\infty}(A \cap [-n, n])$, there is a positive integer n_0 such that

$$m(A \cap [-n_0, n_0]) > 0.$$

We write $B = A \cap [-n_0, n_0]$. Let $x \in B$ and $B_x = \{y \in B: y - x \text{ is rational}\}$. Then $B = \bigcup_{x \in B} B_x$. For $x_1, x_2 \in B$, $B_{x_1} = B_{x_2}$ if $x_1 - x_2$ is rational; otherwise, $B_{x_1} \cap B_{x_2} = \varnothing$. By the Axiom of Choice, there exists a set $E \subset B$ such that E contains exactly one point from each of the distinct sets $\{B_x\}$. We claim that E is not Lebesgue-measurable.

To prove this claim, let $(r_n)_{n=1}^{\infty}$ be the rationals in $[-2n_0, 2n_0]$. Then $E + r_n$ and $E + r_l$ are disjoint (if $n \neq l$); for if $r_n \neq r_l$, $e_n, e_l \in E$ and $e_n + r_n = e_l + r_l$, then $e_n \neq e_l$ and so $e_n - e_l$ is irrational, which is a contradiction. Also $\bigcup_{n=1}^{\infty}(E + r_n) \subset [-3n_0, 3n_0]$ and $B \subset \bigcup_{n=1}^{\infty}(E + r_n)$; for if $x \in B$, then there is some $e \in E$ such that $x \in B_e$ or $x - e = r_n$ (for some n), a rational in $[-2n_0, 2n_0]$. If E is Lebesgue measurable, $E + r_n$ is so for each n, and therefore $0 < m(B) \leq \sum_{n=1}^{\infty} m(E + r_n) = \sum_{n=1}^{\infty} m(E) \leq 6n_0$ (m being translation invariant). If $m(E) = 0$, then $m(B) = 0$, which is not possible. If $m(E) > 0$, then $\sum_{n=1}^{\infty} m(E) = \infty \leq 6n_0$, which is an absurdity. Hence E is not Lebesgue measurable. ∎

Remark 2.4. We have seen that with the Axiom of Choice it can be proved that there are nonmeasurable sets. R. Solovay has shown that if we allow the possibility that there are uncountable sets of cardinality less than c, then it cannot be proved or refuted using the Axiom of Choice that there are nonmeasurable such sets. However, if uncountable measurable sets of reals exist with cardinality less than c, then they must have zero measure. (See Problem 2.4.16.) Let us now point out that if, instead of m, we consider an arbitrary nonzero measure μ that is translation invariant, defined on a σ-algebra containing the Borel sets of R, and finite on finite intervals, then its domain cannot contain the set E. In other words, it is impossible to define a translation-invariant, countably additive nonzero finite measure on the class of all subsets on $[0, 1]$. However, there exists a finitely additive, translation-invariant measure on the class of all subsets of $[0, 1]$ such that the measure of every subinterval in $[0, 1]$ is its length. Such a measure (which is finitely additive and congruence invariant, i.e., two sets that are congruent or isometric have the same measure) also exists on the class of all subsets of $[0, 1] \times [0, 1]$ in R^2. This was first shown by S. Banach.[†] We will sketch the construction on $[0, 1]$ (as an application of the famous Hahn–Banach Theorem) in the chapter on Banach spaces. Since in R^n, in general, the group of isometries becomes increasingly larger with increase in the number of dimensions, it is natural to be less hopeful of finding finitely additive, congruence-invariant measures on the class of all subsets of a general n-dimensional unit cube. Indeed, it has been shown by F. Hausdorff that there does not exist any such measure for $n > 2$. In the context of different extensions of Lebesgue measure to larger classes of sets, we would also mention the works of S. Kakutani and J. C. Oxtoby.[‡] They obtained countably additive, translation-invariant extensions of Lebesgue measure to very large σ-algebras containing properly the class of all Lebesgue-measurable sets.

Actually it can be shown (via the continuum hypothesis)[§] that it is *impossible* to have a finite nonzero measure that is zero for points and is defined on all subsets of a set of cardinality c. This follows immediately from a well-known theorem of S. M. Ulam.[||]

† S. Banach, *Fund. Math.* **4**, 7–33 (1923).

‡ S. Kakutani and J. C. Oxtoby, *Ann. of Math.* **52**(2), 580–590 (1950).

§ The continuum hypothesis is the assertion that each infinite subset of R is either countable or of cardinal number c. P. J. Cohen has shown recently that this hypothesis is independent of the Zermelo–Fraenkel axioms of set theory. See P. J. Cohen, *Proc. Natl. Acad. Sci. USA* **50**, 1143–1148 (1963) and **51**, 105–110 (1964).

|| S. M. Ulam, *Fund. Math.* **16**, 141–150 (1930).

• **Theorem 2.6.** *The Ulam Theorem.* Let Ω be the first uncountable ordinal and $X = [0, \Omega)$. Then a finite measure μ defined for all subsets of X and zero for points must be a zero measure. ▮

Proof.[†] Suppose that μ is a finite measure defined for all subsets of X and zero for points. Let $y \in X$ and $A_y = \{x: x < y\}$. Then A_y is countable and there is a one-to-one correspondence $f(x, y)$ from A_y onto the natural numbers. Let us define

$$B_{x,n} = \{y: x < y, f(x, y) = n\}$$

for each $x \in X$ and each natural number n. Then these sets satisfy

(i) $B_{x,n} \cap B_{z,n} = \emptyset$, if $x \neq z$,
(ii) $X - \bigcup_{n=1}^{\infty} B_{x,n}$ is countable for each $x \in X$.

We establish only (ii). For $x \in X$, $y > x$ implies $f(x, y) = n$ for some n, and hence $\{y: x < y\} \subset \bigcup_{n=1}^{\infty} B_{x,n}$. Since $\{y: y \leq x\}$ is countable, (ii) follows.

By (i) and since $\mu(X) < \infty$, for each natural number n $\mu(B_{x,n}) > 0$ for at most countably many x. Therefore since X is uncountable, there is $x \in X$ such that $\mu(B_{x,n}) = 0$ for all natural numbers n. By (ii), $\mu(X) = 0$ if μ is zero for points. ▮

Next in this section we will introduce and discuss *inner measures* with a view to extending a measure on an algebra $\mathscr{A}$ to a measure on an algebra containing $\mathscr{A}$ and *any* given set E. Through inner measures, we will also obtain a translation-invariant proper extension of Lebesgue measure.

• **Definition 2.5.** Let μ be a measure on an algebra $\mathscr{A}$ and μ^* the induced outer measure as in Theorem 2.4. Then the inner measure μ_* is defined by

$$\mu_*(E) = \sup\{\mu(A) - \mu^*(A - E)\},$$

where the supremum is taken over all sets $A \in \mathscr{A}$ with $\mu^*(A - E) < \infty$. ▮

It follows easily from the definition that

$$\mu_*(E) \leq \mu^*(E), \tag{2.12}$$

$$E \subset F \Rightarrow \mu_*(E) \leq \mu_*(F), \tag{2.13}$$

$$E \in \mathscr{A} \Rightarrow \mu_*(E) = \mu^*(E) = \mu(E). \tag{2.14}$$

[†] Another proof of this theorem is indicated in Problem 3.4.13 in Chapter 3.

But the most interesting and the less obvious properties of the inner measure are perhaps the following.

• **Remarks** *Properties of the Inner Measure.*

2.5. For certain sets the inner measure has a more convenient expression, as in equation (2.15) below.

$$\mu_*(E) = \mu(A) - \mu^*(A - E), \quad \text{whenever } E \subset A,\ A \in \mathscr{A}, \text{ and } \mu^*(A - E) < \infty. \tag{2.15}$$

If $B \in \mathscr{A}$, then for each $C \subset X$, the set $B \cap C \subset B$ and it follows from equation (2.15) that

$$\mu(B) = \mu_*(B \cap C) + \mu^*(B - C). \tag{2.16}$$

Proof of Equation (2.15). Let $E \subset A$, $A \in \mathscr{A}$, and $\mu^*(A - E) < \infty$. For $B \in \mathscr{A}$ and $\mu^*(B - E) < \infty$, using the μ^*-measurability of $A \cup B - A$ and the set equality $A \cup B - E = (A \cup B - A) \cup (A - E)$, we have

$$\mu^*(A \cup B - E) = \mu(A \cup B - A) + \mu^*(A - E).$$

This implies that

$$\begin{aligned}\mu(A) - \mu^*(A - E) &= \mu(A \cup B - A) + \mu(A) - \mu^*(A \cup B - E)\\ &= [\mu(A \cup B - A) + \mu(A \cap B)]\\ &\quad + [\mu(A - B) + \mu^*(B - E)\\ &\quad - \mu^*(A \cup B - E)] - \mu^*(B - E)\\ &\geq \mu(B) - \mu^*(B - E).\end{aligned}$$

[Note that $A \cup B - E \subset (A - B) \cup (B - E)$.] The equality (2.15) now follows easily.

2.6. Every set E has a measurable kernel C. We clarify this below.

$$\text{If } \mathscr{A} \text{ is a } \sigma\text{-algebra, then for each } E \subset X \text{ there exists } C \subset E \text{ and } C \in \mathscr{A} \text{ such that } \mu_*(E) = \mu(C). \tag{2.17}$$

Proof of Equation (2.17). First, let $\mu_*(E) < \infty$. There exist $A_n \in \mathscr{A}$ with $\mu^*(A_n - E) < \infty$ and $\mu(A_n) - \mu^*(A_n - E) > \mu_*(E) - 1/n$. By the definition of μ^*, there exist $B_n \in \mathscr{A}_{\sigma\delta}$ $(= \mathscr{A})$ such that $A_n - E \subset B_n$ and $\mu^*(A_n - E) = \mu(B_n)$. Since $A_n - B_n \subset E$ and $A_n \subset (A_n - B_n) \cup B_n$, we have

$$\mu(A_n - B_n) \geq \mu(A_n) - \mu(B_n) > \mu_*(E) - 1/n.$$

Let $C = \bigcup_{n=1}^{\infty}(A_n - B_n)$. Then $C \subset E$ and $\mu(C) = \mu_*(E)$, by equations (2.12)–(2.14). Clearly C is the measurable kernel of E. In case $\mu_*(E) = \infty$, equation (2.17) follows easily if we replace above the expression $\mu_*(E) - 1/n$ by n.

2.7. For Lebesgue measure m, m_* has the following approximation property: $m_*(A) = \sup\{m(F)\colon F \text{ closed} \subset A\}$.

This follows from Remark 2.6 and Problem 2.3.9(d).

2.8. The inner measure is countably additive in the following sense. For any set E and a disjoint sequence of sets A_n in $\mathscr{A}$,

$$\mu_*\left(E \cap \bigcup_{n=1}^{\infty} A_n\right) = \sum_{n=1}^{\infty} \mu_*(E \cap A_n).$$

Proof of Remark 2.8. We may and do assume that $E \subset \bigcup_{n=1}^{\infty} A_n$. Then $E = \bigcup_{n=1}^{\infty}(E \cap A_n)$. First we show that $\mu_*(E) \leq \sum_{n=1}^{\infty}\mu_*(E \cap A_n)$. Recalling the definition of μ_*, we consider $B \in \mathscr{A}$ with $\mu^*(B - E) < \infty$. Then since $\bigcup_{n=1}^{\infty} A_n$ is μ^* measurable, we have

$$\begin{aligned} \mu(B) &= \mu^*\left[B \cap \left(\bigcup_{n=1}^{\infty} A_n\right)\right] + \mu^*\left(B - \bigcup_{n=1}^{\infty} A_n\right), \\ \mu^*(B - E) &= \mu^*\left[(B - E) \cap \left(\bigcup_{n=1}^{\infty} A_n\right)\right] + \mu^*\left[(B - E) - \left(\bigcup_{n=1}^{\infty} A_n\right)\right]. \end{aligned} \tag{2.18}$$

From equation (2.18), we have

$$\begin{aligned} \mu(B) - \mu^*(B - E) &= \mu^*\left[B \cap \left(\bigcup_{n=1}^{\infty} A_n\right)\right] - \mu^*\left[(B - E) \cap \left(\bigcup_{n=1}^{\infty} A_n\right)\right] \\ &= \sum_{n=1}^{\infty} \mu(B \cap A_n) - \mu^*(B \cap A_n \cap E^c), \text{ by Theorem 2.2.} \\ &\leq \sum_{n=1}^{\infty} \mu_*(E \cap A_n), \quad \text{using the definition of } \mu_*. \end{aligned}$$

It follows that $\mu_*(E) \leq \sum_{n=1}^{\infty}\mu_*(E \cap A_n)$. To prove the converse inequality, we now consider the definition of $\mu_*(E \cap A_n)$ for each n. Let $B_n \in \mathscr{A}$ and $\mu^*(B_n - (E \cap A_n)) < \infty$. Then since $B_n - A_n$ is μ^*-measurable and

$$B_n - (E \cap A_n) = (B_n - A_n) \cup (B_n \cap A_n \cap E^c),$$

we have

$$\mu^*(B_n - (E \cap A_n)) = \mu(B_n - A_n) + \mu^*(B_n \cap A_n \cap E^c)$$

and therefore

$$\begin{aligned}\sum_{n=1}^{k} \mu(B_n) - \mu^*(B_n - (E \cap A_n)) &= \sum_{n=1}^{k} \mu(B_n \cap A_n) - \mu^*(B_n \cap A_n \cap E^c) \\ &= \mu\left[\bigcup_{n=1}^{k} (B_n \cap A_n)\right] \\ &\quad - \mu^*\left[\left(\bigcup_{n=1}^{k} (B_n \cap A_n)\right) \cap E^c\right] \\ &\leq \mu_*(E). \qquad (2.19)\end{aligned}$$

By taking "sup" over all such choices of B_n, we have

$$\sum_{n=1}^{k} \mu_*(E \cap A_n) \leq \mu_*(E), \qquad \text{for all } k. \qquad (2.20)$$

The rest of the proof is clear. ∎

2.9. If $\mu^*(E) < \infty$, then E is μ^*-measurable if and only if $\mu_*(E) = \mu^*(E)$.

Proof of Remark 2.9. First we prove the "if" part. By Remark 2.6, there is a measurable kernel $C \in \mathscr{A}$, $C \subset E$ such that $\mu_*(E) = \mu(C)$. If $\mu^*(E) = \mu_*(E)$, then it follows that $\mu(C) = \mu^*(E) = \mu^*(E \cap C) + \mu^*(E - C)$ implying $\mu^*(E - C) = 0$. Thus $E - C$ is μ^*-measurable and so is $E = C \cup (E - C)$. To prove the "only if" part, let E be μ^*-measurable and $\mu^*(E) < \infty$. Then there is a disjoint sequence of sets A_n in $\mathscr{A}$ such that $E \subset \bigcup_{n=1}^{\infty} A_n$ and $\mu^*(\bigcup_{n=1}^{\infty} A_n) < \infty$. By Remark 2.8, we have

$$\begin{aligned}\mu_*(E) = \sum_{n=1}^{\infty} \mu_*(E \cap A_n) &= \sum_{n=1}^{\infty} [\mu(A_n) - \mu^*(A_n - E)], \text{ by equation (2.15)} \\ &= \sum_{n=1}^{\infty} \mu^*(A_n \cap E) = \mu^*(E). \qquad ∎\end{aligned}$$

We note that in Remark 2.6 it is necessary that $\mathscr{A}$ be a σ-algebra. The reason is that a Lebesgue-measurable set of positive Lebesgue measure, which is nowhere dense in the reals, cannot have a measurable kernel in the algebra generated by the half-open intervals $(a, b]$. We also remark that if μ is σ-finite and E is μ^*-measurable, then $\mu_*(E) = \mu^*(E)$. The converse is of course not true. For, if A is not Lebesgue measurable and $A \subset (0, 1)$, then $E = A \cup (1, \infty)$ is not Lebesgue measurable, despite having the same Lebesgue inner and outer measure.

Now we state and prove the promised extension theorem.

• **Theorem 2.7.** Let μ be a measure on an algebra $\mathscr{A}$ of subsets of X and let E be any subset of X. Let $\mathscr{B}$ be the algebra generated by $\mathscr{A}$ and E. Then if for $B \in \mathscr{B}$

$$\bar{\mu}(B) = \mu^*(B \cap E) + \mu_*(B - E) \tag{2.21}$$

and

$$\underline{\mu}(B) = \mu_*(B \cap E) + \mu^*(B - E), \tag{2.22}$$

then $\bar{\mu}$ and $\underline{\mu}$ are measures on $\mathscr{B}$ such that $\bar{\mu}(A) = \underline{\mu}(A) = \mu(A)$ for $A \in \mathscr{A}$. ▮

Proof. First we observe that

$$\mathscr{B} = \{(A \cap E) \cup (B \cap E^c)\colon A, B \in \mathscr{A}\}.$$

By equation (2.16), $\bar{\mu}(A) = \underline{\mu}(A) = \mu(A)$ for $A \in \mathscr{A}$. We only need to establish that $\bar{\mu}$ and $\underline{\mu}$ are measures on $\mathscr{A}$.

Suppose $D_i = (A_i \cap E) \cup (B_i \cap E^c)$, where A_i, $B_i \in \mathscr{A}$, $D_i \cap D_j = \varnothing$ $(i \neq j)$, and $\bigcup_{i=1}^{\infty} D_i \in \mathscr{B}$. Now

$$\bigcup_{i=1}^{\infty} D_i = \left[\left(\bigcup_{i=1}^{\infty} A_i\right) \cap E\right] \cup \left[\left(\bigcup_{i=1}^{\infty} B_i\right) \cap E^c\right]. \tag{2.23}$$

Clearly for $i \neq j$, $A_i \cap A_j \cap E = \varnothing$ and $B_i \cap B_j \cap E^c = \varnothing$. Let us write $P_1 = A_1$, $Q_1 = B_1$ and for $n > 1$,

$$P_n = A_n - \bigcup_{i=1}^{n-1} A_i, \qquad Q_n = B_n - \bigcup_{i=1}^{n-1} B_i.$$

Then $P_n, Q_n \in \mathscr{A}$ and $P_n \cap E = A_n \cap E$, $Q_n \cap E^c = B_n \cap E^c$; also $P_i \cap P_j = Q_i \cap Q_j = \varnothing$ for $i \neq j$. Therefore, by equation (2.23),

$$\mu^*\left(\bigcup_{n=1}^{\infty} D_n \cap E\right) = \mu^*\left(\bigcup_{n=1}^{\infty} (P_n \cap E)\right) = \sum_{n=1}^{\infty} \mu^*(P_n \cap E) = \sum_{n=1}^{\infty} \mu^*(A_n \cap E). \tag{2.24}$$

Also by Remark 2.8,

$$\mu_*\left(\bigcup_{n=1}^{\infty} D_n \cap E\right) = \sum_{n=1}^{\infty} \mu_*(P_n \cap E) = \sum_{n=1}^{\infty} \mu_*(A_n \cap E). \tag{2.25}$$

Similarly, we have

$$\mu^*\left(\bigcup_{n=1}^{\infty} D_n \cap E^c\right) = \sum_{n=1}^{\infty} \mu^*(B_n \cap E^c) \tag{2.26}$$

and

$$\mu_*\left(\bigcup_{n=1}^{\infty} D_n \cap E^c\right) = \sum_{n=1}^{\infty} \mu_*(B_n \cap E^c). \tag{2.27}$$

From (2.24) and (2.27),

$$\bar{\mu}\left(\bigcup_{n=1}^{\infty} D_n\right) = \sum_{n=1}^{\infty} \bar{\mu}(D_n),$$

and therefore $\bar{\mu}$ is countably additive. Similarly, by equations (2.25) and (2.26) $\underline{\mu}$ is countably additive.

Now we present a simple (but proper) translation-invariant extension of Lebesgue measure that was considered in [46].

Let A be a set $\subset R$ such that both A and $R - A$ have nonempty intersection with every uncountable closed set in R. (Such a set is constructed in Problem 2.4.5.) Such a set is necessarily non-Lebesgue-measurable (see Problem 2.4.5) and has the following properties:

$$E \subset A \text{ is Lebesgue measurable if and only if } m^*(E) = 0. \tag{2.28a}$$

$$E \subset R - A \text{ is Lebesgue measurable if and only if } m^*(E) = 0. \tag{2.28b}$$

The reason for property (2.28a) is that if $E \subset A$ and E is Lebesgue measurable, then

$$m^*(E) = m(E) = \sup\{m(B)\colon B \text{ closed and } B \subset E\};$$

but $B \subset E \subset A \Rightarrow B \cap (R - A) = \varnothing$, and therefore B must be countable. Hence $m(B) = 0$, implying that $m^*(E) = 0$. The same reasoning applies for property (2.28b).

Let $\mathscr{M}$ be the Lebesgue-measurable sets on R and $\mathscr{M}^*$ be the σ-algebra generated by $\mathscr{M}$ and the set A above. Then since the class of sets $\{(E \cap A) \cup (F \cap A^c)\colon E, F \in \mathscr{M}\}$ is a σ-algebra containing A, we have

$$\mathscr{M}^* = \{(E \cap A) \cup (F \cap A^c)\colon E, F \in \mathscr{M}\}. \tag{2.29}$$

Preliminary to proving our extension theorem, we need a lemma.

• **Lemma 2.3.** Let $B \in \mathscr{M}^*$ and let $B = (E \cap A) \cup (F \cap A^c)$, where $E, F \in \mathscr{M}$. Then $m^*(B) + m_*(B) = m(E) + m(F)$. ∎

Proof. Note that

$$B = (E \cap F^c \cap A) \cup (E^c \cap F \cap A^c) \cup (E \cap F).$$

By Remark 2.8, we have

$$m_*(B) = m_*(A \cap E \cap F^c) + m_*(A^c \cap E^c \cap F) + m_*(E \cap F). \quad (2.30)$$

By Remark 2.7 and equation (2.28), it follows that

$$m_*(A \cap E \cap F^c) = m_*(A^c \cap E^c \cap F) = 0.$$

Hence $m_*(B) = m(E \cap F)$. Also, noting that Remark 2.8 remains true if the inner measure is replaced by the outer measure, we have

$$m^*(B) = m^*(A \cap E \cap F^c) + m^*(A^c \cap E^c \cap F) + m^*(E \cap F).$$

It follows from equation (2.16) that

$$m(E \cap F^c) = m_*(E \cap F^c \cap A^c) + m^*(E \cap F^c \cap A) = m^*(E \cap F^c \cap A),$$

by equation (2.28). Similarly, $m^*(A^c \cap E^c \cap F) = m(E^c \cap F)$. Hence $m^*(B) = m(E \cup F)$. The rest is clear. ∎

Theorem 2.8. Let λ be defined on $\mathscr{M}^*$ by

$$\lambda(B) = \tfrac{1}{2}[m^*(B) + m_*(B)]. \quad (2.31)$$

Then λ is a complete measure on $\mathscr{M}^*$ and a proper translation-invariant extension of m. ∎

Proof. Since m^* is translation invariant by Problem 2.3.5, m_* is so also by its definition, and therefore λ is translation invariant. Also since $m^*(B) = m_*(B)$ for Lebesgue-measurable B by Remark 2.9, $\lambda(B) = m(B)$.

We only need to establish that λ is a measure. First note that λ is subadditive since m^* and m_* are both subadditive. Let (B_n) be a disjoint sequence in $\mathscr{M}^*$ such that

$$B_n = (E_n \cap A) \cup (F_n \cap A^c), \qquad E_n, F_n \in \mathscr{M}.$$

Notice that for $n \neq k$, $E_n \cap E_k \subset A^c$ and $F_n \cap F_k \subset A$. Therefore by equation (2.28) $n \neq k$ implies that

$$m(E_n \cap E_k) = 0 = m(F_n \cap F_k).$$

Hence we have

$$m\left(\bigcup_{n=1}^{\infty} E_n\right) = \sum_{n=1}^{\infty} m(E_n)$$

and

$$m\left(\bigcup_{n=1}^{\infty} F_n\right) = \sum_{n=1}^{\infty} m(F_n).$$

Since

$$\bigcup_{n=1}^{\infty} B_n = \left[\left(\bigcup_{n=1}^{\infty} E_n\right) \cap A\right] \cup \left[\left(\bigcup_{n=1}^{\infty} F_n\right) \cap A^c\right],$$

it follows by Lemma 2.3 that

$$\lambda\left(\bigcup_{n=1}^{\infty} B_n\right) = \tfrac{1}{2}\left[m\left(\bigcup_{n=1}^{\infty} E_n\right) + m\left(\bigcup_{n=1}^{\infty} F_n\right)\right] = \sum_{n=1}^{\infty} \tfrac{1}{2}[m(E_n) + m(F_n)]$$
$$= \sum_{n=1}^{\infty} \lambda(B_n).$$

The proof is complete. ∎

Problems

2.4.1. Give an example of a sequence (E_n) of pairwise disjoint sets on R such that $m^*(\bigcup_{n=1}^{\infty} E_n) < \sum_{n=1}^{\infty} m^*(E_n)$.

2.4.2. Give an example of a sequence of sets (E_n) such that $E_n \supset E_{n+1}$, $m^*(E_n) < \infty$, and $m^*(\bigcap_{n=1}^{\infty} E_n) < \lim_{n\to\infty} m^*(E_n)$.

× **2.4.3.** Give an example of a Lebesgue-measurable set that is not a Borel set. {Hint: Take the Cantor set $K \subset [0, 1]$ and the Cantor function L. Let $g(x) = L(x) + x$. Then g is a homeomorphism from $[0, 1]$ onto $[0, 2]$ and $g(K)$ has Lebesgue measure 1. Now there is $E \subset g(K)$, E non-Lebesgue-measurable. Show that $g^{-1}(E)$ is the desired set.}

× **2.4.4.** Show that the Borel measure (the restriction of the Lebesgue measure on the Borel sets of R) is not complete.

2.4.5. *Another Example of a Non-Lebesgue-Measurable Set* [due to F. Bernstein (1908.)] The set $\mathscr{F}$ of all closed (but *uncountable*) subsets of R has cardinality c. (Note that every open set is a finite or countable union of open intervals with rational endpoints.) Assuming the continuum hypothesis, there is a one-to-one correspondence between $\mathscr{F}$ and $[0, \Omega)$, Ω the first uncountable ordinal. Let $(A_\alpha)_{\alpha<\Omega}$ denote a well-ordering of $\mathscr{F}$. For each $\alpha < \Omega$, there exist $a_\alpha, b_\alpha \in R$ such that $a_\alpha \in A_\alpha - \bigcup_{\beta<\alpha}\{a_\beta, b_\beta\}$, $b_\alpha \in A_\alpha - \bigcup_{\beta<\alpha}\{a_\beta, b_\beta\}$, and $a_\alpha \neq b_\alpha$. This is possible since A_α's are all uncountable. Let $A = \{a_\alpha: \alpha < \Omega\}$ and $B = \{b_\alpha: \alpha < \Omega\}$. Then A and B are disjoint and $m^*(A \cap [0, 1]) = m^*(B \cap [0, 1]) = 1$. Therefore A and B are both non-Lebesgue-measurable.

2.4.6. Let E be a Lebesgue-measurable set of positive Lebesgue measure. Show that there are disjoint non-Lebesgue-measurable sets E_1 and E_2 such that $E = E_1 \cup E_2$ and $m(E) = m^*(E_1) = m^*(E_2)$. [Hint: Let A be the set constructed in Problem 2.4.5. Let $E_1 = E \cap A$ and $E_2 = E \cap A^c$ Use equation (2.28).]

2.4.7. *Arbitrary Union of Open Sets of Measure Zero with Measure Zero* (E. Marczewski and R. Sikorski). Let (X, d) be a metric space with a base $\mathscr{B}$ with the property that every finite measure defined for all subsets of a set of cardinality equal to card $\mathscr{B}$ and zero for points is a zero measure. Let μ be a finite (weakly) Borel measure on X. Then the union of any family of open sets of measure zero has measure zero. [Hint: Let $\mathscr{F}$ be a family of open sets of measure zero and let $\{G_\alpha: \alpha \in A\}$ be a well-ordering of $\mathscr{F}$. Let $H_\alpha = G_\alpha - \bigcup_{\beta<\alpha} G_\beta$. Then H_α is an F_σ-set. Notice that $\bigcup_{\alpha \in E} H_\alpha$ is also an F_σ-set for any $E \subset A$; if C_α is a closed subset of H_α and $C_{\alpha,n} = \{x \in C_\alpha: d(x, X - G_\alpha) \geq 1/n\}$, then $C_\alpha = \bigcup_{n=1}^\infty C_{\alpha,n}$ and for each n the set $\bigcup_{\alpha \in E} C_{\alpha,n}$ is closed. Define $\lambda(E) = \mu(\bigcup_{\alpha \in E} H_\alpha)$. Then λ is a finite measure defined for all subsets of E and zero for points. Since card $A \leq$ card $\mathscr{B}$, $\lambda(A) = 0$ (by hypothesis).]

2.4.8. *Thick Subsets of a Measure Space.* Let μ be a measure of a σ-algebra $\mathscr{A}$ of subsets of X. A subset $E \subset X$ is called *thick* if $\mu_*(X - E) = 0$. Suppose E is thick, $\mathscr{A}_E = \{B \cap E: B \in \mathscr{A}\}$, and $\lambda(B \cap E) = \mu(B)$. Prove that λ is a measure on the σ-algebra $\mathscr{A}_E$ of subsets of E.

2.4.9. *Egoroff's Theorem*† *for Families of Functions.* Suppose for each y in $[2, \infty)$, $f(x, y)$ is a real-valued Lebesgue-measurable function on $[0, 1]$ such that $\lim_{y\to\infty} f(x, y) = h(x)$ exists. If for each x in $[0, 1]$, the function $f(x, y)$ is continuous in y, then the above limit is almost uniform. This result is not true in general. Consider the following example due to W. Walter.

Let $E \subset [0, \frac{1}{2})$ be a non-Lebesgue-measurable set as constructed in the proof of Theorem 2.5. Let $Q = [0, 1] \times [2, \infty)$, $Q_n = [0, 1] \times [n, n + 1]$, and the diagonal D_n of $Q_n = \{(x, n + x): 0 \leq x \leq 1\}$. Define $f: Q \to \{0, 1\}$ as follows:

$$f(x, y) = \begin{cases} 1, & \text{if for some } n \geq 2,\ x \in E + \dfrac{1}{n} \text{ and } (x, y) \in D_n; \\ 0, & \text{otherwise.} \end{cases}$$

Then $\lim_{y\to\infty} f(x, y) = 0$, but this limit is not almost uniform.

2.4.10. Let m be the Lebesgue measure on R. Show that there exists a decreasing sequence of thick sets with empty intersection.

† Measurable functions are discussed in Section 3.1. At this point, the reader is expected to verify only the example in this problem.

2.4.11. Let X be a separable metric space, $\mathscr{A}$ the σ-algebra of Borel sets of X and μ be a finite measure on $\mathscr{A}$. A Borel set A is called an atom if for each Borel subset B of A, $\mu(B) = 0$ or $\mu(A - B) = 0$. Prove the following assertions:

(i) $\mathscr{A}$ has no atoms if and only if every singleton has measure zero. [Hint (for the "if" part): Let $E \in \mathscr{A}$ be an atom and $\mu(\{x\}) = 0$ for all $x \in X$. Then $\mu(E) > 0$. Let (B_n) be a base for the relative topology of E. Define D_n to be B_n or $E - B_n$ according as $\mu(B_n) = \mu(E)$ or $\mu(B_n) = 0$. Consider $D = \bigcap_{n=1}^{\infty} D_n$.]

(ii) If there is a thick set whose complement is also a thick set, then $\mathscr{A}$ has no atoms.

(iii) If there is a decreasing sequence of thick sets with empty intersection, then $\mathscr{A}$ has no atoms. [Note: the converses of (ii) and (iii) are also true (and due to S. B. Rao) when X is also complete and has no isolated points.]

2.4.12. Let μ be a semifinite measure on an algebra $\mathscr{A}$. Show that μ_* is a semifinite measure on the smallest σ-algebra $\mathscr{B}$ containing $\mathscr{A}$ and the smallest extension of μ to $\mathscr{B}$.

2.4.13. Let $E_1, E_2 \subset R$ and m^*, m_* be the Lebesgue outer and inner measures, respectively. Show that if $m^*(E_1 \cup E_2) = m^*(E_1) + m^*(E_2)$, then $m_*(E_1 \cup E_2) = m_*(E_1) + m_*(E_2)$. Show also that the converse need not be true.

2.4.14. Let $E \subset R$. Suppose $m^*(E \cap I) \geq \delta m(I)$ for some $\delta > 0$ and all intervals I of R. Show that $m^*(R - E) = 0$ if E is Lebesgue-measurable and $m^*(R - E)$ may be nonzero if E is not so. [Hint: Let A be Lebesgue measurable and $m(A) > 0$. Then there is a non-Lebesgue-measurable subset B of A such that $m^*(B) \geq \frac{1}{2}m(A)$; using this fact observe that there is a non-Lebesgue-measurable subset D of A such that $m^*(D) = m(A)$. Take $E = (R - A) \cup D$.]

2.4.15. *Nonmeasurable Sets Whose Measurable Subsets Are All at Most Countable.* A set $A \subset R$ is said to have property (P) if $A \cap B$ is at most countable for every set B with Lebesgue measure zero. Prove that every uncountable subset of a set with property (P) is non-Lebesgue-measurable. Prove also the following result due to R. E. Dressler and R. B. Kirk: Assuming that the continuum hypothesis holds, there is a partition of the real numbers by an uncountable family of sets (each of which is an uncountable set of Lebesgue measure zero) such that a subset of R has property (P) if and only if it intersects each member of the partition in a set that is at most countable. [Hint: Let $\mathscr{F} = \{A \subset R$: A is an uncountable G_δ set of Lebesgue measure zero. Then the cardinal number of $\mathscr{F}$ is $2^{\aleph_0}$. Assuming the continuum hypothesis, enumerate R as $\{x_\alpha: \alpha < \Omega\}$ and $\mathscr{F}$ as $\{E_\alpha: \alpha < \Omega\}$, where Ω is the first uncountable ordinal. Using transfinite induction, show that there is a family of subsets of R denoted by $\{A_\alpha: \alpha < \Omega\}$ such that

(i) $\alpha < \Omega \Rightarrow E_\alpha - \cup \{A_\beta: \beta \leq \alpha\}$ is at most countable,

(ii) each A_α is an uncountable set of Lebesgue measure zero,

(iii) $\alpha \neq \beta \Rightarrow A_\alpha \cap A_\beta$ is empty,

(iv) $x_\alpha \in \cup \{A_\beta: \beta \leq \alpha\}$.

Then these A_α's provide the desired partition of R.]

2.4.16. Prove that if E is a Lebesgue-measurable set of reals with cardinality less than c, then $m(E) = 0$. [Hint: Let $m(E) > 0$. Assume that E is closed and bounded. Then there are disjoint closed intervals I_0 and I_1 such that $m(E \cap I_0) > 0$ and $m(E \cap I_1) > 0$. Repeat this process to obtain intervals as follows:

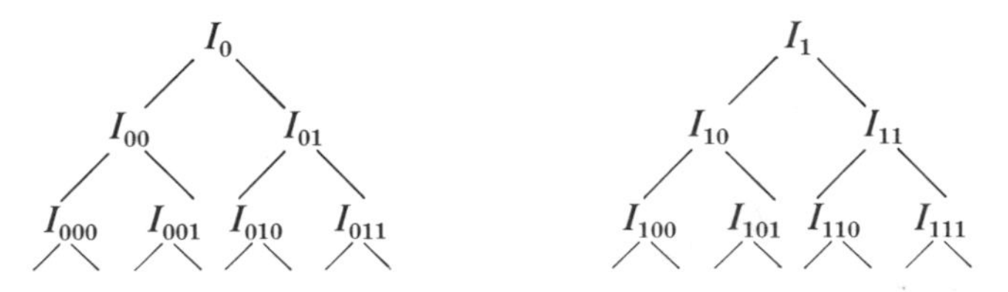

It follows that corresponding to each infinite sequence (a_n) of zeros and ones, we have an infinite sequence of nested intervals $I_{a_1}, I_{a_1a_2}, \ldots$, etc. each of which intersects E and determines at least one point of E.]

2.4.17. *A Property of Non-μ^*-Measurable Sets.* Let μ be a measure on a σ-algebra of subsets of X. If E is a non-μ^*-measurable subset of X, then prove that

$$\inf\{\mu^*(A \cap B): A \supset E \text{ and } B \supset X - E,\ A \text{ and } B \text{ both } \mu^*\text{-measurable}\}$$

is positive.

3

Integration

A. L. Cauchy (1789–1857) was perhaps the first mathematician to give a rigorous definition of an integral as the limit of a sum. But he considered only functions having at most a finite number of discontinuities.

Motivated by the needs of the theory of trigonometric series, already studied by Dirichlet in 1829 in expanding certain monotonic functions f into a series $\sum(a_n \cos nx + b_n \sin nx)$, where

$$a_n = \frac{1}{\pi}\int_0^{2\pi} f(x)\cos nx\,dx \quad \text{and} \quad b_n = \frac{1}{\pi}\int_0^{2\pi} f(x)\sin nx\,dx,$$

B. Riemann (1826–1866) continued the work of Cauchy. He defined the integral in a way similar to Cauchy's, but examined the class of all functions for which the integral could be defined. He discovered that he could even integrate functions with an everywhere dense set of points of discontinuity.

During the time of Cauchy and Riemann, mathematicians were considering mostly integrals for bounded functions. But soon, when unbounded functions started to appear in the theory of trigonometric series, mathematicians turned their attention to the possibility of defining a useful integral for such functions. Harnack (1883) and de la Vallée-Poussin (1894) are among the first mathematicians to take steps in this direction.

Motivated by diverse problems in analysis, investigations in integration theory continued and soon G. Peano (1858–1932) and C. Jordan (1838–1922) connected the concept of integration with the then recent concept of measure, already introduced by G. Cantor (1845–1918) and others. But the decisive step in the theory of integration was taken by H. Lebesgue

(1875–1941) by the discovery of a new theory of the integral contained in his thesis "Intégrale, Longueur, Aire," published in 1902. Mainly motivated by questions of applying the integral, as an effective tool in mathematical analysis, to classical problems such as the determination of curve lengths and areas of planar sets, Lebesgue took up the problem of defining an integral having all the properties that could determine this applicability. He reduced the problem to the problem of defining a countably additive, congruence-invariant measure on the class of all bounded sets on the reals such that the measure of an interval is its length. He defined a measure, now well-known as Lebesgue measure, and showed that this was the unique solution of the problem of measure for a large class of sets, now called Lebesgue-measurable sets.

An important contribution to the theory of integration was also made by W. H. Young (1863–1942). He was mainly influenced by the idea of generalizing the Riemann integral, and he wrote two important papers on this subject.[†] However, the final results of Young's and Lebesgue's contributions are the same.

After Lebesgue had laid the foundations of the modern theory of integration, more work was done later by mathematicians including F. Riesz (1880–1956). A. Denjoy (1884–), J. Radon (1887–1956), and others, to modify and further extend the Lebesgue definition of the integral.

This chapter presents the theory of integration in an abstract setting, containing the Lebesgue theory as a special case. In Section 3.1 we introduce and study the measurable functions, which will serve as the domain of the general integral studied in Section 3.2. In Section 3.3, we study Riemann–Stieltjes integrals from the point of view of measure theory and compare the Riemann and the Lebesgue integrals. In Section 3.4, we present Fubini's Theorem, a cornerstone in modern analysis, unavoidable in the evaluation of most multiple integrals in R^n. Finally, in Section 3.5, the L_p spaces are presented; the Riesz Representation Theorem for bounded linear functionals is, however, deferred till Section 4.3.

Throughout this chapter, $(X, \mathscr{A}, \mu)$ will denote a measure space, that is, a triple where X is a nonempty set, $\mathscr{A}$ is a σ-algebra of subsets of X, and μ is a measure on $\mathscr{A}$. When X is R, $\mathscr{A}$ is the class of Lebesgue-measurable sets, and μ is the Lebesgue measure, then the measure space is referred to as the *Lebesgue measure space*; if here $\mathscr{A}$ is the class of the Borel sets of R, the measure space is called the *Borel measure space*. (Note that in what

[†] W. H. Young, *Phil. Trans. Roy. Soc. London*, 204A, 221–252 (1905); *Proc. London Math. Soc.* **9**(2), 15–20 (1910).

follows m will usually denote the Lebesgue measure whenever X is R or a subset of R.)

Also in this chapter, we will occasionally use the following *notation*: For any extended real-valued function f on a measure space, we will write

$$f^+(x) = \sup\{f(x), 0\}, \qquad f^-(x) = \sup\{-f(x), 0\},$$

and

$$f \vee g = \sup\{f, g\}, \qquad f \wedge g = \inf\{f, g\}.$$

3.1. Measurable Functions

The Riemann integral of classical analysis is defined for any continuous real-valued function on a closed and bounded interval. More precisely, as will be shown in Section 3.3, a bounded function on $[a, b]$ is Riemann integrable if and only if the set of its points of discontinuity has Lebesgue measure zero. The integral, which is the subject of this chapter, will be shown (in Section 3.3) to be an extension of the Riemann integral and will be defined for a much larger class of functions, called *measurable functions*, which are not necessarily bounded.

Measurable functions form the subject of this section. Unlike continuous functions, the class of measurable functions will be found to be closed with respect to additional types of limiting operations; and this makes the integral applicable to a wider class of problems and naturally more useful. Despite being quite general in nature, these functions on the Lebesgue measure space are quite nice in a certain sense. It will be shown in this section that a measurable function is "almost" continuous and in a finite measure space [when $\mu(X) < \infty$], every convergent sequence of measurable functions is "almost" uniformly convergent. The meaning of the word "almost" will be clear later in this section.

Let us begin with the definition of a measurable function.

Definition 3.1. Let $E \in \mathscr{A}$ and let $\bar{R}$ be the extended reals. A function f: $E \to \bar{R}$ is called *measurable* if for each real number α, the set $\{x \in E$: $f(x) > \alpha\} \in \mathscr{A}$. If $\mathscr{A}$ is the class of Lebesgue-measurable subsets (resp. Borel sets) on R $(= X)$, a measurable function f is usually called a *Lebesgue-* (resp. *Borel-*) *measurable* function. ∎

This definition immediately leads to several equivalent definitions in the next proposition.

Proposition 3.1. Let f be as in Definition 3.1. Then the following are equivalent:

(i) f is measurable.

(ii) $\{x \in E: f(x) \geq \alpha\} \in \mathscr{A}$, if $\alpha \in R$.

(iii) $\{x \in E: f(x) < \alpha\} \in \mathscr{A}$, if $\alpha \in R$.

(iv) $\{x \in E: f(x) \leq \alpha\} \in \mathscr{A}$, if $\alpha \in R$.

Moreover, these statements imply

(v) $\{x \in E: f(x) = \alpha\} \in \mathscr{A}$,

for every extended real number α. ∎

Proof. The proof is obvious from the following observations:

$$\{x \in E: f(x) \geq \alpha\} = \bigcap_{n=1}^{\infty} \{x \in E: f(x) > \alpha - 1/n\}; \tag{3.1}$$

$$\{x \in E: f(x) < \alpha\} = E - \{x \in E: f(x) \geq \alpha\}; \tag{3.2}$$

$$\{x \in E: f(x) \leq \alpha\} = \bigcap_{n=1}^{\infty} \{x \in E: f(x) < \alpha + 1/n\}; \tag{3.3}$$

$$\{x \in E: f(x) > \alpha\} = E - \{x \in E: f(x) \leq \alpha\}; \tag{3.4}$$

$$\{x \in E: f(x) = \infty\} = \bigcap_{n=1}^{\infty} (x \in E: f(x) > n\}; \tag{3.5}$$

$$\{x \in E: f(x) = -\infty\} = \bigcap_{n=1}^{\infty} \{x \in E: f(x) < -n\}. \quad ∎ \tag{3.6}$$

Remarks

3.1. If $X = R$ and $\mathscr{A}$ is the class of Lebesgue-measurable sets or the Borel sets on R, then *every* continuous function f on $E \in \mathscr{A}$ is measurable, since $\{x \in E: f(x) > \alpha\}$ is the intersection of E and an open set.

3.2. A function $f: E \to \bar{R}$ is measurable if and only if $f^{-1}(B) \in \mathscr{A}$, for every Borel set $B \subset R$. To see this, it is sufficient to observe that for the measurable function f, the class

$$\{B \subset R: f^{-1}(B) \in \mathscr{A}\}$$

is a σ-algebra containing all intervals of the form (α, ∞).

3.3. The characteristic function $\chi_A(x)$ is a measurable function on X if and only if $A \in \mathscr{A}$.

Proposition 3.2. (i) If f and g are measurable real-valued functions having the same domain, then $f \pm g$, $|f|$, $f \vee g$, $f \wedge g$, and $f \cdot g$ are measurable functions.

(ii) If (f_n) is a sequence of measurable functions (having the same domain), then $\sup_n f_n$, $\inf_n f_n$, $\overline{\lim}_n f_n$, $\underline{\lim}_n f_n$ are all measurable. ∎

Proof. (i) Since for any real number α, there is a rational number r such that

$$f(x) < r < \alpha - g(x)$$

whenever $f(x) + g(x) < \alpha$, we have

$$\{x: f(x) + g(x) < \alpha\} = \bigcup_r [\{x: f(x) < r\} \cap \{x: g(x) < \alpha - r\}].$$

The union is taken over a set of rational numbers. It follows that $f + g$ is measurable. Similarly, $f - g$ is also measurable.

Now

$$f \vee g = \tfrac{1}{2}[f + g + |f - g|] \quad \text{and} \quad f \wedge g = \tfrac{1}{2}[f + g - |f - g|].$$

Also, $|f|$ is measurable whenever f is, since

$$\{x: |f(x)| > \alpha\} = \{x: f(x) > \alpha\} \cup \{x: f(x) < -\alpha\}.$$

This means that $f \vee g$ and $f \wedge g$ are both measurable. Since for $\alpha \geq 0$,

$$\{x: f^2(x) > \alpha\} = \{x: f(x) > \sqrt{\alpha}\} \cup \{x: f(x) < -\sqrt{\alpha}\},$$

and $4fg = (f + g)^2 - (f - g)^2$, it follows that $f \cdot g$ is measurable.

(ii) The proof of (ii) is clear from the following observations:

$$\{x: \sup_n f_n(x) > \alpha\} = \bigcup_n \{x: f_n(x) > \alpha\}, \tag{3.7}$$

$$\inf_n f_n = -\sup_n(-f_n), \tag{3.8}$$

$$\overline{\lim_n} f_n = \inf_k \sup_{n \geq k} f_n, \tag{3.9}$$

$$\underline{\lim_n} f_n = \sup_k \inf_{n \geq k} f_n. \quad ∎ \tag{3.10}$$

Definition 3.2. A property is said to hold *almost everywhere* (or a.e.) if the set of points for which it fails to hold is measurable and has measure zero. ∎

Proposition 3.3. Let $(X, \mathscr{A}, \mu)$ be a complete measure space and $f = g$ a.e. If f is measurable, then g is also measurable. ∎

The proof is left to the reader.

In Chapter 2 (Problems 2.4.3 and 2.4.4) we found that the Borel measure is not complete and there is a set $A \subset K$ (where K is the Cantor set $\subset [0, 1]$) such that A is a Lebesgue-measurable set with Lebesgue measure zero, but not a Borel set. If $(X, \mathscr{A}, \mu)$ is the Borel measure space on $[0, 1]$ then the function

$$f(x) = \begin{cases} 1, & x \in A, \\ 2, & x \in K - A, \\ 0, & x \notin K, \end{cases}$$

is equal to the zero function almost everywhere, but not measurable since

$$\{x\colon f(x) = 1\} = A \notin \mathscr{A}.$$

This shows that completeness is essential in Proposition 3.3.

We will now show how a bounded measurable function can be uniformly approximated by *simple functions*, that is, *measurable real-valued functions that assume only a finite number of real values.*

Proposition 3.4. A simple function f has the form $\sum_{i=1}^{n} \alpha_i \chi_{A_i}$, where $\alpha_i \in R$ and $A_i \in \mathscr{A}$. The sum, product, and difference of two simple functions are simple. ∎

The proof is left to the reader.

Proposition 3.5. Let f be a measurable function. Then f is the pointwise limit of a sequence of simple functions. If f is bounded, then the convergence is uniform. If $f \geq 0$, then the above sequence can be taken as monotonic increasing. ∎

Proof. Let $f^+ = f \vee 0$ and $f^- = (-f) \vee 0$. Then f^+ and f^- are both nonnegative and measurable by Proposition 3.2, and $f = f^+ - f^-$. Therefore, it is no loss of generality to prove the proposition for nonnegative f.

Let f be nonnegative. For each integer $n \geq 1$ and $x \in X$, let

$$f_n(x) = \begin{cases} \dfrac{i-1}{2^n}, & \text{if } \dfrac{i-1}{2^n} \leq f(x) < \dfrac{i}{2^n}, \text{ for } i = 1, 2, \ldots, n2^n, \\ n, & \text{if } f(x) \geq n. \end{cases}$$

Then the f_n's are simple functions and $f_{n+1}(x) \geq f_n(x)$. Also if $f(x) \leq n$, then $0 \leq f(x) - f_n(x) \leq 1/2^n$. Hence $f(x) = \lim_{n\to\infty} f_n(x)$. Clearly, the convergence is uniform when f is bounded. ∎

The next proposition shows that a Lebesgue-measurable function defined on $[a, b]$ is almost continuous in a certain sense. (See also Problems 3.1.11 and 3.1.13.)

Proposition 3.6. Let f be an a.e. real-valued Lebesgue measurable function defined on $[a, b]$. Then given $\varepsilon > 0$, there exists a continuous function g such that $m\{x \in [a, b]: | f(x) - g(x) | \geq \varepsilon\} < \varepsilon$ and $\sup| g(x) | \leq \sup| f(x) |$. ∎

Proof. Since $m(\bigcup_{n=1}^{\infty} \{x: | f(x) | < n\}) = b - a$, there is an N such that $m(\{x: | f(x) | < N\})$ is greater than $b - a - \varepsilon/3$. By Proposition 3.5, there is a simple function h such that $m\{x: | f(x) - h(x) | < \varepsilon/2\} > b - a - \varepsilon/3$. [Note that we are applying Proposition 3.5 to the bounded function $f \cdot \chi_B$, where $B = \{x: | f(x) | < N\}$.] Now let $h = \sum_{k=1}^{n} c_k \chi_{E_k}$, where c_k's are all the distinct values assumed by h on $[a, b]$ and E_k's are pairwise disjoint. Let $F_k \subset E_k$, F_k closed and $m(E_k - F_k) < \varepsilon/3n$. If $F = \bigcup_{k=1}^{n} F_k$, then F is closed and the function g defined on F by $g(x) = c_k$, $x \in F_k$ is continuous on F. Now we can extend g continuously on $[a, b]$ as in Problem 3.1.11(a). Then $m\{x \in [a, b]: | f(x) - g(x) | < \varepsilon/2\}$ is greater than $b - a - 2\varepsilon/3$. ∎

Now we show another important property of measurable functions: the equivalence of a.e. convergence (that is, pointwise convergence everywhere except on a set of measure zero) and almost uniform convergence for a sequence of a.e. real-valued measurable functions.

Definition 3.3. A sequence (f_n) of a.e. real-valued measurable functions is said to *converge almost uniformly* to a measurable function f if for any $\varepsilon > 0$ there exists $E \in \mathscr{A}$ with $\mu(E) < \varepsilon$ such that (f_n) converges to f uniformly on $X - E$. ∎

Remark 3.4. Suppose $f_m \to f$ almost uniformly. Then for each positive integer n, there exists $E_n \in \mathscr{A}$ such that $\mu(E_n) < 1/n$ and $f_m \to f$ uniformly on $X - E_n$. Let $A = \bigcup_{n=1}^{\infty}(X - E_n)$. Then $\mu(X - A) = 0$ and for

$x \in A$, $f_m(x) \to f(x)$. Hence almost uniform convergence implies convergence a.e. The converse is true if $\mu(X) < \infty$, as shown by the next theorem.

Theorem 3.1. (*Egoroff*) Let $\mu(X) < \infty$ and (f_n) be a sequence of a.e. real-valued measurable functions converging a.e. to an a.e. real-valued measurable function f. Then $f_n \to f$ almost uniformly. ▮

Proof. With no loss of generality, we assume that f and the f_n are all real-valued everywhere. For each positive integer k, let

$$A_{n,k} = \bigcap_{m=n}^{\infty} \{x: |f_m(x) - f(x)| < 1/k\}.$$

Since $\lim_{n\to\infty} f_n(x) = f(x)$ a.e., we have

$$\mu\left(\bigcup_{n=1}^{\infty} A_{n,k}\right) = \mu(X).$$

Since $A_{n,k} \subset A_{n+1,k}$, $\lim_{n\to\infty}\mu(A_{n,k}) = \mu(X)$. Now given $\varepsilon > 0$, for each positive integer k there exists n_k such that for $n \geq n_k$,

$$\mu(X - A_{n,k}) < \varepsilon/2^k.$$

Let $A = \bigcap_{k=1}^{\infty} A_{n_k,k}$. Then

$$\mu(X - A) \leq \sum_{k=1}^{\infty} \mu(X - A_{n_k,k}) < \varepsilon.$$

The reader can easily check that f_n converge uniformly to f on A. ▮

Remark 3.5. Theorem 3.1 need not be true if $\mu(X) = \infty$. For example, the sequence $\chi_{(n,\infty)}(x)$ converges to 0 as $n \to \infty$, but not almost uniformly with respect to the Lebesgue measure.

Finally, we introduce another concept of convergence—that of convergence in measure, a weaker concept of convergence. This concept is of basic importance in probability theory.

Definition 3.4. A sequence (f_n) of a.e. real-valued measurable functions is said to converge in measure to a measurable function f if for every $\varepsilon > 0$,

$$\lim_{n\to\infty} \mu\{x: |f_n(x) - f(x)| \geq \varepsilon\} = 0.$$ ▮

Remarks

3.6. If $f_n \to f$ in measure, then f is a.e. real-valued; for, we can find

E with $\mu(E) = 0$ such that the f_n are all real-valued on $X - E$ and $X - E = \bigcup_{n=1}^{\infty} \{x\colon |f_n(x) - f(x)| < \varepsilon\}$.

3.7. If $f_n \to f$ in measure and $f_n \to g$ in measure, then $f = g$ a.e. For, given $\varepsilon > 0$,

$$\{x\colon |f(x) - g(x)| > 2\varepsilon\} \subset \{x\colon |f(x) - f_n(x)| \geq \varepsilon\} \cup \{x\colon |g(x) - f_n(x)| \geq \varepsilon\},$$

and hence $\mu\{x\colon |f(x) - g(x)| > 2\varepsilon\} = 0$. This means that $f = g$ a.e.

3.8. Almost uniform convergence implies convergence in measure; but the converse is not true. (Problem 3.1.7.)

3.9. Almost everywhere convergence implies convergence in measure, when $\mu(X) < \infty$. This follows from Egoroff's Theorem and the previous remark.

Definition 3.5. A sequence (f_n) of a.e. real-valued measurable functions is called a Cauchy sequence in measure, if, for every $\varepsilon > 0$,

$$\mu\{x\colon |f_n(x) - f_m(x)| \geq \varepsilon\} \to 0$$

as $n, m \to \infty$. ∎

Clearly a sequence (f_n), which converges in measure, is a Cauchy sequence in measure. The converse follows from the next result.

Proposition 3.7. Let (f_n) be a Cauchy sequence in measure. Then there is a measurable function f and a subsequence (f_{n_k}) such that $f_n \to f$ in measure and $f_{n_k} \to f$ almost uniformly and hence almost everywhere. ∎

Proof. We choose (n_k) such that $n_{k+1} > n_k$ and

$$\mu\left\{x\colon |f_{n_k}(x) - f_{n_{k+1}}(x)| > \frac{1}{2^k}\right\} < \frac{1}{2^k}.$$

Let

$$A_m = \bigcup_{k=m+1}^{\infty} \left\{x\colon |f_{n_k}(x) - f_{n_{k+1}}(x)| > \frac{1}{2^k}\right\}.$$

Then $\mu(A_m) < 1/2^m$ and on $X - A_m$, the subsequence (f_{n_k}) is uniformly Cauchy. Let $A = \bigcap_{m=1}^{\infty} A_m$. Then $\mu(A) = 0$. Clearly for each m, there exist g_m such that $f_{n_k} \to g_m$ uniformly on $X - A_m$. Since $X - A_m \subset X - A_{m+1}$, $g_m = g_{m+1}$ on $X - A_m$. We define $f(x) = 0$ on A, $= g_m(x)$ on $X - A_m$, $1 \leq m < \infty$. Then $f_{n_k} \to f$ almost uniformly. Hence $f_{n_k} \to f$ a.e. and in measure by Remarks 3.4 and 3.8. Finally, since

$$\{x: |f_n(x) - f(x)| \geq 2\varepsilon\} \subset \{x: |f_n(x) - f_{n_k}(x)| \geq \varepsilon\} \cup \{x: |f_{n_k}(x) - f(x)| \geq \varepsilon\},$$

it follows that $f_n \to f$ in measure. ∎

Problems

3.1.1. Let f be an extended real-valued function such that $\{x: f(x) > \alpha\} \in \mathscr{A}$ for every $\alpha \in D$, a dense set of real numbers. Show that f is measurable.

3.1.2. Give an example of a nonmeasurable function f such that $|f|$ is measurable.

3.1.3. Show that a function of bounded variation on $[a, b]$ is Lebesgue measurable.

× **3.1.4.** Show that $g \circ f$ is measurable whenever f is measurable and g is real-valued and continuous.

3.1.5. Let μ be a measure on the Borel subsets of R such that $\mu(R) = 1$. Prove that $\mu(C + x)$ is an upper semicontinuous function of x for every closed subset C of R. Prove also that for every Borel subset $B \subset R$, $\mu(B + x)$ is a Borel-measurable function of x. [Hint: Notice that if $x_n \uparrow x$ or $x_n \downarrow x$, then $(a + x_n, b + x_n) \triangle (a + x, b + x) \downarrow \varnothing$ and thus, $\mu(I + x)$ is a continuous function of x for every open interval I. For the second part, let $\mathscr{F}$ be the class of all finite disjoint unions of sets of the form $G \cap C$, where G is open and C is closed. Observe that $\mathscr{F}$ is an algebra and for every $B \in \mathscr{F}$, $\mu(B + x)$ is Borel measurable.]

× **3.1.6.** Suppose in a finite measure space, $f_n \to f$ in measure and $g_n \to g$ in measure. Then show that (a) $f_n g_n \to fg$ in measure and (b) if for all x and each n, $f(x) \neq 0$ and $f_n(x) \neq 0$, then $1/f_n \to 1/f$ in measure.

× **3.1.7.** Form the sequence (f_n) on $[0, 1]$ as follows:

$$f_1 = \chi_{[0,1]}, \qquad f_2 = \chi_{[0,1/2]}, \qquad f_3 = \chi_{[1/2,1]},$$

$$f_4 = \chi_{[0,1/3]}, \qquad f_5 = \chi_{[1/3,2/3]}, \qquad f_6 = \chi_{[2/3,1]},$$

etc. Then show that $f_n \to 0$ in measure, but $f_n \not\to 0$ a.e.

3.1.8. Give an example of a continuous function g on $[0, 1]$ and a Lebesgue-measurable function h such that $h \circ g$ is not Lebesgue–measurable. {Hint: Let $f(x) = f_1(x) + x$, where $f_1(x)$ is the Cantor function on $[0, 1]$. Then f is a homeomorphism from $[0, 1]$ onto $[0, 2]$ and $f(K)$, K being the Cantor set, has positive Lebesgue measure. Let E be a nonmeasurable set $\subset f(K)$. Take $g = f^{-1}$ and $h = \chi_{f^{-1}(E)}$.}

3.1.9. (a) Let (f_n) be a sequence of a.e. real-valued measurable functions and let f also be such a function. Prove that the following assertions are equivalent:

(i) $f_n \to f$ almost uniformly;

(ii) for each $\varepsilon > 0$, $\lim_{n\to\infty}\mu(\bigcup_{m=n}^{\infty}\{x: |f_m(x) - f(x)| \geq \varepsilon\}) = 0$;

(iii) for each $\varepsilon > 0$, there exists a natural number $n(\varepsilon)$ such that $\mu(\bigcup_{m=n(\varepsilon)}^{\infty}\{x: |f_m(x) - f(x)| > \varepsilon\}) < \infty$ and $f_n \to f$ a.e.

(b) Let (f_n) and f be as in (a). Suppose that there is an integrable function g such that $|f_n| \leq g$ for all n and $f_n \to f$ a.e. Prove that $f_n \to f$ almost uniformly. (Now do Problem 2.4.9.)

3.1.10. Show that in a finite measure space, for any sequence (f_n) of a.e. real-valued measurable functions, there exist positive numbers a_n such that $(a_n f_n) \to 0$ a.e. [Hint: Choose positive constants b_n such that $\mu(A_n{}^c) < 1/2^{n+1}$, where $A_n = \{x: |f_n(x)| \leq b_n\}$. Take $a_n = 1/nb_n$ and consider the set $\bigcup_{m=1}^{\infty}\bigcap_{n=m}^{\infty} A_n$.]

× **3.1.11.** (a) If f is a continuous real-valued function on a closed set $A \subset R$, then show that there is a continuous extension g of f to R such that

$$\sup_{x\in R} |g(x)| \leq \sup_{x\in A} |f(x)|.$$

[Hint. Writing A^c as the union of disjoint open intervals, define g as f on points of A and linearly on these intervals.]

(b) For any $\varepsilon > 0$ and a simple function f on R, show that there exists a continuous function g and a closed set $A \subset R$ such that $m(R - A) < \varepsilon$ and $g(x) = f(x)$, $x \in A$.

(c) *Lusin's Theorem.* Let f be an a.e. real-valued measurable function on R. Then given $\varepsilon > 0$, there exists a continuous function g on R such that $m\{x \in R: f(x) \neq g(x)\} < \varepsilon$. [Hint: Let (G_i) be a countable basis of open sets in R. Given $\varepsilon > 0$, let $G^{(i)}$ be an open set containing $f^{-1}(G_i)$ such that $m(G^{(i)} - f^{-1}(G_i)) < \varepsilon/2^i$. Let $F_\varepsilon = \bigcup_{i=1}^{\infty}[G^{(i)} - f^{-1}(G_i)]$ and $g = f|_{R-F_\varepsilon}$. Then $g^{-1}(G_i) = G^{(i)} \cap [R - F_\varepsilon]$.]

(d) Prove that a real-valued function f on $[a, b]$ is Lebesgue measurable if and only if there exist a sequence of continuous functions f_n on $[a, b]$ and a sequence of compact subsets $K_n \subset [a, b]$ such that $\lim_{n\to\infty} m(K_n) = b - a$ and for each n, $f = f_n$ on K_n.

3.1.12. Let f be a real-valued function on R and $f(a + b) = f(a) + f(b)$ for all $a, b \in R$. Show that (a) f is continuous if and only if f is continuous at a single point, and (b) f is continuous if f is Lebesgue measurable. [Hint: Use Lusin's Theorem (Problem 3.1.11).]

3.1.13. Prove the following version of Lusin's theorem: Let $(X, \mathscr{A}, \mu)$ be a finite measure space, where X is a locally compact Hausdorff space,

$\mathscr{A}$ the σ-algebra of all Borel sets (the smallest σ-algebra containing the open sets), and for $A \in \mathscr{A}$, $\mu(A) = \sup\{\mu(K): K \subset A, K \text{ compact}\}$. Let f be an a.e. real-valued measurable function. Show that, given $\varepsilon > 0$, there exists a continuous function g vanishing outside a compact set such that $\mu\{x: f(x) \neq g(x)\} < \varepsilon$. [Hint: Assume with no loss of generality, $0 \leq f \leq 1$ and X compact. There exist simple functions f_n, $0 \leq f_n \leq f_{n+1}$, with $\lim_{n\to\infty} f_n = f$. Let $h_n = f_n - f_{n-1}$, $n \geq 2$ and $h_1 = f_1$. Then $f = \sum_{n=1}^{\infty} h_n$ and h_n is of the form $2^{-n} \cdot \chi_{A_n}$, $A_n \in \mathscr{A}$. There exist compact K_n and open V_n such that $K_n \subset A_n \subset V_n$ and $\mu(V_n - K_n) < \varepsilon/2^n$. By Urysohn's Lemma, there is a continuous function g_n, $0 \leq g_n \leq 1$, $g_n(x) = 1$ if $x \in K_n$ and $g_n(x) = 0$ if $x \in X - V_n$. Then $g(x) = \sum_{n=1}^{\infty} 2^{-n} g_n(x)$ is continuous and $g(x) = f(x)$ except on $\cup(V_n - K_n)$.]

3.1.14. Let $\mathscr{A}$ be a *σ-ring* of subsets of X. Then a real-valued function f on X is called *$\mathscr{A}$-measurable* if and only if $\{x: f(x) \neq 0\} \cap f^{-1}(B) \in \mathscr{A}$ for every Borel set $B \subset R$. Prove the following result of Bogdanowicz and McCloskey: f is $\mathscr{A}$-measurable if and only if $\{x: f(x) < -r\} \in \mathscr{A}$ and $\{x: f(x) > r\} \in \mathscr{A}$ for every positive number r.

★ **3.1.15.** Prove the following result due to G. Letac: The constants are the only measurable functions f on $(0, \infty)$ such that for $x, y > 0$, $f(x + y)$ lies in the closed interval joining $f(x)$ and $f(y)$.

★ **3.1.16.** Continuation of Problem 3.1.12: Show that the functions $f: R \to R$ satisfying

(i) $f(x) + f(y) = f(x + y)$

and

(ii) $f(p(x)) = p(f(x))$ for *some* polynomial $p(x)$ of degree ≥ 2 are of the form cx, where $c = 0, 1$, or -1.

★ **3.1.17.** A real-valued function f defined on an open convex set in R^n $(n \geq 1)$ is called *d-convex* $(0 < d < 1)$ if

$$f(dx + (1 - d)y) \leq df(x) + (1 - d)f(y)$$

for all x and y in its domain. Show that a d-convex measurable (with respect to n-dimensional Lebesgue measure) function is continuous and therefore is c-convex *for all* c in $[0, 1]$.

3.1.18. Let $(X, \mathscr{A}, \mu)$ be a measure space such that $\mu(X) = 1$ and f be a real-valued measurable function. Prove that there exists a real number m such that $\mu(f^{-1}(-\infty, m]) \geq 1/2$ and $\mu(f^{-1}[m, \infty)) \geq 1/2$. When is m unique? (Here m is called a median of f.)

3.1.19. Suppose that $h \geq 0$ is a non-Lebesgue-measurable function on $[0, 1]$. Show that there exists a Lebesgue-measurable function f such that $0 \leq f \leq h$ a.e. (m) and also, if $0 \leq g \leq h$ and g Lebesgue measurable, then $g \leq f$ a.e. (m).

3.1.20. *The Recurrence Theorem.* Consider the measure space $(X, \mathscr{A}, \mu)$ and the measure-preserving transformation T of Problem 2.3.25. Let f be a nonnegative measurable real function. Show that for almost all x in $\{x: f(x) > 0\}$, the series $\sum_{n=1}^{\infty} f(T^n(x)) = \infty$.

3.2. Definition and Properties of the Integral

The integral, which will be introduced and studied in this section, will extend the classical Riemann integral to a much wider class of functions, the class of measurable functions. The integral will be first defined for nonnegative measurable functions and then the definition will be extended to an arbitrary measurable function.

In what follows, we will consider only measurable functions and see that the simple functions play a basic role in the definition of the integral. We start this section with a lemma concerning them.

Lemma 3.1. Let $f = \sum_{i=1}^{n} \alpha_i \chi_{A_i} = \sum_{j=1}^{m} \beta_j \chi_{B_j}$, where the α_i and the β_j are real numbers and the A_i and the B_j are in $\mathscr{A}$. Suppose that $\mu(A_i) < \infty$ whenever $\alpha_i \neq 0$ and $\mu(B_j) < \infty$ whenever $\beta_j \neq 0$. Then

$$\sum_{i=1}^{n} \alpha_i \mu(A_i) = \sum_{j=1}^{m} \beta_j \mu(B_j). \qquad \blacksquare$$

Proof. The lemma is trivial if the A_i's as well as the B_j's are disjoint. In the general case, the lemma will be proven if we can write

$$\sum_{i=1}^{n} \alpha_i \chi_{A_i}(x) = \sum_{j=1}^{p} \gamma_j \chi_{C_j}(x),$$

where the C_j's are disjoint and

$$\sum_{i=1}^{n} \alpha_i \mu(A_i) = \sum_{j=1}^{p} \gamma_j \mu(C_j).$$

This is clear if $n = 1$. To use induction on n, we assume for some positive integer q

$$\sum_{i=1}^{n-1} \alpha_i \chi_{A_i}(x) = \sum_{j=1}^{q} \gamma_j \chi_{C_j}(x)$$

and

$$\sum_{i=1}^{n-1} \alpha_i \mu(A_i) = \sum_{j=1}^{q} \gamma_j \mu(C_j),$$

where the C_j's are disjoint. We write

$$\begin{aligned}\sum_{i=1}^{n} \alpha_i \chi_{A_i}(x) &= \sum_{j=1}^{q} \gamma_j \chi_{C_j}(x) + \alpha_n \chi_{A_n}(x)\\ &= \sum_{j=1}^{q} \gamma_j \chi_{C_j - A_n}(x) + \sum_{j=1}^{q} (\gamma_j + \alpha_n) \chi_{C_j \cap A_n}(x) + \alpha_n \chi_{A_n - \cup_{j=1}^{q} C_j}(x).\end{aligned}$$

Then

$$\begin{aligned}\sum_{i=1}^{n} \alpha_i \mu(A_i) &= \sum_{j=1}^{q} \gamma_j \mu(C_j) + \alpha_n \mu(A_n)\\ &= \sum_{j=1}^{q} \gamma_j \mu(C_j - A_n) + \sum_{j=1}^{q} (\gamma_j + \alpha_n) \mu(C_j \cap A_n) + \alpha_n \mu\left(A_n - \bigcup_{j=1}^{q} C_j\right).\end{aligned}$$

The lemma now follows by induction. ∎

This lemma makes the following definition possible.

Definition 3.6. A simple function f is said to be integrable if it can be written as $\sum_{i=1}^{n} \alpha_i \chi_{A_i}$ such that $\mu(A_i) < \infty$ whenever $\alpha_i \neq 0$. If f is integrable or nonnegative, we write $\int f \, d\mu = \sum_{i=1}^{n} \alpha_i \mu(A_i)$. If $E \in \mathscr{A}$, we write $\int_E f \, d\mu = \int f \cdot \chi_E \, d\mu$. ∎

Remarks. Suppose f and g are *both* integrable (or *both* nonnegative) simple functions. Then for $\alpha, \beta \in R$ (α and β are assumed both nonnegative when f and g are assumed so), we have the following:

3.10. $\int (\alpha f + \beta g) \, d\mu = \alpha \int f \, d\mu + \beta \int g \, d\mu$.

3.11. $f \leq g$ a.e. implies $\int f \, d\mu \leq \int g \, d\mu$.

3.12. $\int |f + g| \, d\mu \leq \int |f| \, d\mu + \int |g| \, d\mu$.

3.13. If $A_i \in \mathscr{A}$, $1 \leq i < \infty$ and $A_i \cap A_j$ is empty for $i \neq j$, then

$$\int_{\cup_{i=1}^{\infty} A_i} f \, d\mu = \sum_{i=1}^{\infty} \int_{A_i} f \, d\mu.$$

The reader should become convinced of the validity of these remarks by proving each one of them.

Definition 3.7. Let f be a *nonnegative* measurable function. Let $E \in \mathscr{A}$

and $E \subset$ the domain of f. Then we define

$$\int_E f\,d\mu = \sup\left\{\int_E g\,d\mu\colon 0 \leq g \leq f \text{ and } g \text{ is a simple function}\right\}.$$

If $\int_E f\,d\mu < \infty$, then f is called *integrable* on E. ∎

Remarks

3.14. Definition 3.7 is compatible with Definition 3.6 when f is simple.

3.15. If $\int_E f\,d\mu < \infty$, then $\mu(A) = 0$, where $A = \{x \in E\colon f(x) = \infty\}$. The reason is that for each positive integer n, $0 \leq n \cdot \chi_A \leq f$ and therefore, $n\mu(A) \leq \int_E f\,d\mu$.

3.16. If $\mu(E) < \infty$ and $m \leq f \leq M$, m and M being two nonnegative real numbers, then $m\mu(E) \leq \int_E f\,d\mu \leq M\mu(E)$.

The following proposition contains some basic results on integrals. The proofs are left to the reader.

Proposition 3.8. Let f and g be nonnegative measurable functions whose domains contain $E \in \mathscr{A}$. Then

(i) If $f \leq g$ a.e., then $\int_E f\,d\mu \leq \int_E g\,d\mu$.
(ii) If $f = g$ a.e., then $\int_E f\,d\mu = \int_E g\,d\mu$.
(iii) If $\int_E f\,d\mu = 0$, then $f = 0$ a.e. on E.
(iv) If $\int_E f\,d\mu \leq \int_E g\,d\mu$ for all $E \in \mathscr{A}$, then $f \leq g$ a.e. ∎

To prove (iii), the reader may observe that

$$\{x \in E\colon f(x) > 0\} = \bigcup_{n=1}^{\infty} \{x \in E\colon f(x) > 1/n\};$$

if $A_n = \{x \in E\colon f(x) > 1/n\}$ and $\mu(A_n) > 0$ for some n, then $\int_E f\,d\mu \geq (1/n)\mu(A_n) > 0$, which is a contradiction. To prove (iv), it is sufficient to observe that if the set $E = \{x\colon f(x) > g(x)\}$ has positive measure, then by (i) and (iii)

$$\int_E f\,d\mu > \int_E g\,d\mu,$$

which contradicts the hypothesis in (iv).

An important feature of Definition 3.7 is the fact that limits and integrals can be interchanged for every increasing sequence of functions. This is demonstrated by the next theorem, which also establishes the additivity of the integral in Definition 3.7, which is not evident otherwise.

Theorem 3.2. *The Monotone Convergence Theorem.* If (f_n) is a nondecreasing sequence of nonnegative measurable functions converging to a measurable function f almost everywhere, then for any measurable set E, $\int_E f\,d\mu = \lim_{n\to\infty} \int_E f_n\,d\mu$. ∎

Proof. With no loss of generality, we assume that for all x, $\lim_{n\to\infty} f_n(x) = f(x)$. Since $f_n \leq f_{n+1} \leq f$, we have by Proposition 3.8,

$$\int_E f_n\,d\mu \leq \int_E f_{n+1}\,d\mu \leq \int_E f\,d\mu. \tag{3.11}$$

Let $0 < k < 1$ and g be a simple function $\sum_{i=1}^m c_i \chi_{B_i}$ such that $0 \leq g \leq f$. If

$$A_n = \{x \in E\colon k \cdot g(x) \leq f_n(x)\},$$

then

$$\bigcup_{n=1}^{\infty} A_n = E, \qquad A_n \subset A_{n+1},$$

and therefore $\mu(E) = \lim_{n\to\infty} \mu(A_n)$. Now we have for each positive integer n

$$\int_{A_n} k \cdot g(x)\,d\mu \leq \int_{A_n} f_n(x)\,d\mu \leq \int_E f_n(x)\,d\mu. \tag{3.12}$$

But

$$\int_{A_n} k \cdot g(x)\,d\mu = k \sum_{i=1}^m c_i \mu(B_i \cap A_n);$$

therefore,

$$\lim_{n\to\infty} \int_{A_n} k \cdot g(x)\,d\mu = k \sum_{i=1}^m c_i \mu(B_i \cap E) = k \int_E g(x)\,d\mu, \tag{3.13}$$

which means

$$\int_E k \cdot g(x)\,d\mu \leq \lim_{n\to\infty} \int_E f_n(x)\,d\mu.$$

Since this inequality is true for all k such that $0 < k < 1$, it follows that

$$\int_E g(x)\,d\mu \leq \lim_{n\to\infty} \int_E f_n(x)\,d\mu \tag{3.14}$$

for every simple function g with $0 \leq g \leq f$. The theorem follows now from (3.11) and (3.14). ∎

Corollary 3.1. If $\alpha \geq 0$, $\beta \geq 0$ and f, g are nonnegative measurable functions whose domains contain $E \in \mathscr{A}$,

$$\int_E (\alpha f + \beta g)\, d\mu = \alpha \int_E f\, d\mu + \beta \int_E g\, d\mu. \qquad \blacksquare$$

Proof. By Proposition 3.5, there exist sequences (f_n) and (g_n) of simple functions such that

$$0 \leq f_n \leq f_{n+1}, \qquad 0 \leq g_n \leq g_{n+1};$$

and

$$\lim_{n\to\infty} f_n = f, \qquad \lim_{n\to\infty} g_n = g.$$

By Remark 3.10,

$$\int (\alpha f_n + \beta g_n)\, d\mu = \alpha \int f_n\, d\mu + \beta \int g_n\, d\mu.$$

The corollary now follows by an application of the Monotone Convergence Theorem. $\blacksquare$

The Monotone Convergence Theorem easily leads to the following very useful result.

Lemma 3.2. *Fatou's Lemma.* If (f_n) is a sequence of nonnegative measurable functions whose domains contain $E \in \mathscr{A}$, then

$$\int_E (\underline{\lim}_{n\to\infty} f_n)\, d\mu \leq \underline{\lim}_{n\to\infty} \int_E f_n\, d\mu. \qquad \blacksquare$$

Proof. Let $g_k = \inf\{f_n : n \geq k\}$. Then $0 \leq g_k \leq g_{k+1}$ and $\lim_{k\to\infty} g_k = \underline{\lim}_{n\to\infty} f_n$. By the Monotone Convergence Theorem,

$$\int_E \underline{\lim}_{n\to\infty} f_n\, d\mu = \lim_{k\to\infty} \int_E g_k\, d\mu = \underline{\lim}_{k\to\infty} \int_E g_k\, d\mu \leq \underline{\lim}_{k\to\infty} \int_E f_k\, d\mu. \qquad \blacksquare$$

Lemma 3.2 yields easily the following extension of the Monotone Convergence Theorem.

Corollary 3.2. Let (f_n) be a sequence of nonnegative measurable functions that converge a.e. to a measurable function f, such that $f_n \leq f$

for all n. Then

$$\int_E f\,d\mu = \lim_{n\to\infty} \int_E f_n\,d\mu,$$

where $E \in \mathscr{A}$ is contained in the domain of each f_n and f. ∎

Now we define the integral for an arbitrary measurable function.

Definition 3.8. Let f be any measurable function whose domain contains $E \in \mathscr{A}$. Then if at least one of the integrals $\int_E f^+\,d\mu$ and $\int_E f^-\,d\mu$ is finite, we define

$$\int_E f\,d\mu = \int_E f^+\,d\mu - \int_E f^-\,d\mu.$$

f is called integrable on E if $|\int_E f\,d\mu| < \infty$. (Note that f is integrable on E if and only if f^+ and f^- are both integrable on E, which is true if and only if $|f|$ is integrable on E.) ∎

If $(X, \mathscr{A}, \mu)$ is the Lebesgue measure space on R, then the integral defined above is called the *Lebesgue integral*. In the next section, we will show that the Lebesgue integral is a proper extension of the classical Riemann integral. The reader can easily see that $\chi_A(x)$ is integrable whenever $\mu(A) < \infty$; this means that $\chi_{\{\text{rationals}\}}(x)$ is Lebesgue integrable, though not Riemann integrable. One should, of course, be careful enough not to overlook the fact that there are many "nice" functions that are not Lebesgue integrable. For instance, if μ is the Lebesgue measure on $X = (0, 1)$, then $1/x$, despite being a continuous function on X, is not Lebesgue integrable; the reason is that, for each positive integer k, the simple function

$$\sum_{n=1}^{k} n\chi_{(1/(n+1),1/n]}(x) \le \frac{1}{x},$$

and therefore

$$\int_0^1 \frac{1}{x}\,d\mu \ge \int_0^1 \left[\sum_{n=1}^{k} n\chi_{(1/(n+1),1/n]}(x)\right] d\mu$$

$$= \sum_{n=1}^{k} \frac{1}{n+1},$$

so that

$$\int_0^1 \frac{1}{x}\,d\mu = \infty.$$

Despite the existence of many such nonintegrable functions, the class of integrable functions is indeed very large. Clearly any bounded measurable function f vanishing outside a set A of finite measure is integrable, since $|f| \leq k\chi_A$ (k being the bound of f) and therefore $\int |f| \, d\mu \leq k\mu(A) < \infty$. Some of the basic properties of the integral are contained in the next few remarks.

Remarks

3.17. If f is integrable on $E \in \mathscr{A}$ and if $E = \bigcup_{n=1}^{\infty} E_n$, $E_n \cap E_m = \emptyset$ if $n \neq m$, and $E_n \in \mathscr{A}$, then

$$\int_E f \, d\mu = \sum_{n=1}^{\infty} \int_{E_n} f \, d\mu.$$

This follows easily by writing $f = f^+ - f^-$ and then applying the Monotone Convergence Theorem.

3.18. If f and g are integrable on $E \in \mathscr{A}$ and $f = g$ a.e., then

$$\int_E f \, d\mu = \int_E g \, d\mu.$$

3.19. If f is integrable on $E \in \mathscr{A}$, then $A = \{x \in E: f(x) \neq 0\}$ has σ-finite measure; for, if $A_n = \{x \in E: |f(x)| > 1/n\}$ then $A = \bigcup_{n=1}^{\infty} A_n$ and $\int (1/n)\chi_{A_n} \, d\mu \leq \int_E |f| \, d\mu$, which means that $\mu(A_n) < \infty$ for each positive integer n.

3.20. The integral is linear, i.e., if f and g are integrable on $E \in \mathscr{A}$ and α, β are real numbers, then

$$\int_E (\alpha f + \beta g) \, d\mu = \alpha \int_E f \, d\mu + \beta \int_E g \, d\mu.$$

To show this, one needs to observe the following: For $\alpha \geq 0$,

$$\int_E (\alpha f) \, d\mu = \int_E (\alpha f)^+ \, d\mu - \int_E (\alpha f)^- \, d\mu$$

$$= \alpha \int_E f^+ \, d\mu - \alpha \int_E f^- \, d\mu = \alpha \int_E f \, d\mu. \tag{3.15}$$

For $\alpha < 0$,

$$\int_E (\alpha f)\,d\mu = \int_E (\alpha f)^+\,d\mu - \int_E (\alpha f)^-\,d\mu$$
$$= \int_E (-\alpha) f^-\,d\mu - \int_E (-\alpha) f^+\,d\mu = \alpha \int_E f\,d\mu. \quad (3.16)$$

If $f \geq 0$, $g \geq 0$, then an application of the Monotone Convergence Theorem yields easily (Corollary 3.1)

$$\int_E (f+g)\,d\mu = \int_E f\,d\mu + \int_E g\,d\mu. \quad (3.17)$$

In the general case, let

$$E_1 = \{x \in E\colon f(x) \geq 0,\ g(x) \geq 0\},$$
$$E_2 = \{x \in E\colon f(x) < 0,\ g(x) < 0\},$$
$$E_3 = \{x \in E\colon f(x) \geq 0,\ g(x) < 0,\ f(x) + g(x) < 0\},$$
$$E_4 = \{x \in E\colon f(x) < 0,\ g(x) \geq 0,\ f(x) + g(x) < 0\},$$
$$E_5 = \{x \in E\colon f(x) \geq 0,\ g(x) < 0,\ f(x) + g(x) \geq 0\},$$
$$E_6 = \{x \in E\colon f(x) < 0,\ g(x) \geq 0,\ f(x) + g(x) \geq 0\}.$$

The reader can easily show by using equation (3.17), that

$$\int_{E_i} (f+g)\,d\mu = \int_{E_i} f\,d\mu + \int_{E_i} g\,d\mu, \qquad 1 \leq i \leq 6. \quad (3.18)$$

The linearity of the integral now follows easily from equations (3.15), (3.16), and (3.18).

3.21. If f and g are integrable on $E \in \mathscr{A}$, and $f \leq g$ a.e., then

$$\int_E f\,d\mu \leq \int_E g\,d\mu.$$

Indeed, since $|f|$ and $|g|$, being integrable, are finite a.e.,

$$f \leq g \text{ a.e.} \Rightarrow f^+ + g^- \leq g^+ + f^- \text{ a.e.}$$

Then by Proposition 3.8,

$$\int (f^+ + g^-)\,d\mu \leq \int (g^+ + f^-)\,d\mu.$$

which means that

$$\int f^+ \, d\mu - \int f^- \, d\mu \leq \int g^+ \, d\mu - \int g^- \, d\mu.$$

3.22. If f is integrable on $E \in \mathscr{A}$ and g is a measurable function such that on E, $|g| \leq f$, then g is integrable.

To see this, we need to observe that $g^+ \leq f$ and $g^- \leq f$ so that

$$0 \leq \int_E g^+ \, d\mu \leq \int_E f \, d\mu \qquad \text{and} \qquad 0 \leq \int_E g^- \, d\mu \leq \int_E f \, d\mu,$$

and therefore g^+ and g^- are both integrable.

Now we present the most important theorem of this section, the Dominated Convergence Theorem, providing a very useful criterion for the interchange of limit and integrals.

Theorem 3.3. *The Dominated Convergence Theorem.*† If (f_n) is a sequence of measurable functions converging a.e. to a measurable function f such that $|f_n| \leq g$ a.e. on $E \in \mathscr{A}$, where g is an integrable function on E, then

$$\int_E f \, d\mu = \lim_{n\to\infty} \int_E f_n \, d\mu. \qquad \blacksquare$$

Proof. We will apply Fatou's Lemma to the sequence $g + f_n$ and the sequence $g - f_n$. Clearly, $g \pm f_n \geq 0$ for each n. Therefore, by Fatou's Lemma,

$$\int_E (g + f) \, d\mu \leq \varliminf_{n\to\infty} \int_E (g + f_n) \, d\mu, \tag{3.19}$$

$$\int_E (g - f) \, d\mu \leq \varliminf_{n\to\infty} \int_E (g - f_n) \, d\mu. \tag{3.20}$$

From (3.20), and the integrability of f, we have

$$\int_E g \, d\mu - \int_E f \, d\mu \leq \int_E g \, d\mu - \varlimsup_{n\to\infty} \int_E f_n \, d\mu. \tag{3.21}$$

† This theorem is also known as the *Lebesgue Convergence Theorem*.

From (3.19) and (3.21), we have

$$\overline{\lim_{n\to\infty}} \int_E f_n \, d\mu \leq \int_E f \, d\mu \leq \underline{\lim_{n\to\infty}} \int_E f_n \, d\mu.$$

The theorem now follows easily. ∎

We present a few examples to show that convergence a.e. or even uniform convergence need not imply that the integral and the limit can be interchanged.

Examples

3.1. If $(X, \mathscr{A}, \mu)$ is the Lebesgue measure space on [0, 1] and $f_n = n\chi_{[0,1/n]}$, then $f_n \to 0$ a.e. and in measure; but $\int_0^1 f_n \, d\mu = 1 \neq 0$.

3.2. If $(X, \mathscr{A}, \mu)$ is the Lebesgue measure space on R and $f_n = (1/n)\chi_{[0,n]}$, then $f_n \to 0$ uniformly on R; but $\int_R f_n \, d\mu = 1 \neq 0$.

3.3. If μ is the counting measure on X, the positive integers, and $\mathscr{A} = 2^X$, then the sequence

$$f_n(k) = \begin{cases} 1/k, & 1 \leq k \leq n \\ 0, & k > n \end{cases}$$

converges uniformly to $f(k) = 1/k$, $k \geq 1$. Note that the f_n are all integrable, whereas f is not integrable since $\int f \, d\mu = \sum_{k=1}^{\infty} 1/k = \infty$. However, the reader may easily become convinced of the following:

If $(X, \mathscr{A}, \mu)$ is any measure space with $\mu(X) < \infty$, then uniform convergence of a sequence of integrable functions f_n to a function f implies that f is integrable and $\lim_{n\to\infty} \int_E f_n \, d\mu = \int_E f \, d\mu$ for all $E \in \mathscr{A}$.

Before we close this section, let us mention an often useful, but very simple, generalization of the Dominated Convergence Theorem.

Theorem 3.4. Let (f_n) and (g_n) be two sequences of measurable functions which converge a.e. to the measurable functions f and g, respectively. Suppose $|f_n| \leq g_n$ and

$$\lim_{n\to\infty} \int_E g_n \, d\mu = \int_E g \, d\mu < \infty.$$

Then

$$\lim_{n\to\infty} \int_E f_n \, d\mu = \int_E f \, d\mu.$$ ∎

The proof follows immediately by applying Fatou's Lemma to the sequences $(g_n - f_n)$ and $(g_n + f_n)$.

Corollary 3.3. Let (f_n) be a sequence of integrable functions such that $f_n \to f$ a.e. with f integrable. Then $\int_E |f - f_n| \, d\mu \to 0$ if and only if

$$\int_E |f_n| \, d\mu \to \int_E |f| \, d\mu,$$

with $E \in \mathscr{A}$. ∎

Proof. The "only if" part is easy. We prove the "if" part. Suppose $\int_E |f_n| \, d\mu \to \int_E |f| \, d\mu$. Then since $|f_n - f| \le |f_n| + |f|$ and

$$\lim_{n \to \infty} \int_E (|f_n| + |f|) \, d\mu = 2 \int_E |f| \, d\mu < \infty,$$

the corollary follows from Theorem 3.4. ∎

Finally we point out that the Lebesgue integral $\int f \, dm$ is often written as $\int f(x) \, dx$. This notation is consistent with the fact that a Riemann-integrable function on a closed and bounded interval $[a, b] \subset R$ is Lebesgue integrable and the integrals are equal. This will be discussed at length in the next section.

Problems

× **3.2.1.** Show that for an integrable function f, for every positive number ε there is a positive number δ such that $\int_A |f| \, d\mu < \varepsilon$ whenever $\mu(A) < \delta$.

× **3.2.2.** Show that for a Lebesgue-integrable function f, the function $g(x) = \int_{-\infty}^{x} f \, dm$ is continuous. Is f^2 integrable when f is? Does the converse hold?

× **3.2.3.** Let f be integrable on $E \in \mathscr{A}$. Then given $\varepsilon > 0$, there is a simple function g vanishing outside a set of finite measure such that $\int_E |f - g| \, d\mu < \varepsilon$.

× **3.2.4.** Suppose that $\lambda(E) = \int_E f \, d\mu$, where E is a measurable set and f is a nonnegative measurable function. Show that λ is a measure and $\int g \, d\lambda = \int fg \, d\mu$ for each nonnegative measurable function g. (Hint: Prove the result first for simple functions and then use the Monotone Convergence Theorem.)

× **3.2.5.** Prove the Monotone Convergence Theorem, Corollary 3.2, and the Dominated Convergence Theorem by replacing almost everywhere convergence by convergence in measure.

× **3.2.6.** *Translation Invariance of the Lebesgue Integral.* Show that for any Lebesgue-integrable function f on R and any real number t, (i) $\int f\,dm = \int f_t\,dm$, $f_t(x) = f(x+t)$ and (ii) $\int f\,dm = \int f_-\,dm$, $f_-(x) = f(-x)$. (Hint: Prove the results first for simple functions.)

3.2.7. *Uniformly Integrable Functions.* Let $(X, \mathscr{A}, \mu)$ be a finite measure space and (f_n) be a sequence of real-valued measurable functions. The f_n are called *uniformly integrable* if $\lim_{k\to\infty}\sup_n \int_{A_{nk}} |f_n|\,d\mu = 0$, where $A_{nk} = \{x: |f_n(x)| > k\}$. Prove the following assertions:

(i) Let $\beta > 0$. Then there exists a positive integer N and a set $A \in \mathscr{A}$ such that $\mu(A^c) < \beta$ and $f_k(x) \leq \lim_n\sup f_n(x)$, whenever $k \geq N$ and $x \in A$.

(ii) Let $|f_n| \leq g$, g integrable. Then the f_n are uniformly integrable.

(iii) The f_n are uniformly integrable if and only if

(a) $$\sup_n \int |f_n|\,d\mu < \infty$$

and

(b) given $\varepsilon > 0$, there exists a $\beta > 0$ such that $\int_A |f_n|\,d\mu < \varepsilon$ for every n, whenever $\mu(A) < \beta$.

(iv) Suppose that the f_n are uniformly integrable. Then

(a) $$\int (\lim_n\inf f_n)\,d\mu \leq \lim_n\inf \int f_n\,d\mu \leq \lim_n\sup \int f_n\,d\mu \leq \int (\lim_n\sup f_n)\,d\mu;$$

(b) if $f_n \to f$ a.e. or in measure, then f is integrable and

$$\lim_{n\to\infty} \int |f_n - f|\,d\mu = 0.$$

(v) Suppose that $\lim_n\sup f_n$ as well as each f_n is integrable. Then the f_n are uniformly integrable if and only if for each $E \in \mathscr{A}$,

$$\lim_n\sup \int_E f_n\,d\mu \leq \int_E (\lim_n\sup f_n)\,d\mu.$$

× **3.2.8.** Use Proposition 3.6 to show that, given $\varepsilon > 0$ and any Lebesgue-integrable function f on E (Lebesgue measurable), there is a continuous function g on R vanishing outside a finite interval such that $\int_E |f - g|\,dm < \varepsilon$.

3.2.9. Show that $f(x) = (\sin x)/x$ on $[0, \infty)$, $f(0)$ being 1, is not Lebesgue integrable. (Notice that $\lim_{n\to\infty} \int_0^n [(\sin x)/x]\,dx$ exists.)

3.2.10. Let (f_n) be a sequence of integrable functions. Prove that, if

$$\sum_{n=1}^{\infty} \int |f_n|\,d\mu < \infty,$$

then

$$\sum_{n=1}^{\infty} f_n(x)$$

is convergent a.e. to an integrable function f and

$$\int f\,d\mu = \sum_{n=1}^{\infty} \int f_n\,d\mu.$$

3.2.11. Let $\mu(X) < \infty$. Show that a sequence (f_n) of a.e. real-valued measurable functions is convergent in measure to zero if and only if

$$\int \frac{|f_n|}{1 + |f_n|}\,d\mu \to 0, \qquad \text{as } n \to \infty.$$

3.2.12. Let g be a measurable function such that $|\int g \cdot f\,d\mu| < \infty$ for every integrable function f. Show that g is bounded a.e., whenever μ is σ-finite. (This result is false if μ is not σ-finite.)

3.2.13. Find a nonnegative Lebesgue-integrable function f such that for any real number a and positive integers n and k,

$$m(\{x: f(x) \geq k\} \cap (a, a + 1/n)) > 0.$$

3.2.14. Let f be a bounded Lebesgue-measurable function on $[0, 1]$. Show that

$$\left[\int_0^1 f(x)\,dx\right]^2 \leq \int_0^1 f^2(x)\,dx,$$

using the inequality $b \leq (1 + b^2)/2$.

3.2.15. Prove the following result known as the Steinhaus Lemma. The difference set $D(E) = \{x - y: x \in E,\ y \in E\}$ of a Lebesgue-measurable set E with $m(E) > 0$ contains an open interval. [Hint: Let $f(x) = \int_a^b \chi_{E_1}(y)\chi_{E_1}(x + y)\,dy$, where $E_1 = E \cap [a, b]$ and $m(E_1) > 0$. Then $f(0) > 0$ and f is continuous at 0.] Use this lemma to find the Lebesgue measure of a proper Lebesgue-measurable subgroup of the additive group of reals. (Does a similar result hold for the difference set $D(E, F) = \{x - y: x \in E, y \in F\}$ where $m(E) > 0$ and $m(F) > 0$? If you fail to find a solution at this point, see later Problem 3.4.12.)

3.2.16. Let $\mu(X) < \infty$ and f: $X \times [0, 1] \to R$. Suppose that (i) for each $t \in [0, 1]$, $f(x, t)$ is integrable, and (ii) the partial derivative $\partial f(x, t)/\partial t$ exists and is uniformly bounded for all $x \in X$. Show that $\int f(x, t)\,d\mu$ is differentiable on (0, 1) and

$$\frac{d}{dt}\int f(x, t)\, d\mu = \int \frac{\partial f(x, t)}{\partial t}\, d\mu.$$

3.2.17. Let (f_n) be a sequence of integrable functions such that for some integrable function f, $\lim_{n\to\infty} \int | f_n - f |\, d\mu = 0$. Show that

$$\lim_{n\to\infty} \int_{A_n} f_n\, d\mu = \int_A f\, d\mu \text{ if } \lim_{n\to\infty} \mu(A_n \triangle A) = 0.$$

3.2.18. (a) Prove the Riemann–Lebesgue theorem: Show that

$$\lim_{n\to\infty} \int_{-\infty}^{\infty} f(x) \cos nx\, dx = \lim_{n\to\infty} \int_{-\infty}^{\infty} f(x) \sin nx\, dx = 0$$

for every Lebesgue-integrable function f. [Hint: First prove the theorem when f is a step function. Then use Problem 3.2.3 above and Problem 2.3.9 (Chapter 2), (i)-(c).]

(b) *Cantor–Lebesgue Theorem.* Let $A_n(x) = c_n e^{inx} + c_{-n} e^{-inx}$, where the c_n and c_{-n} are real. If $A_n(x) \to 0$ as $n \to \infty$ for each $x \in E$, a Lebesgue-measurable set with $m(E) > 0$, then prove that $\lim_{n\to\infty} c_n = \lim_{n\to\infty} c_{-n} = 0$. [Hint: Let $b_n = c_n{}^2 + c_{-n}^2$. Use (a) to show that one may assume the b_n to be bounded and also $m(E)$ to be finite. Notice that then $\int_E | A_n(x) |^2\, dx = b_n \cdot m(E) + r_n$, where $r_n \to 0$.]

3.2.19. Given any Lebesgue-integrable function f and $\varepsilon > 0$, show that there exists an infinitely differentiable bounded function g vanishing outside a finite interval such that $\int_{-\infty}^{\infty} | f - g |\, dm < \varepsilon$. [Hint: First, by Problem 3.2.3, there is a simple function f_1 vanishing outside an interval $[a, b]$ such that $\int_{-\infty}^{\infty} | f - f_1 |\, dm < \varepsilon/3$. By Problem 2.3.9 there exists a step function f_2 such that $\int_{-\infty}^{\infty} | f_1 - f_2 |\, dm < \varepsilon/3$. Therefore f can be assumed to be a step function.]

3.2.20. (*J. Gillis*). Suppose for each $\lambda \in \Lambda$, an infinite set, there is a Lebesgue-measurable subset $A_\lambda \subset [0, 1]$ such that $m(A_\lambda) \geq \beta > 0$. Show that:

(a) Given $\varepsilon > 0$ and any positive integer p, there exist $\lambda_1, \lambda_2, \ldots, \lambda_p$ in Λ such that $m(\bigcap_{i=1}^{p} A_{\lambda_i}) \geq \beta^p - \varepsilon$. [Hint for $p = 2$: Use Problem 3.2.14 to show

$$\left[\sum_{i=1}^{n} m(A_i)\right]^2 \leq \sum_{i=1}^{n} m(A_i) + 2 \sum_{i<j\leq n} m(A_i \cap A_j).]$$

(b) If Λ is uncountable, then there is an uncountable subset $\Lambda_1 \subset \Lambda$ such that for any $\lambda, \lambda' \in \Lambda_1$, $m(A_\lambda \cap A_{\lambda'}) \geq \beta - \varepsilon$. [Hint: The set $\{A_\lambda: \lambda \in \Lambda\}$ has a condensation point C with respect to the pseudometric d, where $d(A, B) = m(A \triangle B)$, i.e., every open set containing C contains uncountably many of the A_λ.]

★ **3.2.21.** *An Extension of the Riemann–Lebesgue Theorem* (Kestelman). Suppose f is a Lebesgue-integrable function on $(0, \infty)$, and, for each positive λ, $I(\lambda)$ is a subinterval of $(0, \infty)$. Then $\lim_{\lambda\to\infty} \int_{I(\lambda)} f(t) \cos \lambda t \, dt = 0$. This result is false if we assume only that $I(\lambda)$ is a finite union of intervals. [Hint: Note that

$$s_\lambda = \int_{a_\lambda}^{b_\lambda} f(t) \cos \lambda t \, dt = -\int_{a_\lambda - \pi/\lambda}^{b_\lambda - \pi/\lambda} f\left(t + \frac{\pi}{\lambda}\right) \cos \lambda t \, dt,$$

and so

$$2 \mid s_\lambda \mid \leq \int_{a_\lambda}^{b_\lambda} \left| f(t) - f\left(t + \frac{\pi}{\lambda}\right)\right| dt + \int_{a_\lambda}^{a_\lambda + \pi/\lambda} \mid f(t) \mid dt + \int_{b_\lambda}^{b_\lambda+\pi/\lambda} \mid f(t) \mid dt.]$$

3.2.22. Suppose (f_n) is a sequence of continuous real-valued functions on $[0, 1]$ such that $f_1(x) \geq f_2(x) \geq \cdots \geq 0$ for all $x \in [0, 1]$. Suppose also that the only *continuous* function f such that $f_n(x) \geq f(x) \geq 0$ for all $x \in [0, 1]$ and all n is the zero function. Show by example that $\int_0^1 f_n(x) \, dx$ need *not* have limit 0 as $n \to \infty$. [Hint: Let K be a Cantor set $\subset [0, 1]$ with $m(K) > 0$, m being the Lebesgue measure. Then since $\chi_K(x)$ is upper semicontinuous, there exists a sequence of continuous functions f_n with $f_n(x) \geq f_{n+1}(x)$ for all n, $\lim_{n\to\infty} f_n(x) = \chi_K(x)$, and $\lim_{n\to\infty} \int_0^1 f_n(x) \, dx = m(K) > 0$.]

3.2.23. *Limit of the Derivatives as the Derivative of the Limit* (*without Uniform Convergence*). Suppose f_n is a sequence of continuously differentiable functions on (a, b) such that $\lim_{n\to\infty} f_n(x) = f(x)$ and $\lim_{n\to\infty} f_n'(x) = \Phi(x)$. Then if Φ and f' are continuous, prove that $f' = \Phi$. {Hint: Let $[c, d] \subset (a, b)$ and let $A_n = \{x \in [c, d]: \mid f_k'(x) - \Phi(x) \mid \leq 1 \text{ for all } k \geq n\}$. Since $[c, d]$ is of the second category, A_n is dense in some subinterval of $[c, d]$ for some n, and hence (f_n') is uniformly bounded on some $[s, t] \subset [c, d]$. Now use the Dominated Convergence Theorem on $[s, x]$, $s < x < t$.}

3.2.24. *Integration of Complex-Valued Functions.* Let u and v be the real and imaginary parts respectively of a complex-valued function f. Then f is called measurable (integrable) if both u and v are measurable (integrable). If f is integrable, we write: $\int f \, d\mu = \int u \, d\mu + i \int v \, d\mu$. Show that the integrable complex-valued functions form a vector space over the complex numbers. Also show that a measurable complex-valued function f is integrable if and only if $|f|$ is integrable, and then $|\int f\,d\mu| \leq \int |f|\,d\mu$. [Hint: $R(\overline{\int f\,d\mu} \cdot f) \leq |\int f\,d\mu| \cdot |f|$ and then integrate both sides. Here $R(g)$ is the real part of g and the bar denotes the conjugate.]

3.2.25. Let f be a bounded Lebesgue-measurable function on [0, 1]. Show that $\lim_{n\to\infty} \int_{1/n}^{1} |f(x - 1/n) - f(x)|\,dx = 0$. Is it true that $\lim_{n\to\infty} f(x - 1/n) = f(x)$ a.e.?

3.2.26. Prove *Scheffe's theorem.* Let (f_n) and f be real measurable functions such that for each n, $\int f_n\,d\mu = \int f\,d\mu$ and f integrable. Suppose that $f_n \to f$ a.e. Then $\lim_{n\to\infty} \int |f_n - f|\,d\mu = 0$. [Hint: Let $g_n = f - f_n$. Then $0 \leq g_n^+ \leq f$ and $g_n^+ \to 0$ a.e.]

3.2.27. *An Application of the Dominated Convergence Theorem.* Here is an example of a limit point of a countable set in a compact Hausdorff space that is not a subsequential limit. Let X denote the topological space of all functions f: $[0, 1] \to [0, 1]$ such that an open neighborhood of f is of the form $\{g \in X: |g(t_i) - f(t_i)| < r_i,\ i = 1, 2, \ldots, n\}$, where $r_i > 0$, $t_i \in [0, 1]$ for each i. Then X can be identified with the product space $\prod_{t\in[0,1]} I_t$, $I_t = [0, 1]$. Let A be a finite set of rationals s_i such that $0 < s_1 < \cdots < s_n < 1$ and let $f_A(x)$ be defined by

$$f_A(x) = \begin{cases} 2s_i'[x - s_i], & \text{if } x \in [s_i, s_i^0], \\ 2s_i'[s_{i+1} - x], & \text{if } x \in [s_i^0, s_{i+1}], \end{cases}$$

where $0 \leq i \leq n$, $s_0 = 0$, $s_{n+1} = 1$, $s_i' = 2/[s_{i+1} - s_i]$, and $s_i^0 = [s_i + s_{i+1}]/2$. Let Y be the class of all such functions. Use Theorem 3.3 to show that the function 0 is a limit point of Y, but not a subsequential limit of Y.

3.2.28. Let E be a Lebesgue-measurable subset of the reals. Find $\lim_{n\to\infty} \int_E \cos^2 nx\,dx$.

3.2.29. (*Jensen's Inequality*). Let $(X, \mathscr{A}, \mu)$ be a measure space such that $\mu(X) = 1$, and let f be an integrable function. Let F be a convex function on (a, b), where Range $F \subset (a, b)$. Prove the following assertions:

(i) $F(\int f\,d\mu) \leq \int F(f)\,d\mu$.

(ii) Let $F(x) = e^x$, $X = \{a_1, a_2, \ldots, a_n\}$ and $\mu(a_i) = 1/n$ for each i. Use (i) to prove that the geometric mean of any n positive numbers is less than or equal to their arithmetic mean.

(iii) If for some real function F on R and every bounded Lebesgue-measurable function f, $F(\int_0^1 f(x)\,dx) \leq \int_0^1 F(f(x))\,dx$, then F is convex. [Hint: Let $v = \int f\,d\mu$. Let $a < u < v < w < b$. Use Problem 1.5.22 to show that

$$\frac{F(v) - F(u)}{v - u} \leq M \leq \frac{F(w) - F(v)}{w - v},$$

where $M = \sup\{[F(v)] - F(u)]/[v - u]: a < u < v\}$. Thus, if $a < y < b$, then $F(y) \geq F(v) + M(y - v)$.]

3.2.30. (*The Vitali–Caratheodory Theorem.*) Let μ be a Borel measure on R and f be a real-valued integrable function on R. Prove that given $\varepsilon > 0$, there exist functions f_1 and f_2 on R such that $f_1 \leq f \leq f_2$, f_1 is upper semicontinuous and bounded above, f_2 is lower semicontinuous and bounded below, and $\int (f_2 - f_1)\, d\mu < \varepsilon$. [This result remains true when μ is a measure on a locally compact Hausdorff space such that $\mu(K) < \infty$ for each compact set K.]
[Hint: Let $f \geq 0$. There are simple functions $h_n \uparrow f$, $h_0 = 0$. Write $g_n = h_n - h_{n-1}$. Then $f = \sum_{n=1}^{\infty} g_n$. Write $f = \sum_{n=1}^{\infty} c_n \chi_{A_n}$, where $c_n > 0$ and $\mu(A_n) < \infty$. By Problem 2.3.12, there are compact sets K_n and open sets G_n such that $K_n \subset A_n \subset G_n$ and $\mu(G_n - K_n) < \varepsilon/2^{n+1}$. Consider $f_1 = \sum_{n=1}^{N} c_n \chi_{K_n}$ and $f_2 = \sum_{n=1}^{\infty} c_n \chi_{G_n}$ with N sufficiently large.]

3.2.31. Let g be Lebesgue integrable on $[a, b]$ and f be a nonnegative increasing function on $[a, b]$. Prove that there is a point $c \in [a, b]$ such that

$$\int_a^b f(x)g(x)\, dx = f(b) \cdot \int_c^b g(x)\, dx.$$

[Hint: First note that given $\varepsilon > 0$, there is a nonnegative increasing step function h such that $0 \leq f - h < \varepsilon$ on $[a, b]$. Solve the problem for h first.]

3.3. Lebesgue–Stieltjes Measure and the Riemann–Stieltjes Integral

In this section we will study the Riemann–Stieltjes integral from the point of view of measure and integration theory presented thus far. We will show that a bounded function f defined on $[a, b]$, $-\infty < a < b < \infty$ is Riemann–Stieltjes integrable with respect to a right-continuous monotonic increasing function g on $[a, b]$ if and only if f is continuous almost everywhere with respect to the Lebesgue–Stieltjes measure induced by g. In particular, we will show that every Riemann-integrable function is continuous almost everywhere with respect to the Lebesgue measure (hence Riemann-integrable functions are Lebesgue integrable), and the integrals are equal.

Let us recall our discussion of Lebesgue–Stieltjes measures that we presented briefly at the end of Section 2.3 in Chapter 2. There we found that the Lebesgue–Stieltjes measure μ induced by a monotonic increasing right-continuous function f is a complete measure on a σ-algebra containing all the Borel sets of R such that $\mu((a, b]) = f(b) - f(a)$. Actually, there the measure μ was obtained by considering the outer measure μ^* induced by

$$\mu^*(A) = \inf\left\{\sum_{i=1}^{\infty} f(b_i) - f(a_i):\ A \subset \bigcup_{i=1}^{\infty} (a_i, b_i]\right\} \tag{3.22}$$

and considering the completion of its restriction on all the Borel sets, which are μ^*-measurable. Since the Borel sets are μ^*-measurable, μ^* is a metric outer measure by Theorem 2.3 of Chapter 2. It should be noticed that the condition $\mu((a, b]) = f(b) - f(a)$ requires the function f to be right continuous since $\lim_{h_n \to 0+} \mu((a, b + h_n]) = \mu((a, b])$. A natural question then is: What happens when we consider equation (3.22) for an *arbitrary* monotonic increasing function f? Of course, μ^* will be a metric outer measure as before. [This can be checked directly by the reader, or the reader can follow the proof of part (ii) of Proposition 3.9 below.] Therefore μ^* will again give us a measure μ on the class of all Borel sets by Theorem 2.3; but in this case the reader can easily verify that $\mu((a, b]) \leq f(b) - \lim_{h \to 0+} f(a + h)$. Should we then call the completion of μ the Lebesgue–Stieltjes measure induced by f? This is possible. However, a more satisfactory definition can be obtained by considering open intervals instead of half-open intervals as follows:

$$\mu^*(A) = \inf\left\{\sum_{i=1}^{\infty} f(b_i) - f(a_i):\ A \subset \bigcup_{i=1}^{\infty} (a_i, b_i)\right\}. \tag{3.23}$$

In what follows we will show that the outer measure in equation (3.23) induced by f (not necessarily right continuous) coincides with the outer measure in equation (3.23) induced by the right-continuous monotonic increasing function f_2 defined by

$$f_2(x) = \lim_{h \to 0+} f(x + h) \equiv f(x+).$$

In the case of a right-continuous monotonic increasing function, equations (3.22) and (3.23) induce the same outer measure. (See Remark 3.23.)

Proposition 3.9. Let f be a monotonic increasing function and μ^* be as in equation (3.23). Then

(i) $\mu^*(\{a\}) = f(a+) - f(a-)$;

(ii) μ^* is a metric outer measure;

(iii) $\mu^*([a, b]) = f(b+) - f(a-)$,

$\mu^*((a, b]) = f(b+) - f(a+)$,
$\mu^*([a, b)) = f(b-) - f(a-)$,
$\mu^*((a, b)) = f(b-) - f(a+)$;

(iv) $\mu^*(A) = \inf\{\mu^*(0)\colon A \subset 0 \text{ open}\}$;

(v) the outer measure induced as in equation (3.23) by f_2 or f_1, where $f_2(x) = f(x+)$ and $f_1(x) = f(x-)$, coincides with μ^*. ∎

Proof. We omit the proof of (i). For (ii) let $A, B \subset R$ and $d(A, B) > 0$. Then there is a positive integer m such that $d(A, B) > 1/m$. We must prove that for any $\varepsilon > 0$

$$\mu^*(A) + \mu^*(B) \leq \mu^*(A \cup B) + 2\varepsilon. \tag{3.24}$$

We may and do assume that $\mu^*(A \cup B) < \infty$. Then there exist open intervals (a_i, b_i) such that $A \cup B \subset \bigcup_{i=1}^{\infty}(a_i, b_i)$ and

$$\sum_{i=1}^{\infty} f(b_i) - f(a_i) < \mu^*(A \cup B) + \varepsilon. \tag{3.25}$$

We know that the points of discontinuity of a monotonic increasing function are at most countable, and therefore the points of continuity of f are dense in every open interval. Therefore, for each i we can find subintervals (c_{ij}, d_{ij}), $1 \leq j \leq n_i$, such that

$$| d_{ij} - c_{ij} | < 1/m \qquad \text{for each } i, j, \tag{3.26}$$

$$(a_i, b_i) \subset \bigcup_{j=1}^{n_i} (c_{ij}, d_{ij}) \qquad \text{for each } i, \tag{3.27}$$

$$f(b_i) - f(a_i) + \varepsilon/2^i \geq \sum_{j=1}^{n_i} f(d_{ij}) - f(c_{ij}) \qquad \text{for each } i. \tag{3.28}$$

[To obtain (3.26)–(3.28), we can argue briefly as follows: Let $0 < 2s < 1/m$ and k_i be a positive integer such that $b_i - a_i < k_i s$. Choose $c_{i1} = a_i$. Let d_{i1} be a point of continuity of f such that $c_{i1} < d_{i1} \leq b_i$ and $s < d_{i1} - c_{i1} < 2s$. Choose c_{i2} such that $c_{i1} < c_{i2} < d_{i1}$, $s < c_{i2} - c_{i1}$ and $f(d_{i1}) - f(c_{i2}) < \varepsilon/k_i 2^i$. In this way, we construct the intervals (c_{ij}, d_{ij}). If p_i is the smallest positive integer such that $b_i - c_{ip_i} < 2s$, then we take $d_{ip_i} = b_i$.] Now since

$$A \cup B \subset \bigcup_{i=1}^{\infty} \bigcup_{j=1}^{n_i} (c_{ij}, d_{ij})$$

and

$$| d_{ij} - c_{ij} | < 1/m \qquad \text{for all } i, j,$$

each subinterval (c_{ij}, d_{ij}) can intersect only one of A and B, not both. Therefore, some of these intervals will cover A while the rest will cover B. Since $\sum_{i=1}^{\infty} \sum_{j=1}^{n_i} f(d_{ij}) - f(c_{ij})$ is a convergent series of nonnegative terms, every rearrangement will also converge to the same sum. It is now clear from (3.25) that

$$\begin{aligned} \mu^*(A) + \mu^*(B) &\leq \sum_{i=1}^{\infty} \sum_{j=1}^{n_i} f(d_{ij}) - f(c_{ij}) \\ &\leq \sum_{i=1}^{\infty} f(b_i) - f(a_i) + \varepsilon \\ &< \mu^*(A \cup B) + 2\varepsilon. \end{aligned}$$

Since $\varepsilon > 0$ is arbitrary, $\mu^*(A \cup B) = \mu^*(A) + \mu^*(B)$. This proves (ii).

To prove (iii), we notice that $[a, b] \subset (a - 1/n, b + 1/n)$ for each n, and therefore

$$\begin{aligned} \mu^*([a, b]) &\leq \lim_{n \to \infty} f(b + 1/n) - f(a - 1/n) \\ &= f(b+) - f(a-). \end{aligned}$$

For the converse inequality, let $[a, b] \subset \bigcup_{i=1}^{\infty} (a_i, b_i)$. Then since $[a, b]$ is compact, there exists N such that $[a, b] \subset \bigcup_{i=1}^{N} (a_i, b_i)$. It is easy to see that

$$f(b+) - f(a-) \leq \sum_{i=1}^{N} f(b_i) - f(a_i).$$

This proves that $f(b+) - f(a-) = \mu^*([a, b])$. The rest of the proof of this proposition is left to the reader. ∎

Remark 3.23. For right-continuous, monotonic increasing functions f, equations (3.22) and (3.23) above yield the *same* outer measure μ^*.

Proposition 3.9 now leads to the following definitions.

Definition 3.9. Let f be a monotonic increasing function and μ^* be the outer measure induced by f as in equation (3.23). Then the completion μ_f of the restriction of μ^* on the Borel sets is called the *Lebesgue–Stieltjes measure* induced by f. (If f is defined on a finite closed interval $[a, b]$ only, then μ_f is defined as before, restricting μ^* to subsets of $[a, b]$.) ∎

Definition 3.10. Let g be any μ_f-measurable function on R, that is, $g^{-1}(B)$ belongs to the domain of μ_f for every Borel subset B of R. Then $\int g\, d\mu_f$, when it exists, is called the *Lebesgue–Stieltjes integral* of g with respect to the monotonic increasing function f. ∎

We now consider briefly the classical Riemann–Stieltjes integral via measure theory, and discuss its connections with the Lebesgue–Stieltjes integral defined above. In order to get nicer and less complicated results in this context, we assume that the function f is right continuous.

Let g be a *bounded* real-valued function on the closed and bounded interval $[a, b]$, and let f be a *right-continuous, monotonic increasing* function on $[a, b]$.

For a partition P_n: $a = x_0 < x_1 < \cdots < x_n = b$, let us define

$$M_i = \sup\{g(x):\ x_{i-1} \le x \le x_i,\ 1 \le i \le n\}, \tag{3.29}$$

$$m_i = \inf\{g(x):\ x_{i-1} \le x \le x_i,\ 1 \le i \le n\}, \tag{3.30}$$

$$g_n = \sum_{i=1}^{n} M_i \chi_{(x_{i-1}, x_i]}, \tag{3.31}$$

$$h_n = \sum_{i=1}^{n} m_i \chi_{(x_{i-1}, x_i]}, \tag{3.32}$$

the upper sum

$$U(P_n) = \sum_{i=1}^{n} M_i[f(x_i) - f(x_{i-1})] = \int g_n\, d\mu_f, \tag{3.33}$$

the lower sum

$$L(P_n) = \sum_{i=1}^{n} m_i[f(x_i) - f(x_{i-1})] = \int h_n\, d\mu_f. \tag{3.34}$$

Note that the notation x_i above should really be $x_{i(n)}$; however, we have decided to use x_i for notational simplicity.

Let us consider a sequence of partitions $P_1, P_2, \ldots$ of $[a, b]$ such that $P_{k+1} \supset P_k$ for each k, and the norm of P_k (the length of the largest subinterval of P_k) tends to zero as $k \to \infty$. Then we have

$$g_n \ge g_{n+1} \ge \cdots \ge g \ge \cdots \ge h_{n+1} \ge h_n. \tag{3.35}$$

Suppose $s(x) = \lim_{n\to\infty} g_n(x)$ and $t(x) = \lim_{n\to\infty} h_n(x)$. Then by the Dominated Convergence Theorem we have

$$\int s \, d\mu_f = \lim_{n\to\infty} \int g_n \, d\mu_f \geq \lim_{n\to\infty} \int h_n \, d\mu_f = \int t \, d\mu_f. \tag{3.36}$$

We are now in a position to define the Riemann–Stieltjes integral properly.

Definition 3.11. If $\int s \, d\mu_f$ and $\int t \, d\mu_f$ above are equal and have a common value α that is independent of the particular sequence of partitions P_k chosen, then g is called *Riemann–Stieltjes integrable* with respect to f on $[a, b]$; and we write

$$(R) \int_a^b g \, df = \alpha$$

as the Riemann–Stieltjes integral of g with respect to f on $[a, b]$. ∎

The next theorem will now give a necessary and sufficient condition of Riemann–Stieltjes integrability in terms of μ_f.

Theorem 3.5. Let g be a bounded real-valued function on $[a, b]$ and f be a right continuous monotonic increasing function on $[a, b]$. Then g is Riemann–Stieltjes integrable with respect to f if and only if g is continuous a.e. (μ_f). Furthermore, if $(R) \int_a^b g \, df$ exists, then g is integrable with respect to μ_f, and $(R) \int_a^b g \, df = \int g \, d\mu_f$. ∎

Proof. Suppose $(R) \int_a^b g \, df$ exists. Consider a sequence of partitions P_m, $P_{m+1} \supset P_m$ and the corresponding functions g_m and h_m, as in (3.35). Then from (3.36), $\int [s(x) - t(x)] \, d\mu_f(x) = 0$. By Proposition 3.8(iii), $s(x) = t(x) = g(x)$ a.e. (μ_f). Let $x \in [a, b]$, $x \notin \bigcup_{n=1}^{\infty} P_n$ and $s(x) = t(x) = g(x)$. Then given $\varepsilon > 0$, there exists a positive integer n such that $g_n(x) - h_n(x) < \varepsilon$. Since x is an interior point of one of the intervals $(x_{i-1}, x_i]$, where $x_i \in P_n$, $0 \leq i \leq n$, there exists $\delta > 0$ such that $(x - \delta, x + \delta) \subset (x_{i-1}, x_i]$ for some i. Then it is clear that

$$| x - y | < \delta \Rightarrow | g(x) - g(y) | \leq g_n(x) - h_n(x) < \varepsilon.$$

It follows that g is continuous at x. Thus the set of discontinuities of g is a subset of $\bigcup_{n=1}^{\infty} (P_n \cup E)$, where $\mu_f(E) = 0$. Now considering a different sequence of partitions whose intervals will contain the points of P_n as interior points and noting that $\mu_f(\{a\}) = 0$, it follows that g is continuous a.e. (μ_f).

Conversely, suppose that g is continuous a.e. (μ_f). Let x $(\neq a)$ be a point of continuity of g. Then given $\varepsilon > 0$, there exists a $\delta > 0$ such that $\sup g(y) - \inf g(y) < \varepsilon$ for $y \in (x - \delta, x + \delta)$. Let $P_n(\subset P_{n+1})$ be a sequence of partitions of $[a, b]$ such that the norm of P_n tends to 0 as $n \to \infty$. Then for some positive integer N, there exist x_{i-1}, x_i in P_N such that $x \in (x_{i-1}, x_i] \subset (x - \delta, x + \delta)$. It follows that $s(x) - t(x) \leq g_N(x) - h_N(x) < \varepsilon$. Since ε is arbitrary, we have $s = t = g$ a.e. (μ_f). It is clear that g is μ_f-measurable (since s is) and $\int g \, d\mu_f$ exists (since g is bounded). The rest is clear. ∎

Corollary 3.4. A bounded function g on $[a, b]$ is Riemann integrable if and only if g is continuous a.e. with respect to the Lebesgue measure on $[a, b]$. Moreover, every Riemann-integrable function is Lebesgue integrable, and the integrals, when they exist, are equal. ∎

Proof. In Theorem 3.5, if $f(x) = x$, then μ_f becomes the Lebesgue measure on $[a, b]$ and $(R) \int_a^b g \, df$ becomes the standard Riemann integral $\int_a^b g(x) \, dx$. The corollary now follows easily. ∎

We note that the definition for Riemann–Stieltjes integral used above is equivalent to the standard definition in terms of the equality of the upper and lower integrals

$$\bar{\int_a^b} g \, df \quad \text{and} \quad \underline{\int_a^b} g \, df$$

when f is a continuous function.

Let us consider now a slightly more restricted definition originally considered by Young. Suppose f is a monotonic increasing function on $[a, b]$. If independent of the choice of partitions $a = x_1 < x_2 < \cdots < x_{n-1} < x_n = b$ and the choice of points $y_i \in [x_i, x_{i+1}]$, the sums

$$\sum_{i=1}^{n-1} g(y_i)[f(x_{i+1}) - f(x_i)]$$

converge to a unique and finite limit $\int_a^b g \, df$ as the length of the greatest subinterval approaches zero (and $n \to \infty$), then g is said to be Riemann–Stieltjes integrable with respect to f on $[a, b]$.

Now we state the following theorem of Young characterizing the above Riemann–Stieltjes integrability of g with respect to f in terms of the variation of f at the points of discontinuity of g. We omit the proof; for the details, the serious reader can consult page 133 of Young's paper.[†]

Theorem 3.6. (*Young*). In order that a bounded function g be Riemann–Stieltjes integrable with respect to a monotonic increasing function f on $[a, b]$, it is necessary and sufficient that for every $\varepsilon > 0$ it be possible to include the set of discontinuities of g in a set of intervals $(I_j)_{j=1}^{\infty}$ such that $\sum_{j=1}^{\infty} w(f, I_j) < \varepsilon$, where $w(f, I_j) = \sup_{x \in I_j} f(x) - \inf_{x \in I_j} f(x)$. ∎

Problems

{In Problems 3.3.1–3.3.5, f is an absolutely continuous monotonic increasing function. Assume that f is absolutely continuous on $[a, b]$ if and only if f' is Lebesgue integrable on each $[c, d] \subset [a, b]$ and $\int_c^d f'(x)\,dx = f(d) - f(c)$. This will be discussed at length in the next chapter.}

3.3.1. Show that $(R) \int_a^b df = \int_a^b f'(x)\,dx$.

3.3.2. If E is Lebesgue measurable and $m(E) = 0$ (m = Lebesgue measure), then show that E is Lebesgue–Stieltjes measurable and $\mu_f(E) = 0$. Conversely, if $\mu_f(E) = 0$, then $f'(x) = 0$ a.e. (Lebesgue measure) on E.

3.3.3. If E is any Lebesgue-measurable set $\subset [a, b]$, then show that $\mu_f(E) = \int_E f'(x)\,dx$.

3.3.4. If E is any μ_f-measurable set, then show that $E - \{x : f'(x) = 0\}$ is Lebesgue measurable.

3.3.5. Show that a function g is μ_f-measurable if and only if $g \cdot f'$ is Lebesgue measurable.

3.3.6. Prove the following theorem: Let f be an absolutely continuous monotonic increasing function on $[a, b]$. Suppose that *either* g is integrable with respect to μ_f *or* $g \cdot f'$ is Lebesgue integrable. Then both integrals exist and

$$\int_a^b g\,d\mu_f = \int_a^b g(x) f'(x)\,dx.$$

3.3.7. Let f be a continuous monotonic increasing function on $[a, b]$ with $f(a) = c$, $f(b) = d$. Then show that for any nonnegative Borel-mea-

[†] W. H. Young, Integration with respect to a function of bounded variation, *Proc. London Math. Soc.* **13**(2), 109–150 (1914).

surable function g on $[c, d]$,

$$\int_a^b g \circ f \, d\mu_f = \int_c^d g(y) \, dy.$$

(Hint: First consider the case when g is a simple function.)

3.3.8. Prove that a real-valued function defined on $[a, b]$ is Riemann integrable if it has a finite limit at each point of $[a, b]$. [Hint: Let $f = g + h$, where $g(x) = \lim_{y \to x} f(y)$; show that the sets $\{x: h(x) \geq 1/n\}$ and $\{x: h(x) \leq -1/n\}$ are all finite so that $h(x) = 0$, a.e. and therefore f is continuous a.e.]

3.3.9. *Integration by Parts.* Let f and g be right-continuous, monotonic increasing functions on R with no common discontinuity point in $(a, b]$. Then prove that

$$\int_{(a,b]} f \, d\mu_g + \int_{(a,b]} g \, d\mu_f = f(b)g(b) - f(a)g(a).$$

3.3.10. *Another Necessary and Sufficient Condition for Riemann Integrability* (La Vita). Prove that a bounded real-valued function f on $[a, b]$ is Riemann integrable if and only if f has a finite right-hand limit a.e. {Hint: Suppose f has a finite right-hand limit at x and the oscillation of f at x, i.e.,

$$\lim_{\delta \to 0+} [\sup_{|x-y|<\delta} f(y) - \inf_{|x-y|<\delta} f(y)]$$

is greater than $1/n$. Then there is an interval $(x, x + \delta)$ where the oscillation of f is less than or equal to $1/n$. It follows that the set of discontinuity points of f where f has a finite right-hand limit is at most countable.}

3.3.11. *A Bounded Derivative Which Is Not Riemann Integrable* (Goffman). Let $H \subset [0, 1]$ be the union of pairwise disjoint open intervals (I_n) (forming the complement of a Cantor set) of total length $\frac{1}{2}$. Let $J_n \subset I_n$ be a closed subinterval in the center of I_n such that $l(J_n) = [l(I_n)]^2$, $l =$ length. Define f $(0 \leq f \leq 1)$ on each J_n to be continuous, 1 at the center and 0 at the end points. Let $f = 0$ in the complement of the J_n. Then f is not Riemann integrable, but $f(x) = F'(x)$, $F(x) = \int_0^x f(t) \, dt$. (Hint: If $x \notin H$, then $l([x, y] \cap J_n) \leq 16\{l([x, y] \cap I_n)\}^2$. Use this to show that $F'(x) = 0$ for $x \notin H$.)

3.3.12. Let f be a bounded real-valued Lebesgue measurable function on $[a, b]$. Then there are continuous functions f_n, compact subsets $K_n \subset [a, b]$ such that $f = f_n$ on K_n and $\lim_{n \to \infty} m(K_n) = b - a$. [See 3.1.11(d).] Show that

$$\int_{[a,b]} f \, dm = \lim_{n \to \infty} (R) \int_a^b f_n(x) \, dx.$$

Is this result true for Lebesgue-measurable real functions that are not bounded?

3.3.13. Let $(X, \mathscr{A}, \mu)$ be a measure space such that $\mu(X) = 1$. Let f and g be two measurable real functions such that $0 \leq f \leq 1$, $0 \leq g \leq 1$. Suppose that f and g are independent, that is, $\mu(f^{-1}(B) \cap g^{-1}(C)) = \mu(f^{-1}(B)) \cdot \mu(g^{-1}(C))$, whenever B and C are Borel subsets of $[0, 1]$. Let F be the distribution of f, i.e., $F(x) = \mu(f^{-1}(-\infty, x])$ for real x. Similarly, let G be the distribution of g. Show that the distribution of $f + g$ is given by $H(x) = (R) \int_0^1 F(x - y)\, dG(y)$ and for any Borel subset $B \subset [0, 1]$, $\mu_H(B) = \int \mu_F(B - y)\mu_G(dy)$.

3.3.14. Let F be a nondecreasing right continuous function on R such that $\lim_{x\to-\infty} F(x) = 0$ and $\lim_{x\to\infty} F(x) = 1$. Suppose that $\int |x|\, d\mu_F(x) < \infty$. Prove that

$$\int x\, d\mu_F(x) = \int_0^\infty [1 - F(x)]\, dx - \int_{-\infty}^0 F(x)\, dx.$$

Prove also that if the integrals on the right are finite, then so is the integral on the left and both sides are equal.

3.4. Product Measures and Fubini's Theorem

Lebesgue theory in $R^n (n > 1)$ is in most respects completely analogous to the theory in the one-dimensional case. Nevertheless the evaluation of an integral in R^n is usually accomplished by a succession of n one-dimensional integrals. This method of evaluation is based on a result known as Fubini's Theorem, named after G. C. Fubini (1879–1943). It is one of the most useful results in analysis. The theory of Fourier transforms, which has numerous applications in analysis and many applied areas, owes much to this theorem. Fubini's Theorem is applicable often in various contexts in probability theory. Its presentation requires some information regarding product measures, one of the principal concerns of this section. Further considerations of product measures in a topological setting are given in Chapter 5.

First, we consider several examples illustrating a few iterated integrals.

Examples

3.4. Let $I^2 = (0, 1) \times (0, 1)$ and

$$f(x, y) = \frac{x^2 - y^2}{(x^2 + y^2)^2}, \qquad (x, y) \in I^2.$$

Then

$$\int_0^1 f(x, y)\, dy = \frac{1}{1+x^2},$$

and therefore

$$\int_0^1 \left[\int_0^1 f(x, y)\, dy\right] dx = \frac{\pi}{4}.$$

But

$$\int_0^1 \left[\int_0^1 f(x, y)\, dx\right] dy = -\frac{\pi}{4}.$$

Notice that for $0 < x < 1$

$$\int_0^x f(x, y)\, dy = \frac{1}{2x}.$$

It follows that

$$\int_0^1 | f(x, y) |\, dy \geq \frac{1}{2x}$$

and

$$\int_0^1 \left[\int_0^1 | f(x, y) |\, dy\right] dx \geq \frac{1}{2} \int_0^1 \frac{dx}{x} = \infty.$$

This will mean, as will be apparent later, that f is not Lebesgue integrable on I^2. Notice that the iterated integrals are unequal in this example.

3.5. Let $I_1{}^2 = [0, 1] \times [0, 2]$ and $f(x, y) = x^2 + xy$. In this case

$$\int_0^2 \left[\int_0^1 f(x, y)\, dx\right] dy = \frac{5}{3}$$

and

$$\int_0^1 \left[\int_0^2 f(x, y)\, dy\right] dx = \frac{5}{3}.$$

Note that here $f(x, y)$ is a nonnegative bounded continuous function on $I_1{}^2$ and hence integrable with respect to the Lebesgue measure on $I_1{}^2$.

3.6. Let $f(x, y) = xy/(x^2 + y^2)^2$, whenever $(x, y) \neq (0, 0)$. Then

$$\int_0^\infty f(x, y)\, dy = \frac{1}{2x}$$

and

$$\int_{-\infty}^0 f(x, y)\, dy = -\frac{1}{2x}.$$

Therefore,

$$\int_{-\infty}^{\infty}\left[\int_{-\infty}^{\infty} f(x, y)\, dy\right] dx = 0.$$

Similarly,

$$\int_{-\infty}^{\infty}\left[\int_{-\infty}^{\infty} f(x, y)\, dx\right] dy = 0.$$

But

$$\int_{-\infty}^{\infty} | f(x, y) |\, dy = \frac{1}{| x |},$$

and therefore

$$\int_{-\infty}^{\infty}\left[\int_{-\infty}^{\infty} | f(x, y) |\, dy\right] dx = \int_{-\infty}^{\infty} \frac{1}{| x |}\, dx = \infty.$$

In this case also, as in Example 3.4, f is not Lebesgue integrable.

3.7. Let $I = [0, 1] \times [0, 1]$ and μ_1, μ_2 be the Lebesgue and the counting measures, respectively, on $[0, 1]$. Let $f(x, y) = \chi_{\triangle}(x, y)$, where

$$\triangle = \{(x, x)\colon\ 0 \leq x \leq 1\}.$$

Then

$$\int_0^1\left[\int_0^1 f(x, y)\, d\mu_1(x)\right] d\mu_2(y) = 0$$

and

$$\int_0^1\left[\int_0^1 f(x, y)\, d\mu_2(y)\right] d\mu_1(x) = 1.$$

Note that in this case f is nonnegative and the iterated integrals are unequal.

The above four examples clearly show that the iterated integrals of many "nice" functions need not always be equal. One way to find a useful criterion for the equality of the iterated integrals is through the introduction of product measures. First we need the following lemma.

Lemma 3.3. Let $\mathscr{F}$ be a *semialgebra* of subsets of X (that is, a collection $\mathscr{F}$ closed with respect to finite intersections and such that the complement of any set in $\mathscr{F}$ is a finite disjoint union of sets in $\mathscr{F}$), and let μ be a nonnegative set function defined on $\mathscr{F} \cup \{\varnothing\}$ with $\mu(\varnothing) = 0$. Then μ has a unique extension to a measure on the algebra generated by $\mathscr{F} \cup \{\varnothing\}$, provided that the following two conditions are satisfied:

(i) If $A \in \mathscr{F}$ and $A = \bigcup_{i=1}^{n} A_i$, where A_i's are a finite disjoint collection in $\mathscr{F}$, then

$$\mu(A) = \sum_{i=1}^{n} \mu(A_i).$$

(ii) If $A \in \mathscr{F}$ and $A = \bigcup_{i=1}^{\infty} A_i$, where A_i's are an at most countable collection of disjoint sets in $\mathscr{F}$, then

$$\mu(A) \leq \sum_{i=1}^{\infty} \mu(A_i). \qquad \blacksquare$$

Proof. First, the class $\mathscr{A}$ consisting of the empty set and all finite disjoint unions of sets in $\mathscr{F}$ is the algebra generated by $\mathscr{F} \cup \{\varnothing\}$. Second, the condition (i) implies that if a set A is the union of each of two finite disjoint collections $\{A_1, A_2, \ldots, A_n\}$ and $\{B_1, B_2, \ldots, B_m\}$ in $\mathscr{F}$, then

$$\sum_{i=1}^{n} \mu(A_i) = \sum_{j=1}^{m} \mu(B_j).$$

Defining $\lambda(A) = \sum_{i=1}^{n} \mu(A_i)$, the reader can verify through the condition (ii) that λ is countably additive and the unique extension to $\mathscr{A}$. $\blacksquare$

We are now ready to introduce the product measure. One way to do this is by using the Carathéodory Extension Theorem. We will present this method, although a second method of obtaining the product measure is inherent in Theorem 3.9.

Let $(X, \mathscr{A}, \mu)$ and $(Y, \mathscr{B}, \nu)$ be two measure spaces and $X \times Y$ be the Cartesian product of X and Y. If $A \subset X$, $B \subset Y$, then $A \times B$ is called a rectangle (a measurable rectangle when $A \in \mathscr{A}$ and $B \in \mathscr{B}$). The reader can easily verify that the class $\mathscr{R}$ of all measurable rectangles forms a semi-algebra.

We define the set function λ on $\mathscr{R}$ by $\lambda(A \times B) = \mu(A) \cdot \nu(B)$. We have in mind the idea of extending λ to a measure on the algebra generated by $\mathscr{R}$ and then easily to a complete measure on a σ-algebra containing $\mathscr{R}$ by using the Extension Theorem (Theorem 2.4). To do this the following lemma is necessary.

Lemma 3.4. Let $(A_i \times B_i)_{i=1}^{\infty}$ be an at most countable disjoint collection of measurable rectangles whose union is a measurable rectangle $A \times B$. Then

$$\lambda(A \times B) = \sum_{i=1}^{\infty} \lambda(A_i \times B_i). \qquad \blacksquare$$

Proof. Since $A\times B=\bigcup_{i=1}^\infty(A_i\times B_i)$ and these rectangles are pairwise disjoint,

$$\chi_A(x)\chi_B(y)=\chi_{A\times B}(x,y)=\sum_{i=1}^\infty \chi_{A_i\times B_i}(x,y)=\sum_{i=1}^\infty \chi_{A_i}(x)\chi_{B_i}(y).$$

Integrating with respect to ν, we have: $\chi_A(x)\nu(B)=\sum_{i=1}^\infty\chi_{A_i}(x)\nu(B_i)$. The lemma now follows by integrating with respect to μ. ∎

Now by Lemma 3.3 and Lemma 3.4, λ has a unique extension to a measure λ_0 on the algebra generated by the measurable rectangles. By Theorem 2.4, the outer measure $\lambda_0{}^*$ is a measure on the class $\mathscr{M}$ of $\lambda_0{}^*$-measurable sets and an extension of λ_0 to the σ-algebra $\mathscr{M}$. This extension, a complete measure on $\mathscr{M}$, is called the product of μ and ν, and denoted by $\mu\times\nu$.

Remark 3.24. $\mu\times\nu$ is defined on a σ-algebra containing $\mathscr{A}\times\mathscr{B}$, the smallest σ-algebra containing $\mathscr{R}$. The measure space $(X\times Y, \mathscr{A}\times\mathscr{B}, \mu\times\nu)$ can be incomplete even if μ and ν are complete. For example, if $A\subset X$, $A\notin\mathscr{A}$ and $\nu(B)=0$ with B nonempty, then $A\times B\notin\mathscr{A}\times\mathscr{B}$; but $A\times B\subset X\times B$ and $\mu\times\nu(X\times B)=0$. Furthermore, when μ and ν are both σ-finite, $\mu\times\nu$ is the unique extension of λ on $\mathscr{R}$ to a measure on $\mathscr{A}\times\mathscr{B}$. The motivation behind considering $\mu\times\nu$ on $\mathscr{M}$ rather than on $\mathscr{A}\times\mathscr{B}$ is that $\mu\times\nu$ on $\mathscr{M}$ is exactly the Lebesgue measure on R^2 when $\mu=\nu$ is the Lebesgue measure on R.

The next few lemmas will lead to Fubini's Theorem, the main result of this section, and will also describe the structure of the sets in $\mathscr{M}$.

First, a notation: For $E\subset X\times Y$, let

$$E_x=\{y\in Y\colon (x,y)\in E\}$$

and

$$E^y=\{x\in X\colon (x,y)\in E\}.$$

Then it is easy to verify the following:

(i) $(E^c)_x=(E_x)^c$,

(ii) $(\cup E_\alpha)_x=\cup(E_\alpha)_x$, for any collection $\{E_\alpha\}$.

Lemma 3.5. If $E\in\mathscr{A}\times\mathscr{B}$, then

(i) $E_x\in\mathscr{B}$ for all $x\in X$,

and

(ii) $E^y \in \mathscr{A}$ for all $y \in Y$. ∎

Proof. We will prove only (i). Let

$$\mathscr{F} = \{E \in \mathscr{A} \times \mathscr{B}\colon E_x \in \mathscr{B} \text{ for all } x \in X\}.$$

Then $\mathscr{F}$ contains $\mathscr{R}$. The reader can easily verify that $\mathscr{F}$ is a σ-algebra. Hence $\mathscr{F} \supset \mathscr{A} \times \mathscr{B}$. ∎

Lemma 3.6. Let f be an extended real-valued $\mathscr{A} \times \mathscr{B}$ measurable function on $X \times Y$. Then

(i) f_x, where $f_x(y) = f(x, y)$, is $\mathscr{B}$-measurable for all $x \in X$,

and

(ii) f^y, where $f^y(x) = f(x, y)$, is $\mathscr{A}$-measurable for all $y \in Y$. ∎

Proof. This lemma follows easily from Lemma 3.5 when f is a simple function. Since there exist simple functions $f_n(x, y)$ such that $\lim_{n\to\infty} f_n(x, y) = f(x, y)$, the rest of the proof is clear. ∎

Lemma 3.7. Suppose that μ and ν are complete. Let E be a set in $\mathscr{R}_{\sigma\delta}$ (that is, of the form $\bigcap_{i=1}^{\infty} \bigcup_{j=1}^{\infty} E_{ij}$ where $E_{ij} \in \mathscr{R}$) with $\mu \times \nu(E) < \infty$. Then

(i) the function $\Gamma_E(x) = \nu(E_x)$ is $\mathscr{A}$-measurable,

(ii) the function $\Gamma^E(y) = \mu(E^y)$ is $\mathscr{B}$-measurable,

and

(iii) $\mu \times \nu(E) = \int \mu(E^y)\nu(dy) = \int \nu(E_x)\mu(dx)$. ∎

Proof. This lemma is clearly true for $E \in \mathscr{R}$. Now suppose $E \in \mathscr{R}_\sigma$. Then we can write $E = \bigcup_{j=1}^{\infty} E_j$, $E_j \in \mathscr{R}$ and $E_i \cap E_k = \varnothing$ $(i \neq k)$. Since $\nu(E_x) = \sum_{j=1}^{\infty} \nu((E_j)_x)$ and by the Monotone Convergence Theorem,

$$\begin{aligned}\int \nu(E_x)\mu(dx) &= \sum_{j=1}^{\infty} \int \nu((E_j)_x)\mu(dx) \\ &= \sum_{j=1}^{\infty} \mu \times \nu(E_j) = \mu \times \nu(E)\end{aligned}$$

and similarly

$$\mu\times\nu(E)=\int\mu(E^y)\nu(dy),$$

the lemma is true for all $E\in\mathscr{R}_\sigma$.

Finally, let $E\in\mathscr{R}_{\sigma\delta}$ and $\mu\times\nu(E)<\infty$. Since $\mu\times\nu$ is an outer measure, by definition there are sets in $\mathscr{R}_\sigma$ that contain E and have finite $\mu\times\nu$-measure. Thus, we can assume that $E=\bigcap_{j=1}^{\infty}E_j$, $E_j\in\mathscr{R}_\sigma$, $E_j\supset E_{j+1}$, and $\mu\times\nu(E_1)<\infty$. Now $\mu\times\nu(E_1)=\int\nu\big((E_1)_x\big)\mu(dx)<\infty$, and therefore $\nu\big((E_1)_x\big)<\infty$ a.e. (μ). Hence we have

$$\nu(E_x)=\lim_{j\to\infty}\nu\big((E_j)_x\big)\text{ a.e. }(\mu).$$

Since μ is complete, Γ_E is measurable; and, by the Dominated Convergence Theorem,

$$\begin{aligned}\int\nu(E_x)\mu(dx)&=\lim_{j\to\infty}\int\nu\big((E_j)_x\big)\mu(dx)\\&=\lim_{j\to\infty}\mu\times\nu(E_j)\\&=\mu\times\nu(E),\end{aligned}$$

and similarly

$$\mu\times\nu(E)=\int\mu(E^y)\nu(dy).$$ ∎

It may be noted that Lemma 3.7 uses the completeness of μ and ν. Nevertheless, completeness is not needed if μ and ν are assumed σ-finite, as the next lemma will demonstrate. First we give an example showing that Lemma 3.7 need *not* be valid for all $E\in\mathscr{R}_{\sigma\delta}$.

Example 3.8. Let $X=Y=[0,1]$, $\mathscr{A}=\mathscr{B}=$ Lebesgue-measurable subsets of [0, 1], and $\mu=\nu$ be the measure defined by

$$\mu\{x\}=\begin{cases}2, & x\in A,\\ 1, & x\notin A,\end{cases}$$

where A is a nonmeasurable subset of [0, 1]. Then μ and ν are complete, and the set $D=\{(x,y)\in X\times Y\colon x=y\}\in\mathscr{R}_{\sigma\delta}$, but $\nu(D_x)=\chi_A(x)+1$, so Γ_D is nonmeasurable.

Lemma 3.8. The conclusion for Lemma 3.7 remains true (without the assumption of completeness) for *all* $E \in \mathscr{A} \times \mathscr{B}$, if μ and ν are both σ-finite. ∎

Proof. We will briefly outline the proof. Observe that as in Lemma 3.7 $\mathscr{R}_\sigma \subset \mathscr{F}$, the class of sets for which the lemma holds. The reader can verify that $\mathscr{F}$ is a monotone class by first assuming that μ and ν are both finite. This means that $\mathscr{F} \supset \mathscr{A} \times \mathscr{B}$. The σ-finite case then follows easily. ∎

Proposition 3.10. Let μ and ν be complete. Let $E \in \mathscr{M}$ and $\mu \times \nu(E) < \infty$. Then for almost all x, $E_x \in \mathscr{B}$ and $\Gamma_E(x)$ [$= \nu(E_x)$] is defined and $\mathscr{A}$-measurable. In addition

$$\mu \times \nu(E) = \int \nu(E_x)\mu(dx).$$

The same is true for $\Gamma^E(y)$ and $\mu \times \nu(E) = \int \mu(E^y)\nu(dy)$. ∎

Proof. By the definition of $\mu \times \nu$, there exists $F \in \mathscr{R}_{\sigma\delta}$, $F \supset E$ and $\mu \times \nu(F - E) = 0$. Hence if $G = F - E$, then $\mu \times \nu(G) = 0$ and

$$\mu \times \nu(E) = \mu \times \nu(F) - \mu \times \nu(G).$$

Because of Lemma 3.7, it is enough to prove the proposition for G. To do this, let $H \in \mathscr{R}_{\sigma\delta}$, $G \subset H$, and $\mu \times \nu(H) = 0$. Then by Lemma 3.7

$$\mu \times \nu(H) = \int \nu(H_x)\mu(dx) = \int \mu(H^y)\nu(dy).$$

Hence $\nu(H_x) = 0$ a.e. (μ) and $\mu(H^y) = 0$ a.e. (ν). Since $G_x \subset H_x$, $G^y \subset H^y$, and μ, ν are complete, the proposition is clear. ∎

Theorem 3.7. *Fubini's Theorem.* Let μ and ν be complete and f be integrable with respect to $\mu \times \nu$ on $\mathscr{M}$. Then

(i) $\int f(x, y)\nu(dy)$ and $\int f(x, y)\mu(dx)$
are integrable functions of x and y, respectively, and

(ii) $\int f\, d(\mu \times \nu) = \int\int f\, d\mu\, d\nu = \int\int f\, d\nu\, d\mu$. ∎

Proof. By Proposition 3.10 this theorem is clearly true for $\chi_E(x, y)$, where $E \in \mathscr{M}$ and $\mu \times \nu(E) < \infty$. Suppose then that f is nonnegative. Since f is integrable, by Proposition 3.5 f is the limit of an increasing sequence of simple functions f_n, each vanishing outside a set of finite measure. The theorem then follows for this f by an application of the Monotone Convergence

Theorem. Since for any function f, $f=f^+-f^-$, where $f^+=\sup\{f(x), 0\}$ and $f^-=\sup\{-f(x), 0\}$, the theorem follows easily. ∎

Theorem 3.8. *Tonelli's Theorem.* Let μ and ν be complete and σ-finite. Let f be a nonnegative $\mathscr{M}$-measurable function on $X\times Y$. Then

(i) For almost all x, $f_x(y)=f(x, y)$ is $\mathscr{B}$-measurable and for almost all y, $f^y(x)=f(x, y)$ is $\mathscr{A}$-measurable;
(ii) $\int f(x, y)\, d\nu(y)$ and $\int f(x, y)\, d\mu(x)$ are both measurable functions of x and y, respectively;
(iii) $\int f\, d(\mu\times\nu)=\iint f\, d\mu\, d\nu=\iint f\, d\nu\, d\mu$. ∎

Proof. Since μ and ν are σ-finite, $\mu\times\nu$ is σ-finite; and therefore by Proposition 3.5 every nonnegative f is the limit of an increasing sequence of simple functions, each vanishing outside a set of finite measure. The proof now follows as in Theorem 3.7. ∎

Remark 3.25. We note that the integrability condition in Theorem 3.7 and the nonnegativeness (as well as the σ-finiteness) condition in Theorem 3.8 are essential, as Examples 3.4 and 3.7 demonstrate. The completeness assumption for μ and ν has been necessary in the proofs of Fubini's and Tonelli's Theorems (Theorems 3.7 and 3.8). But these theorems above have been stated for $\mathscr{M}$-measurable functions. As we will see shortly in this section, the completeness assumption is unnecessary in the above two theorems if f is $\mathscr{A}\times\mathscr{B}$ measurable. (Recall that $\mathscr{M}\supset\mathscr{A}\times\mathscr{B}$. This inclusion may be proper.)

Product measures are very useful in probability theory. One simple instance in which they can arise in probability theory is the following. Suppose we make two observations, one resulting in a point x in X and the other in a point y in Y. Suppose also that the probability $P(x, B)$ that the second observation belongs to B given that the first observation is x, is a probability measure on $\mathscr{B}$ for each x in X, and is $\mathscr{A}$-measurable for each B in $\mathscr{B}$. Now if μ is a probability measure such that $\mu(A)$ is the probability that the first observation belongs to A, then intuitively the probability that the first observation results in a point of A and the second in a point of B should be given by $\int_A P(x, B)\mu(dx)$. The next theorem will show that this probability is actually a product measure in $\mathscr{A}\times\mathscr{B}$. This theorem is also a generalization of Lemma 3.8.

• **Theorem 3.9.** Let $(X, \mathscr{A}, \mu)$ be a σ-finite measure space and $(Y, \mathscr{B})$ be a measurable space. (Here $\mathscr{A}$ and $\mathscr{B}$ are σ-algebras.) Suppose that $P(x, B)$

is an extended real-valued function on $X \times \mathscr{B}$ such that (i) for each x in X, $P(x, B)$ is a measure on $\mathscr{B}$ and for each B in $\mathscr{B}$, $P(x, B)$ is an $\mathscr{A}$-measurable function; (ii) $Y = \bigcup_{n=1}^{\infty} B_n$ and for each x in X, $P(x, B_n) \leq k_n < \infty$. Then there is a unique σ-finite measure Q on $\mathscr{A} \times \mathscr{B}$ such that

$$\text{(i)} \quad Q(A \times B) = \int_A P(x, B)\mu(dx), \qquad A \in \mathscr{A},\ B \in \mathscr{B},$$

and

$$\text{(ii)} \quad Q(E) = \int P(x, E_x)\mu(dx), \qquad E \in \mathscr{A} \times \mathscr{B}.$$

■

Proof. With no loss of generality, we can assume that the sets B_n are pairwise disjoint. Then let $P_n(x, B) = P(x, B \cap B_n)$. If we can show the existence of a unique measure Q_n on $\mathscr{A} \times \mathscr{B}$ such that

$$Q_n(E) = \int P_n(x, E_x)\mu(dx), \qquad E \in \mathscr{A} \times \mathscr{B},$$

for each n, then the measure Q given by $Q(E) = \sum Q_n(E)$ will be the desired product measure of the theorem. Therefore, it is no loss of generality to assume that the measures $P(x, B)$ are finite.

We observe that for each $E \in \mathscr{A} \times \mathscr{B}$, the function $P(x, E_x)$ is $\mathscr{A}$-measurable. To see this, let

$$\mathscr{F} = \{E \in \mathscr{A} \times \mathscr{B}\colon P(x, E_x) \text{ in } \mathscr{A}\text{-measurable}\}.$$

Then the reader can easily verify that $\mathscr{F}$ contains finite disjoint union of measurable rectangles, and $\mathscr{F}$ is a monotone class. Thus, $\mathscr{F} = \mathscr{A} \times \mathscr{B}$. Since $P(x, E_x)$ is $\mathscr{A}$-measurable, we can define

$$Q(E) = \int P(x, E_x)\mu(dx).$$

Now it follows easily that Q is a measure, and $Q(A \times B) = \int_A P(x, B)\mu(dx)$.

Finally for the uniqueness proof, we observe that another measure Q_0 satisfying the properties of Q has to be σ-finite, will coincide with Q on $\mathscr{R}$, and will, by the Carathéodory Extension Theorem, coincide with Q on $\mathscr{A} \times \mathscr{B}$. ■

Now we will state the σ-finite version of the Fubini–Tonelli Theorem, the most important theorem in this section.

Theorem 3.10. *The Fubini–Tonelli Theorem* (*σ-Finite Version*). Let $(X, \mathscr{A}, \mu)$ and $(Y, \mathscr{B}, \nu)$ be two σ-finite measure spaces. Suppose f is an extended real-valued $\mathscr{A} \times \mathscr{B}$-measurable function on $X \times Y$ such that f is either nonnegative or integrable. Then the iterated integrals of f are defined and

$$\int f\,d(\mu\times\nu) = \iint f\,d\mu\,d\nu = \iint f\,d\nu\,d\mu. \qquad \blacksquare$$

Proof. Using Lemma 3.8, the theorem follows immediately for simple functions. The proof then follows as in Theorem 3.7 and 3.8. ∎

Remarks

3.26. Interestingly enough, Fubini's Theorem (Theorem 3.7) has a category analog, proved in 1932 by Kuratowski and Ulam, which can be stated as follows:

If E is a plane set ($\subset R^2$) of first category, then E_x is a linear set ($\subset R$) of first category for all x outside a set of the first category. If E is a nowhere dense subset of the plane, then E_x is a nowhere dense linear subset for all x outside a set of first category.

Actually this theorem can be reduced to Fubini's Theorem using the following fact: For any set E of first category in the plane, there exists a product homeomorphism h of the plane onto itself such that $m_2(h(E)) = 0$, m_2 being the Lebesgue measure on R^2. For proofs of these facts the interested reader can consult *Measure and Category* by J. C. Oxtoby.†

3.27. The requirement that f be $\mathscr{A}\times\mathscr{B}$-measurable (or $\mathscr{M}$-measurable) is *crucial* in Fubini's Theorem as well as Tonelli's Theorem. Even measurability of all the sections f_x and f^y is not enough for the validity of the above theorems. For instance, let $(X, \mathscr{A}, \mu) = (Y, \mathscr{B}, \nu)$ be the Lebesgue measure space on $[0, 1]$. By the continuum hypothesis and Proposition 1.6 there is a one-to-one mapping Φ of $[0, 1]$ onto a well-ordered set (the set of all ordinals $< \Omega$, the first uncountable ordinal) such that $\Phi(x)$ has at most countably many predecessors in this set for each $x \in [0, 1]$. Let $Q = \{(x, y) \in [0, 1]\times[0, 1]\colon \Phi(x) < \Phi(y)\}$. Then for $x \in [0, 1]$, Q_x is the complement of a countable set and for $y \in [0, 1]$, Q^y is countable. Let $f = \chi_Q$. Then

$$\int_0^1\int_0^1 f\,dy\,dx = 1, \qquad \int_0^1\int_0^1 f\,dx\,dy = 0.$$

3.28. Note that $\mu\times\nu$ need not be semifinite even when μ and ν are both semifinite. For instance, consider Example 3.7. There $\mu\times\nu(\triangle) = \infty$,

† J. C. Oxtoby, *Measure and Category*, Springer-Verlag, pp. 56–60 (1971).

since otherwise by Fubini's Theorem the iterated integrals will be equal. If $A \subset \triangle$, $A \in \mathscr{A}\times\mathscr{B}$ and $\mu\times\nu(A) < \infty$, then by Fubini's Theorem

$$\mu\times\nu(A) = \iint \chi_A(x, y)\, d\mu\, d\nu = 0.$$

This shows that $\mu\times\nu$ is not semifinite.

Actually semifiniteness of the product measure is inseparably connected with the validity of Tonelli's Theorem. *By the validity of Tonelli's Theorem we mean* the following: For every nonnegative extended real-valued $\mathscr{A}\times\mathscr{B}$-measurable function $f(x, y)$ whenever one of the iterated integrals $\iint f\, d\mu\, d\nu$ or $\iint f\, d\nu\, d\mu$ is well defined and finite, the other one is also; and $\int f\, d(\mu\times\nu) = \iint f\, d\mu\, d\nu = \iint f\, d\nu\, d\mu$. Tonelli's Theorem is, of course, valid when μ and ν are both σ-finite. But even when μ is an arbitrary measure, Tonelli's Theorem can be valid provided ν is the counting measure on a countable set.

• **Proposition 3.11.** Let μ, ν be complete and $\mu\times\nu$ be semifinite. Then Tonelli's Theorem is valid. ▮

Proof. Suppose $f(x, y)$ is a nonnegative $\mathscr{A}\times\mathscr{B}$-measurable function such that $\iint f(x, y)\, d\mu\, d\nu$ is well-defined and equal to some nonnegative number k. Then $\{(x, y): f(x, y) \neq 0\}$ has σ-finite $\mu\times\nu$ measure. If not, we can find a positive integer n such that $\mu\times\nu(A_n) = \infty$, where $A_n = \{(x, y): f(x, y) > 1/n\}$. Since $\mu\times\nu$ is semifinite, there exists $B \subset A_n$, $B \in \mathscr{A}\times\mathscr{B}$ and $2kn < \mu\times\nu(B) < \infty$. Then, by Fubini's Theorem,

$$\begin{aligned} 2k < \frac{1}{n}\mu\times\nu(B) &= \iint \frac{1}{n}\cdot \chi_B(x, y)\, d\mu\, d\nu \\ &\leq \iint f(x, y)\, d\mu\, d\nu, \end{aligned}$$

which is a contradiction. Therefore, the support of f (that is, $\{x: f(x) \neq 0\}$) is σ-finite, and f is the limit of an increasing sequence of simple functions each vanishing outside a set of finite measure. The proof now follows as in Theorem 3.7. ▮

Actually, much more than Proposition 3.11 is true.

• **Theorem 3.11.** Suppose μ and ν are complete measures. Then the following conditions are equivalent:

(a) $\mu \times \nu$ is semifinite on $\mathscr{A} \times \mathscr{B}$ (respectively, $\mathscr{M}$);

(b) μ and ν are semifinite and Tonelli's Theorem is valid for $\mathscr{A} \times \mathscr{B}$ (respectively, $\mathscr{M}$) measurable functions. ∎

The proof is omitted. The interested reader is referred to [35].

Problems

× **3.4.1.** Let $X = Y = [0, 1]$, and $\mathscr{A} = \mathscr{B}$ is a σ-algebra containing the open sets. Show that $\mathscr{A} \times \mathscr{B}$ contains all the open sets of $X \times Y$.

× **3.4.2.** Let f and g be integrable on $(X, \mathscr{A}, \mu)$ and $(Y, \mathscr{B}, \nu)$, respectively. Show that $f(x) \cdot g(y)$ is integrable on $(X \times Y, \mathscr{A} \times \mathscr{B}, \mu \times \nu)$ and

$$\int fg \, d(\mu \times \nu) = \int f \, d\mu \cdot \int g \, d\nu.$$

It is interesting to notice that if $\mu(X) = \nu(Y) = 1$ here, then for any two positive measurable functions f and g on X and Y, respectively, such that $fg \geq 1$,

$$\int f \, d\mu \cdot \int g \, d\nu \geq 1.$$

3.4.3. Show that

$$\lim_{n \to \infty} \int_0^n \frac{\sin x}{x} \, dx = \frac{\pi}{2}.$$

[Hint: Note that

$$\int_0^n \frac{\sin x}{x} \, dx = \int_0^n \left(\int_0^\infty e^{-xt} \, dt \right) \sin x \, dx.$$

and then use Fubini's Theorem.]

3.4.4. Show that completeness of the measures is essential in Theorem 3.7. [Hint: Consider $\mu = \nu =$ the Borel measure on R and the function $\chi_{A \times [0,1]}(x, y)$ where A is a non-Borel Lebesgue-measurable set with finite positive Lebesgue measure.]

3.4.5. Show that a nonnegative real-valued function f on R is Lebesgue measurable if and only if the set $E = \{(x, y) \colon 0 \leq y \leq f(x)\}$ is product measurable. (Hint: Consider simple functions first and then pass to the limit, for the "only if" part. To prove the "if" part, observe that

$$\{(x, y): f(x) > c, y > 0\} = \bigcup_{n=1}^{\infty} \left\{(x, y): \left(x, \frac{y}{n} + c\right) \in E,\ y > 0\right\}.)$$

Using the set E and also the set $D = \{(x, y): 0 \leq y < f(x)\}$, show that the graph of a measurable function is measurable. Note that the graph may be product measurable even for a nonmeasurable function.

3.4.6. Let f be Lebesgue integrable in Problem 3.4.5. Show that $\mu \times \mu(E) = \int f(x)\, dx$, where μ is the Lebesgue measure on R. Thus, the integral of f is the area under the curve $y = f(x)$.

3.4.7. Prove that a real-valued function f on R^2 is Borel measurable (i.e., $f^{-1}(H)$ is a Borel set for any open set H), if each section f_x is Borel measurable on R and each section f^y is continuous on R. [Hint: Consider the functions

$$\sum_{i=-n^2}^{i=n^2} f\left(\frac{i-1}{n}, y\right)\chi_{[(i-1)/n, i/n)}(x).\ ]$$

3.4.8. × (a) Show that the set $\{(x, y): x - y \in E\}$ is $\mu \times \mu$ measurable on R^2, where μ is the Lebesgue measure on R and E is a Lebesgue-measurable set on R. (Hint: Consider first when E is open and then, when E is a G_δ-set or a set of measure zero.) Use this to show that $f(x - y)$ is $\mu \times \mu$ measurable for a Lebesgue-measurable function f.

(b) Let f and g be complex-valued Lebesgue-integrable functions on R. Use Tonelli's theorem to show that $f * g(x) = \int f(x - y)g(y)\, dy$ is a well-defined Lebesgue-integrable function on R.

(c) Let f and g be as in (b). We define the Fourier transform $\hat{f}$ of f by $\hat{f}(t) = \int e^{-itx} f(x)\, dx$. [Sometimes by analogy with the Fourier series and also for the purpose of getting simpler forms for certain results,

$$\hat{f}(t) = \frac{1}{2\pi} \int e^{-itx} f(x)\, dx.$$

defines the Fourier transform.] Show that

(i) $\hat{f}$ is a complex-valued continuous function on R vanishing at ∞;

(ii) $\widehat{f * g} = \hat{f} \cdot \hat{g}$;

(iii) $\overline{\hat{f}} = \widehat{f^*}$, $f^*(x) = \overline{f(-x)}$ (bar = conjugate);

(iv) the set of all Fourier transforms of complex-valued Lebesgue-integrable functions is dense (relative to uniform convergence) in the set of all complex-valued continuous functions vanishing at ∞;

(v) for any two finite Borel measures μ and ν on R, the equality $\int e^{itx}\,d\mu(x) = \int e^{itx}\,d\nu(x)$ for all real t implies that $\mu = \nu$;

(vi) $\hat{f} = \hat{g}$ implies that $f = g$ a.e.

[Hint: (iv) Use the Stone–Weierstrass theorem and see also Remark 1.92. (v) Note that $\int \hat{f}(-t)\,d\mu(t) = \int \hat{f}(-t)\,d\nu(t)$. Hence using (iv), $\int f\,d\mu = \int f\,d\nu$ for all f in $C_0(R)$. Use Problem 2.3.12 to conclude that $\mu = \nu$. (vi) Suppose first that f and g are real-valued. Write $f = f^+ - f^-$ and $g = g^+ - g^-$ (usual notation). Let

$$\mu(A) = \int_A (f^+ + g^-)\,dx \quad \text{and} \quad \nu(A) = \int_A (g^+ + f^-)\,dx.$$

Then $\int e^{itx}\,d\mu(x) = \int e^{itx}\,d\nu(x)$. Use (v).]

3.4.9. Let X be an uncountable set and $\mathscr{A}$ be the smallest σ-algebra of subsets of $X \times X$ containing all sets of the form $A \times B$, where $A \subset X$, $B \subset X$. When does $\mathscr{A}$ contain all subsets of $X \times X$? (Hint: (1) Assume card $X > c$. Let $D = \{(x, y): x = y\} \subset X \times X$. Suppose $D \in \mathscr{A}$. Then there is a class $\mathscr{F}$ consisting of an at most countable number of rectangles $(A_i \times B_i)$ such that $D \in \sigma(\mathscr{F})$. Let the A_i's generate the σ-algebra $\mathscr{B}$. Then by Proposition 1.10 of Chapter 1, card $\mathscr{B} \leq c$. But for each $y \in X$, $D^y \in \mathscr{B}$, which is a contradiction. (2) Assume card $X = c$. By assuming the continuum hypothesis, identify X with $[0, \Omega)$ and $[0, 1]$, where Ω is the first uncountable ordinal. Observe the following:

(i) For any function f from $[0, \Omega)$ (or from one of its subsets) to $[0, 1]$, the graph of $f \in \mathscr{A}$.

(ii) For $B \subset C_1 = \{(x, y) \in [0, \Omega) \times [0, \Omega): y \leq x\}$ or $B \subset C_2 = \{(x, y) \in [0, \Omega) \times [0, \Omega): y > x\}$, $B \in \mathscr{A}$.

Therefore $\mathscr{A}$ consists of all subsets of $X \times X$.)

The first case of this result is due to P. R. Halmos, and the second one is due to B. V. Rao.

3.4.10. *An Example of a Set of Positive Product Measure Not Containing Any Measurable Rectangle of Positive Product Measure.* Let m be the Lebesgue measure on R and C be a Cantor set $\subset [0, 1]$ with $m(C) > 0$, as in Problem 2.3.7(b) of Chapter 2. Let $S = \{(x, y): x - y \in C\} \subset [0, 1] \times [0, 1]$. Show that S is a compact subset of R^2, $m \times m(S) > 0$, and S is the desired set. [Hint: If $A \times B \subset S$, then $A - B \subset C$; since $m(A) > 0$ and $m(B) > 0$ imply $A - B$ has nonempty interior (see Problem 3.2.15), $m \times m(A \times B) = 0$.]

3.4.11. *Integration by Parts Using Fubini's Theorem.* Let f and g be two monotonic increasing, bounded, right-continuous functions on $[a, b]$.

Let $f^*(x) = f(x) - f(x-)$ and $g^*(x) = g(x) - g(x-)$, and let (d_i) be the set of common points of discontinuity of f and g. Prove that

$$\int_{(a,b]} f\,d\mu_g + \int_{(a,b]} g\,d\mu_f = f(b)g(b) - f(a)g(a) + \sum_{a<d_i\leq b} f^*(d_i)g^*(d_i).$$

[Hint: Let $A = \{(x, y)\colon\ a < x \leq b,\ y \leq x\}$, $B = \{(x, y)\colon\ a < x \leq b,\ y \leq a\}$, and $C = \{(x, y)\colon\ y \leq x \leq b,\ a < y \leq b\}$, so that $A = B \cup C$ and $B \cap C = \varnothing$. By Fubini's Theorem,

$$\int_{(a,b]} f\,d\mu_g = \int_A \int d\mu_f\,d\mu_g = \int_B \int d\mu_g\,d\mu_f + \int_C \int d\mu_g\,d\mu_f.]$$

3.4.12. *Extension of Steinhaus Lemma: An Application of Fubini's Theorem.* Let A and B be Lebesgue-measurable subsets of the reals such that $m(A) > 0$ and $m(B) > 0$. Prove that the difference set $D(A, B) = \{x\colon x = y - z,\ y \in A,\ z \in B\}$ contains an open interval. [Hint: We may assume A and B to be closed and bounded. Use Fubini's theorem on $\chi_A(y - x) \cdot \chi_B(y)$ to prove that $m\big(B \cap (A + x)\big) > 0$ for some real x. Let $C = B \cap (A + x)$. Use Problem 3.2.15 on the set C.]

3.4.13. *Another Proof of the Ulam Theorem Using Fubini's Theorem.* Let μ be a measure defined on the class 2^X of all subsets of $X = [0, \Omega)$, Ω being the first uncountable ordinal such that $\mu(\{x\}) = 0$ for every singleton $\{x\}$. Then μ is the zero measure. [Hint: By Problem 3.4.9, the product σ-algebra $2^X \times 2^X = 2^{X\times X}$. Let $X_1 = \{(\alpha, \beta)\colon\ \alpha \leq \beta,\ \alpha$ and β in $X\}$ and $X_2 = \{(\alpha, \beta)\colon\ \alpha > \beta,\ \alpha$ and β in $X\}$. By Fubini's Theorem, $\mu \times \mu(X_1) = \mu \times \mu(X_2) = 0$.]

× **3.4.14.** *Uniqueness of Translation-Invariant Measures*: Let μ and ν be two σ-finite translation-invariant measures defined on the Borel sets of R^n such that for *some* Borel set $A \subset R^n$ and some constant c, $0 < \mu(A) = c \cdot \nu(A) < \infty$. Then $\mu = c \cdot \nu$. Can R^n be replaced by any separable metric space with a vector space structure where the addition is continuous? [Hint: Apply Theorem 3.10 to the iterated integral

$$\iint \chi_B(x)\chi_A(x + y)\mu(dx)\nu(dy).]$$

3.4.15. *Interchange of the Order of Partial Differentiation*: Let f be a continuous real function defined on an open rectangle V of R^2 such that (i) $\partial f/\partial x$ exists and is continuous on V; (ii) for some a, $(d/dy)[f(a, y)]$ exists at all points (a, y) in V; and (iii) $(\partial/\partial y)(\partial f/\partial x)$ exists and is continuous on V. Then $\partial f/\partial y$ and $(\partial/\partial x)(\partial f/\partial y)$ exist on V; moreover, $(\partial/\partial x)(\partial f/\partial y) = (\partial/\partial y)$

$(\partial f/\partial x)$ on V. [Hint: Use Fubini's Theorem and the fundamental theorem of calculus to show that

$$f(x, y) - f(a, y) - f(x, b) + f(a, b) = \int_b^y \int_a^x \frac{\partial}{\partial y}\left(\frac{\partial f}{\partial x}\right) dx\, dy.$$

Notice that for each x, the inner integral defines a continuous function of y.]

3.4.16. *Measures Which Coincide on the Translates of a Set.* Suppose μ and ν are any two Borel measures on R such that $\mu(R) = \nu(R) = 1$. Let B be a Borel set with finite positive Lebesgue measure. Suppose that $\hat{\chi}_B(t) = \int e^{-itx}\chi_B(x)\, dx$ and the set $\{t: \hat{\chi}_B(t) \neq 0\}$ is dense in R. Also, let $\mu(B + t) = \nu(B + t)$ for all t in R. Show that $\mu = \nu$. [Hint: If $f(x) = \int \chi_B(y - x)\, d\mu(y)$, then $\hat{f}(t) = \hat{\mu}(t) \cdot \hat{\chi}_B(-t)$, where $\hat{\mu}(t) = \int e^{-itx}\, d\mu(x)$. It follows that $\hat{\mu}(t) = \hat{\nu}(t)$. Now use Problem 3.4.8c (v).] This result remains true without the hypothesis on $\hat{\chi}_B(t)$, and is due to N. A. Sapogov. See also [4]. Let us point out one more way we can use Fubini's Theorem in this context. Consider the integral

$$g(z) = \int_B e^{-izx}\, dx,$$

where z is a complex variable and B is a bounded Borel set $\subset R$ with positive Lebesgue measure. Note that if C is any closed path given by $z = \gamma(t)$, $t \in [a, b] \subset R$, where $\gamma(t)$ is continuously differentiable, then

$$\int_C g(z)\, dz = \int_a^b g(\gamma(t))\gamma'(t)\, dt.$$

Thus, we can use Fubini's theorem to show that

$$\int_C g(z)\, dz = \int_B \left[\int_C e^{-izx}\, dz\right] dx = 0,$$

by Cauchy's theorem in complex analysis. It follows by Morera's theorem in complex analysis that g is an analytic function of z. Therefore, g cannot be zero on any interval $I \subset R$. This means that $\{t: \hat{\chi}_B(t) \neq 0\}$ is dense in R.

3.4.17. Let $p \geq 1$ and $f(x, y)$ be any product-measurable function on $[0, 1] \times [0, 1]$. Let us write

$$\| g \|_p = \left(\int_0^1 | g(y) |^p\, dy\right)^{1/p}.$$

It will be proven in the next section that

(i) $\| g + h \|_p \leq \| g \|_p + \| h \|_p$,
(ii) $\int_0^1 | g(y) |\, dy \leq \| g \|_p$.

Use these facts to prove that

$$\left\| \int | f(x, y) | \, dy \right\|_p \leq \int \| f(x, y) \|_p \, dx.$$

[Hint: Assume that f is bounded. Note that $L_p = \{g: \| g \|_p < \infty\}$ is a separable pseudometric space with the metric $d(g, h) = \| g - h \|_p$. Then there exist functions $f_n(x, y) = \sum_i \chi_{A_i}(x) g_i(y)$, $g_i \in L_p$ such that for almost all x, $\lim_{n\to\infty} \| f - f_n \| = 0$. Note that the inequality in the problem holds for f_n and by (ii) above,

$$\lim_{n\to\infty} \left\| \int | f(x, y) | \, dy - \int | f_n(x, y) | \, dy \right\|_p = 0.]$$

3.4.18. Let $F(x, t)$ be a real-valued function on R^2. Find appropriate conditions on F to justify the following steps in the proof of $(d/dt) \int_u^v F(x, t) \, dx = \int_u^v (\partial/\partial t)[F(x, t)] \, dx$. The steps are

$$\begin{aligned} \int_u^v \frac{\partial}{\partial t} [F(x, t)] \, dx &= \frac{d}{dt} \int_a^t ds \left[\int_u^v \frac{\partial}{\partial s} [F(x, s)] \, dx \right] \\ &= \frac{d}{dt} \int_u^v dx \left[\int_a^t \frac{\partial}{\partial s} [F(x, s)] \, ds \right] \\ &= \frac{d}{dt} \int_u^v F(x, t) \, dx. \end{aligned}$$

3.4.19. Suppose that $(X, \mathscr{A}, \mu)$ is a nonatomic measure space and $(Y, \mathscr{B}, \nu)$ is an arbitrary measure space such that there is a one-to-one map $f\colon X \to Y$ mapping measurable sets into measurable sets. Prove the following assertions:

(i) The set $D = \{(x, f(x))\colon x \in X\}$ is $\mathscr{M}$-measurable.

(ii) If $A \subset D$ and $\mu \times \nu(A) < \infty$, then $\mu \times \nu(A) = 0$.

3.4.20. Let $(X, \mathscr{A}, \mu)$ and $(Y, \mathscr{B}, \nu)$ be two complete and nonatomic measure spaces. Suppose also that there is a bijection $f\colon X \to Y$ taking and carrying back measurable sets into measurable sets such that

(a) $\mu(A) = \infty \Rightarrow \nu(f(A)) > 0$;

(b) $\nu(B) = \infty \Rightarrow \mu(f^{-1}(B)) > 0$.

Then show that Tonelli's Theorem is valid for $\mathscr{M}$-measurable functions and points (singleton sets) are measurable if and only if μ and ν are both σ-finite. [Hint: Use 3.4.19.]

3.4.21. Let μ and ν be two finite Borel measures on R. By Problem 3.1.5, we can define for any Borel subset $B \subset R$,

$$\mu * \nu(B) = \int \mu(B - x) \nu(dx).$$

Show that $\mu * \nu$ is a finite Borel measure on R and for every bounded Borel measurable function f on R,

$$\int f \, d\mu * \nu = \iint f(x + y) \, d\mu(x) \, d\nu(y).$$

Prove also the following assertions:

(i) If $\hat{\mu}(t) = \int e^{itx} \, d\mu(x)$, then $\widehat{\mu * \nu} = \hat{\mu} \cdot \hat{\nu}$.

(ii) If $\mu(R) = 1$ and for each $t \in R$, $| \hat{\mu}(t) |^2 = 1$, then μ is discrete.

3.4.22. Consider the Lebesgue measure on $(0, \infty)$. Let f be a strictly positive Lebesgue integrable function. Let $F(x) = (1/x) \int_0^x f(t) \, dt$, $0 < x < \infty$. Prove the following assertions:

(i) F is not Lebesgue integrable.

(ii) Let $1 < p < \infty$. If $| f |^p$ is Lebesgue integrable, then $| F |^p$ is Lebesgue integrable.

[Hint: Consider $\int_0^\infty F(x) \, dx$ and use Theorem 3.10.]

3.5. The L_p Spaces

Let $(X, \mathscr{A}, \mu)$ be a measure space and p a positive real number. By the L_p space, we mean the set of all real- or complex-valued measurable functions f such that $| f |^p$ is integrable. Let us recall (from linear algebra) that a linear (or vector) space V over a field F of scalars is an Abelian group under addition (+), together with a scalar multiplication from $F \times V$ into V such that

(i) $\alpha(x + y) = \alpha x + \alpha y$,

(ii) $(\alpha + \beta)x = \alpha x + \beta x$,

(iii) $(\alpha\beta)x = \alpha(\beta x)$,

(iv) $1x = x$,

for all α, β in F and x, y in V. (Here 1 denotes the multiplicative identity in F and 0 the additive identity in V.) In what follows, F will always be taken as either the field of reals or the field of complex numbers. Since for α, β in F

$$| \alpha f + \beta g |^p \leq 2^p(| \alpha |^p | f |^p + | \beta |^p | g |^p),$$

it is clear that L_p is a linear space. For $f \in L_p$, let

$$\| f \|_p = \left(\int | f |^p \, d\mu \right)^{1/p}.$$

It is then easy to see that, for a scalar α and $f \in L_p$,

(i) $\| f \|_p \geq 0$,
(ii) $\| f \|_p = 0 \Leftrightarrow f = 0$ a.e. (μ),
(iii) $\| \alpha f \|_p = | \alpha | \, \| f \|_p$.

It will be shown later in Proposition 3.14 that for $p \geq 1$, the function $\| \cdot \|_p$ satisfies a fourth property, namely,

(iv) $\| f + g \|_p \leq \| f \|_p + \| g \|_p$, $\quad f$ and g in L_p.

A function from a linear space into the reals with the above four properties is usually called a pseudonorm. Identifying functions in L_p that are equal a.e., the second property above reads as

(ii)′ $\| f \|_p = 0 \Leftrightarrow f = 0$.

In this situation, the function $\| \cdot \|_p$, $p \geq 1$, with properties (i), (ii)′, (iii), and (iv) is called a norm. The reader should notice that when we identify functions that differ only on a set of zero measure, we are really replacing the L_p space by a new space $\mathscr{L}_p$, whose elements are equivalence classes of functions in L_p with respect to the relation: $f \sim g \Leftrightarrow f = g$ a.e. Thus, if $\tilde{f} \in \mathscr{L}_p$ is the equivalence class containing f, then we define $\| \tilde{f} \|_p = \| g \|_p$, where g is any member of the equivalence class $\tilde{f}$. The space $\mathscr{L}_p$ is also linear, since

$$\tilde{f} + \tilde{g} = (f + g)^{\sim} \quad \text{and} \quad \alpha \tilde{f} = (\alpha f)^{\sim}.$$

From now on, without any further mention, we will write L_p to mean $\mathscr{L}_p$. Thus, $L_p (p \geq 1)$ is a normed linear space, which is also a metric space with the metric d defined by $d(f, g) = \| f - g \|_p$. One of the main aims in this section is to show that this metric is complete.

The reason we restrict ourselves to values of $p \geq 1$ is that for $p < 1$, the function $\| \cdot \|_p$ generally does not have the triangle property (iv) above. For example, if $p = 1/2$ and the measure space is the Lebesgue measure space on $[0, 1]$, then for $f = 4.\chi_{[0,1/2]}$, $g = 4.\chi_{[1/2,1]}$, we have

$$\| f + g \|_{1/2} = 4, \qquad \| f \|_{1/2} + \| g \|_{1/2} = 2.$$

But for $0 < p < 1$ and $f, g \in L_p$, if we define

$$d(f, g) = \int | f - g |^p \, d\mu,$$

then (L_p, d) becomes a metric linear space (i.e., a linear space with a metric topology where vector addition and scalar multiplication are continuous mappings). This is due to the fact that

$$x^p + y^p \geq (x + y)^p, \qquad 0 < p < 1, \qquad x \geq 0, \qquad y \geq 0.$$

M. M. Day† showed that even this consideration of L_p (as a metric linear space for $0 < p < 1$) fails to be useful in a certain sense. He proved that in most measure spaces any continuous linear scalar-valued function on the L_p spaces, $0 < p < 1$, is identically zero. Many important results in functional analysis depend on the existence of nontrivial continuous linear functionals (i.e., scalar-valued functions) and therefore are not available for these L_p spaces.

Let us now present the proofs of the triangle property and the completeness property of the L_p norm for $p \geq 1$. Though general normed linear spaces will not be treated in this volume, it will be convenient for us to present the first result in this section for such spaces.

Definition 3.12. A linear space over a field F of scalars (F is either the field of reals or the field of complex numbers) is called a normed linear space if to each $x \in X$ is associated a nonnegative real number $\| x \|$, called the norm of x such that

(i) $\| x \| = 0$, if and only if $x = 0$;
(ii) $\| \alpha x \| = | \alpha | \, \| x \|$, for all $\alpha \in F$;
(iii) $\| x + y \| \leq \| x \| + \| y \|$, for all $x, y \in X$.

A normed linear space X is called a Banach space if it is complete in the topology induced by the metric d defined by $d(x, y) = \| x - y \|$. ∎

Examples

3.9. R^n is a Banach space over the field of reals under usual vector addition and scalar multiplication, if we define the norm by $\| (a_1, a_2, \ldots, a_n) \| = \sum_{i=1}^{n} | a_i |$.

3.10. The space $C_1[0, 1]$ of complex-valued continuous functions on $[0, 1]$ is a Banach space over F, if we define $\| f \| = \sup_{0 \leq x \leq 1} | f(x) |$.

3.11. The set of all polynomials on $[0, 1]$ as a subspace of $C[0, 1]$ (with the same sup norm as in Example 3.10) is a normed linear space, but not a Banach space. The reason is, of course, that nonpolynomial continuous functions can be uniformly approximated in $[0, 1]$ by polynomials. (See Corollary 1.1.)

3.12. The space $l_p (p \geq 1)$ of all sequences $x = (x_i)$, $x_i \in F$ and

† M. M. Day, *Bull. Amer. Math. Soc.* **46**, 816–823 (1940).

$\sum | x_i |^p < \infty$, is a Banach space under natural addition and scalar multiplication and norm defined by $\| x \|_p = (\sum_{i=1}^{\infty} x_i^p)^{1/p}$. Note that l_p is a special case of L_p when X is the natural numbers and the measure of every point is 1. The proof that l_p spaces are Banach spaces is not trivial and will not be presented here. However, it will follow from the corresponding fact about the L_p spaces.

Definition 3.13. A series $\sum_{i=1}^{\infty} x_i$ in a normed linear space X over F is called summable if there is an $x \in X$ such that $\| \sum_{i=1}^{n} x_i - x \|$ goes to zero as $n \to \infty$. For a summable series, we write

$$\sum_{i=1}^{\infty} x_i = \lim_{n\to\infty} \sum_{i=1}^{n} x_i.$$

The series $\sum_{i=1}^{\infty} x_i$ is called absolutely summable if $\sum_{i=1}^{\infty} \| x_i \| < \infty$. ∎

We know that an absolutely summable series of real numbers is summable. This is a consequence of the completeness of the reals. The converse is also true. In fact, this result holds for normed linear spaces as the following proposition shows.

Proposition 3.12. Every absolutely summable series in a normed linear space X is summable if and only if X is complete. ∎

Proof. For the "if" part, let X be complete. Suppose that for each n, $x_n \in X$ and $\sum_{n=1}^{\infty} \| x_n \| < \infty$. Let $y_k = \sum_{n=1}^{k} x_n$. Then

$$\| y_{k+p} - y_k \| \leq \sum_{n=k+1}^{k+p} \| x_n \|,$$

which converges to zero as $k \to \infty$. Thus, (y_k) is a Cauchy sequence in X. Since X is complete, there is an x in X such that $\lim_{k\to\infty} y_k = x$. This proves the "if" part.

For the "only if" part, suppose that every absolutely summable series in X is summable. Let (x_n) be a Cauchy sequence in X. Then there is an increasing sequence (n_k) such that for $n > m \geq n_k$, $\| x_n - x_m \| < 1/2^k$. Let $z_1 = x_{n_1}$ and $z_k = x_{n_k} - x_{n_{k-1}}$ for $k \geq 2$. Then $\sum_{k=1}^{\infty} \| z_k \| < \infty$. Therefore, there is a z in X such that

$$\lim_{m\to\infty} x_{n_m} = \lim_{m\to\infty} \sum_{k=1}^{m} z_k = z.$$

Since (x_n) is Cauchy, $x_n \to z$ as $n \to \infty$. ∎

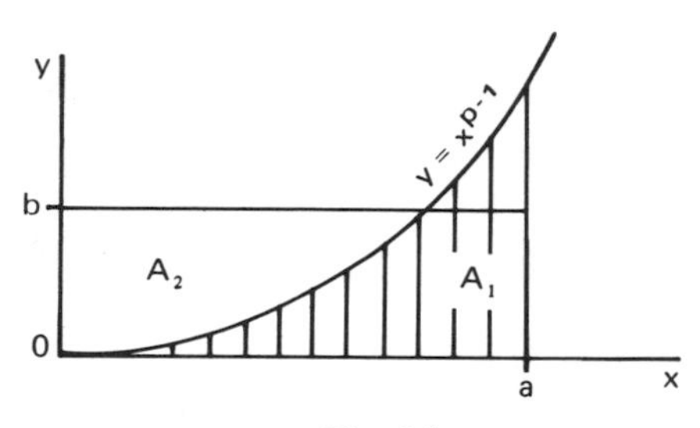

Fig. 3.1

Lemma 3.9. Let a and b be nonnegative reals and p, q be reals with $p > 1$, $q > 1$ and $1/p + 1/q = 1$. Then

$$ab \leq a^p/p + b^q/q,$$

with equality if and only if $a^p = b^q$. ▮

Proof. Suppose $a > b$. The function $y = x^{p-1}$ is sketched in Figure 3.1. Clearly, from the picture, $a \cdot b \leq A_1 + A_2$, where

$$A_1 = \int_0^a x^{p-1}\,dx = \frac{a^p}{p}$$

and

$$A_2 = \int_0^b x\,dy = \int_0^b y^{q-1}\,dy = \frac{b^q}{q}.$$

Hence $a \cdot b \leq a^p/p + b^q/q$. The case $a \leq b$ is taken care of similarly. Also, from the picture it is clear that $ab = A_1 + A_2$ if and only if $a^{p-1} = b$, i.e., $a^p = b^q$. ▮

Proposition 3.13. *Hölder Inequality.* Let p and q be real numbers with $p > 1$, $q > 1$ and $1/p + 1/q = 1$. If $f \in L_p, g \in L_q$, then $f \cdot g \in L_1$ and $\int |f \cdot g|\,d\mu \leq \|f\|_p \cdot \|g\|_q$, with equality if and only if $\alpha |f|^p = \beta |g|^q$ a.e. for some constants α and β, not both zero. ▮

Proof. Let $f_1 = f/\|f\|_p$ and $g_1 = g/\|g\|_q$. (The proposition is trivial when either $\|f\|_p$ or $\|g\|_q = 0$; so we may assume $\|f\|_p > 0$ and $\|g\|_q > 0$.) Now $\|f_1\|_p = \|g_1\|_q = 1$. Taking $a = |f_1|$, $b = |g_1|$ in Lemma 3.9, we have

$$|f_1 g_1| \leq \frac{|f_1|^p}{p} + \frac{|g_1|^q}{q}.$$

Therefore,

$$\int |f_1 g_1| \, d\mu \leq \frac{1}{p} \cdot \| f_1 \|_p^p + \frac{1}{q} \cdot \| g_1 \|_q^q,$$

which means

$$\int |fg| \, d\mu \leq \| f \|_p \, \| g \|_q.$$

The rest is left to the reader. ∎

Proposition 3.14. *Minkowski Inequality.* If f and g are in L_p, $p \geq 1$, then

$$\| f + g \|_p \leq \| f \|_p + \| g \|_p. \qquad ∎$$

Proof. The case $p = 1$ is trivial. Hence we assume $p > 1$. Then

$$\int |f + g|^p \, d\mu \leq \int |f + g|^{p-1} |f| \, d\mu + \int |f + g|^{p-1} |g| \, d\mu.$$

By the Hölder Inequality (taking $q = p/(p - 1)$),

$$\int |f + g|^{p-1} |f| \, d\mu \leq \| f \|_p \cdot \| \, |f + g|^{p-1} \, \|_q$$

and

$$\int |f + g|^{p-1} |g| \, d\mu \leq \| g \|_p \, \| \, |f + g|^{p-1} \, \|_q.$$

Since

$$\| \, |f + g|^{p-1} \, \|_q = (\| f + g \|_p)^{p-1},$$

the proposition follows from the last two inequalities. ∎

Theorem 3.12. (*Riesz–Fischer*). For $p \geq 1$, $(L_p, \| \cdot \|_p)$ is a Banach space. ∎

Proof. We only need to show that $(L_p, \| \cdot \|_p)$ is complete. We will use Proposition 3.12. Let $(f_n)_{n=1}^{\infty}$ be a sequence in L_p with $\sum_{n=1}^{\infty} \| f_n \|_p = a < \infty$. We must show that $\sum_{n=1}^{\infty} f_n$ is summable in L_p.

Set $g_n(x) = \sum_{k=1}^{n} | f_k(x) |$. Hence, by the Minkowski Inequality,

$$\| g_n \|_p \leq \sum_{k=1}^{n} \| f_k \|_p \leq a$$

or

$$\int g_n{}^p \, d\mu \leq a^p.$$

Let $\lim_{n\to\infty} g_n(x) = g(x)$. By Fatou's Lemma,

$$\int g^p \, d\mu \leq a^p.$$

This means that $g(x)$ is finite a.e. Now since $\sum_{k=1}^{\infty} | f_k(x) | < \infty$ a.e., there is a function $h(x)$ with $| h(x) |$ finite a.e. and $\sum_{k=1}^{\infty} f_k(x) = h(x)$. Since

$$\left| \sum_{k=1}^{n} f_k(x) - h(x) \right|^p \leq 2^p [g(x)]^p,$$

by the Lebesgue Convergence Theorem

$$\left\| \sum_{k=1}^{n} f_k(x) - h(x) \right\|_p \to 0, \qquad \text{as } n \to \infty.$$

This completes the proof. ∎

So far we have considered p to be a real number ≥ 1. We now take $p = \infty$ and denote by L_∞ the space of all measurable functions that are bounded except possibly on a set of measure zero. As before, we identify functions in L_∞ that are equal a.e. and L_∞ becomes a complete normed linear space if we define

$$\|f\|_\infty = \text{ess sup}\,|f|, \qquad \text{where ess sup}\,|f| = \inf\{M\colon \mu\{x\colon |f(x)| > M\} = 0\}.$$

(See Problem 3.5.1.) The Hölder Inequality, as well as the Riesz–Fischer theorem, remains true when $p = \infty$ and $q = 1$.

When the measure space X is the set of positive integers with each integer having measure 1, the L_p spaces are, as mentioned in Example 3.12, called the l_p spaces. Thus, we define the l_p space $(1 \leq p < \infty)$ as the class of all sequences $x = (x_n)_{n=1}^{\infty}$, $x_n \in F$ with $\sum_{n=1}^{\infty} | x_n |^p < \infty$. Then l_p, like L_p, is a Banach space under the norm $\| x \|_p = (\sum | x_n |^p)^{1/p}$. Also, we define the l_∞ space as the class of all bounded sequences $x = (x_n)_{n=1}^{\infty}$, $x_n \in F$. Again l_∞, like L_∞, is a Banach space under the norm $\| x \|_\infty = \sup_n | x_n |$.

Finally, in this section we study an important theorem on the convergence of certain arithmetic means of L_1 functions.

• The Individual Ergodic Theorem

Many interesting problems in mathematics and physics arise naturally in the context of L_p, $p \geq 1$ spaces. The problem of the convergence of certain arithmetic means is one such problem in ergodic theory. To be more specific, we need the following concept.

Let $(X, \mathscr{A}, \mu)$ be a measure space. Then a mapping $\Phi: X \to X$ is called *measure preserving* if

(i) Φ is bijective;
(ii) $A \in \mathscr{A} \Leftrightarrow \Phi(A) \in \mathscr{A}$;
(iii) $\mu(\Phi(A)) = \mu(A), \qquad A \in \mathscr{A}$.

A simple example of a measure-preserving mapping is $\Phi(x) = x + t$, where the measure is the Lebesgue measure on R.

Let T be a linear operator on the linear space of measurable functions defined by

$$T(f)(x) = f(\Phi(x)),$$

where Φ is a measure-preserving mapping on the measure space $(X, \mathscr{A}, \mu)$. It can be verified easily that T is one-to-one and onto.

An important concern of ergodic theory is to find when the averages

$$\frac{1}{n} \sum_{k=0}^{n-1} T^k(f)$$

converge for various classes of measurable functions f. The individual ergodic theorem solves this problem for L_1 functions with respect to pointwise convergence. This theorem was first proven by G. D. Birkhoff in 1931. We present below F. Riesz' proof of this theorem.

Theorem 3.13. *The Individual Ergodic Theorem.* Let $(X, \mathscr{A}, \mu)$ be a finite measure space and Φ a measure-preserving mapping on X. Then for $f \in L_1$, the sequence

$$T_n(f)(x) = \frac{1}{n} \sum_{k=0}^{n-1} f(\Phi^k(x))$$

converges pointwise a.e. ∎

Proof. The proof will be given in two steps. It is no loss of generality to assume that f is real valued.

Step I. In this step, we show that

$$\int_E f\,d\mu \geq 0,$$

where

$$E = \left\{x\colon \sum_{k=0}^{n} f(\Phi^k(x)) \geq 0, \text{ for some } n = 0, 1, 2, \ldots\right\}.$$

To show this, we consider the set

$$E_m = \left\{x\colon \sum_{k=0}^{n} f(\Phi^k(x)) \geq 0, \text{ for some } n \leq m\right\}.$$

Then it is enough to show that

$$\int_{E_m} f\,d\mu \geq 0, \qquad m = 1, 2, \ldots .$$

Let m be a fixed positive integer and n be any given positive integer. For $k \leq n$, we write

$$F_k = \{x\colon f(\Phi^k(x)) + \cdots + f(\Phi^{k+i-1}(x)) \geq 0, \quad \text{for some } i \leq m\}.$$

For $k > n$, we write

$$F_k = \{x\colon f(\Phi^k(x)) + \cdots + f(\Phi^{k+i-1}(x)) \geq 0, \quad \text{for some } i \text{ such that } k+i-1 \leq n+m-1\}.$$

Then it is clear that for each x

$$\sum_{k=0}^{n+m-1} \chi_{F_k}(x) f(\Phi^k(x)) \geq 0.$$

The reason is that, in the above summation, if a particular term (say, the kth term) is negative, then there is a positive integer $i(k)$ such that $k + i(k) - 1 \leq n + m - 1$ and

$$\sum_{j=k}^{k+i(k)-1} \chi_{F_j}(x) f(\Phi^j(x)) \geq 0.$$

Thus the above summation is the sum of a number of nonnegative subsums. Since Φ is bijective, $F_0 = E_m$ and $\Phi(F_k) = F_{k-1}$, $k \leq n - 1$. This means that

$$n \int_{E_m} f\,d\mu + \sum_{k=n}^{n+m-1} \int_{F_k} f \circ \Phi^k\,d\mu \geq 0.$$

It follows that

$$n \int_{E_m} f \, d\mu + m \int | f | \, d\mu \geq 0.$$

Since this inequality is valid for all positive integers n, it follows that $\int_{E_m} f \, d\mu \geq 0$. Thus we have proven step I.

Step II. Let r, s be rationals and

$$E_{rs} = \left\{ x\colon \liminf_{n\to\infty} T_n(f)(x) < r < s < \limsup_{n\to\infty} T_n(f)(x) \right\}.$$

The theorem will be proven if $\mu(E_{rs}) = 0$ for all rationals r and s. It is clear that $\Phi(E_{rs}) = E_{rs}$. Therefore, replacing X by E_{rs} and applying step I, we have

$$\int_{E_{rs}} (f - s) \, d\mu \geq 0, \qquad \int_{E_{rs}} (r - f) \, d\mu \geq 0.$$

This means that $\int_{E_{rs}} (r - s) \, d\mu \geq 0$, proving that $\mu(E_{rs}) = 0$.

Problems

× **3.5.1.** Prove that L_∞ is a Banach space under the norm $\| f \|_\infty = \text{ess sup} | f |$.

× **3.5.2.** Let $f \in L_p$, $1 \leq p < \infty$. Then prove the following.

(a) There exists a sequence $f_n \in L_p$, each vanishing outside a set of finite measure, $| f_n | \leq | f |$ for all n and $\| f_n - f \|_p \to 0$ as $n \to \infty$.

(b) Given $\varepsilon > 0$, there exists $g \in L_p$ such that $| g |$ is bounded, $| g | \leq | f |$ and $\| g - f \|_p < \varepsilon$.

(c) Given ε positive, there exists a simple function g in L_p such that $| g | \leq | f |$ and $\| g - f \|_p \leq \varepsilon$.

× **3.5.3.** In a finite-measure space, every L_p function is in L_q whenever $1 \leq q \leq p \leq \infty$. Prove this and then show that this is false in an infinite-measure space. (In this context, see also Problem 3.5.16.)

3.5.4. Show that in any measure space for $f \in L_1 \cap L_\infty$, the limit of $\| f \|_p$ is $\| f \|_\infty$ as $p \to \infty$. (Hint: $\int | f |^{p+1} \, d\mu \leq \| f \|_\infty^p \int | f | \, d\mu$.)

3.5.5. Let (f_n) be a sequence of L_p $(1 \leq p < \infty)$ functions, each vanishing outside a set E_n, where (E_n) is a sequence of pairwise disjoint measurable sets. Show that $f = \sum_{n=1}^\infty f_n$ is in L_p if and only if $\sum_{n=1}^\infty \| f_n \|_p{}^p < \infty$, and in this case $\| f - \sum_{n=1}^k f_n \|_p \to 0$ as $k \to \infty$.

× **3.5.6.** Let g be a real-valued integrable function in a finite-measure space such that for some constant M and all simple functions h, $| \int gh \, d\mu |$

$\leq M \| h \|_p$, where $1 \leq p < \infty$. Show that g is in L_q, where $1/p + 1/q = 1$. [Hint: Let $p > 1$ and (g_n) be a sequence of nonnegative simple functions which increase to $| g |^q$. Set $h_n = (g_n)^{1/p} \cdot g/| g |$ when $g \neq 0$, and $= 0$ when $g = 0$. Then $g_n \leq h_n g$ and $\int g_n \, d\mu \leq M^q$.]

3.5.7. Let $1 \leq p < \infty$ and let (f_n) be a sequence in L_p converging a.e. to a function f in L_p. Show that $\| f_n - f \|_p \to 0$ as $n \to \infty$ if and only if $\| f_n \|_p \to \| f \|_p$ as $n \to \infty$. (Hint: The "if" part is nontrivial. Use Theorem 3.4.)

3.5.8. Let $f \in L_p$ and (f_n) be a sequence in L_p, where $1 < p < \infty$, and $f_n \to f$ a.e. Suppose also that for some constant M, $\| f_n \|_p \leq M$ for all n. Then show that for each function $g \in L_q$, $1/p + 1/q = 1$,

$$\int f \cdot g \, d\mu = \lim_{n \to \infty} \int f_n \cdot g \, d\mu.$$

This result is false for $p = 1$. Give an example. [Hint: For this problem use the Hölder inequality and reduce the problem to one in a finite-measure space, then use Egoroff's Theorem.]

3.5.9. Let $1 \leq p \leq \infty$ and $1/p + 1/q = 1$. Let μ be a σ-finite measure and let g be a measurable function such that $f \cdot g \in L_1$ for every $f \in L_p$. Prove that $g \in L_q$.

3.5.10. Let $\| f_n - f \|_p \to 0$ as $n \to \infty$, $1 \leq p < \infty$, where (f_n) is a sequence in L_p and $f \in L_p$. Suppose (g_n) is a sequence of measurable functions such that $| g_n | \leq M$ for some constant M, for all n and $g_n \to g$ a.e. Show that $\| g_n f_n - g f \|_p \to 0$ as $n \to \infty$.

3.5.11. *A Generalization of the Lebesgue Convergence Theorem* (W. R. Wade). Let $(X, \mathscr{A}, \mu)$ be a complete finite-measure space, and let (f_n) be a sequence of measurable functions such that for some r with $0 < r \leq \infty$, $\| f_n \|_r \leq M$ for all n. If $\lim_{n \to \infty} f_n(x) = f(x)$ and $0 < p < r$, then $\lim_{n \to \infty} \| f_n - f \|_p = 0$. This result is false if $p = r$ or if $\mu(X) = \infty$. (Hint: Let $0 < r < \infty$. Use Fatou's Lemma to show that f^p is integrable for $0 < p < r$; then use Egoroff's Theorem and the Hölder Inequality.)

3.5.12. *An Infinite-Measure Version of Problem 3.5.11* (W. R. Wade). Let $(X, \mathscr{A}, \mu)$ be a complete measure space. Suppose as in Problem 3.5.11 that $0 < r \leq \infty$, $\| f_n \|_r \leq M$ for all n, and $\lim_{n \to \infty} f_n(x) = f(x)$ a.e. If $0 < p < r$ and $1/q + p/r = 1$, then $\lim_{n \to \infty} \| f_n \cdot g - f \cdot g \|_p = 0$ for every $g \in L_q$.

3.5.13. *A Stone-Weierstrass Theorem for L_2-Valued Real Functions.* Let $(X, \mathscr{A}, \mu)$ be a finite-measure space with $\mu(X) = 1$, and let $x: T \to L_2(\mu)$ be a mapping where

(i) T is a compact set in some topological space

and

(ii) $\lim_{s\to t} \int |x_t - x_s|^2 \, d\mu = 0$, $x_t \equiv x(t)$.

Such a map is called *mean-square-continuous* (m.s.c.). A family of uniformly bounded [i.e., $|x_t(\cdot)| \le M$ for all t] m.s.c. maps $[x_t(\cdot)]$ is called an *algebra* if it is closed under pointwise multiplication and finite linear combinations. An algebra $\mathscr{W}$ is said to separate points of T if, given $t_1, t_2 \in T$, there exists $x_t(\cdot) \in \mathscr{W}$ such that $|x_{t_1}(\cdot) - x_{t_2}(\cdot)| = 1$ a.e. Prove the following result: let $\mathscr{W}$ be an algebra of uniformly bounded m.s.c. maps on T containing all bounded constant maps [i.e., $x_t(\cdot) = a(\cdot)$ for some $a(\cdot) \in L_2$, $|a(\cdot)| \le M$] and separating points of T. Then given any m.s.c. map $y_t(\cdot)$, there exists a sequence $x_t^{(n)}(\cdot) \in \mathscr{W}$ converging in mean square to $y_t(\cdot)$ uniformly on T. (Hint: The proof can follow the same lines as used in that of Theorem 1.28.)

3.5.14. *Necessary and Sufficient Conditions for Convergence in L_p.* Let (f_n) be a sequence in $L_p(X, \mathscr{A}, \mu)$, $1 \le p < \infty$ and $\lim_{n\to\infty} f_n = f$ a.e., where f is a real-valued measurable function. Prove that the following are equivalent:

(a) $f \in L_p$ and $\lim_{n\to\infty} \| f_n - f \|_p = 0$.

(b) Whenever (A_n) is a decreasing sequence of sets in $\mathscr{A}$ with $\mu(\bigcap_{n=1}^\infty A_n) = 0$, then $\lim_{k\to\infty} \int_{A_k} |f_n|^p \, d\mu = 0$ uniformly in n.

(c) (i) For each $\varepsilon > 0$, there exists $A_\varepsilon \in \mathscr{A}$ such that $\mu(A_\varepsilon) < \infty$ and $\int_{X-A_\varepsilon} |f_n|^p \, d\mu < \varepsilon$ for all n; and

(ii) for each $\varepsilon > 0$, there exists a $\delta > 0$ such that $\mu(A) < \delta$ implies $\int_A |f_n|^p \, d\mu < \varepsilon$ for all n.

(d) The condition (i) in (c) holds and $\lim_{k\to\infty} \int_{A_{nk}} |f_n|^p \, d\mu = 0$ uniformly in n, where $A_{nk} = \{x: |f_n(x)|^p \ge k\}$.

[Hint: Show $(a) \Leftrightarrow (c)$; then show $(a) \Rightarrow (b) \Rightarrow (d) \Rightarrow (c)$.]

3.5.15. Let $(X, \mathscr{A}, \mu)$ be a measure space and a, b positive real numbers such that $a + b = 1$. Prove that if f and g are integrable, then $|f|^a \cdot |g|^b$ is integrable and $\int |f|^a \cdot |g|^b \, d\mu \le \| f \|_1^a \cdot \| g \|_1^b$. Extend this result to n functions.

3.5.16. *On the Inclusion $L_p \subset L_q$.* Let $(X, \mathscr{A}, \mu)$ be a measure space and p, q positive real numbers such that $0 < p < q$. Prove the following assertions due to B. Subramanian:

(i) If $L_1 \subset L_2$, then $L_1 \subset L_\infty$.

(ii) Let $\mu(X) < \infty$ and $L_1 \subset L_2$. Then any collection of disjoint measurable sets of positive measure is finite.

(iii) Let $\mu(X) < \infty$. Suppose that any collection of disjoint measurable sets of positive measure is finite. Then $L_p = L_q = L_\infty$.

(iv) Let $\mu(X) = \infty$. Then $L_p \subset L_q$ if and only if for any sequence (E_n) of disjoint measurable sets of positive measure, the sequence $(\mu(E_n))$ is bounded away from zero.

× **3.5.17.** Consider the Lebesgue measure space on R. Prove that if $f \in L_p$ $(1 \leq p < \infty)$ and $\varepsilon > 0$, then there is a continuous function g vanishing outside a finite interval such that $\| f - g \|_p < \varepsilon$. Use this to prove that

$$\lim_{t \downarrow 0} \| f_t - f \|_p = \lim_{t \uparrow 0} \| f_t - f \|_p = 0,$$

where $f_t(x) = f(x + t)$. (Let us mention that the first part of this problem holds for locally compact Hausdorff spaces and regular measures. See Problem 5.3.12.)

× **3.5.18.** Let $(X, \mathscr{A}, \mu)$ be any semifinite measure space. Let $g \in L_\infty$ and for each $f \in L_1$, $\Phi_g(f) = \int fg \, d\mu$. (Here the scalar field F is taken to be the reals.) Show that

$$\| g \|_\infty = \sup\{| \Phi_g(f) | : \| f \|_1 \leq 1\}.$$

[Hint: If $A = \{x : | g(x) | > \| g \|_\infty - \varepsilon\}$, then $\mu(A) > 0$. Let $E \subset A$ be such that $0 < \mu(E) < \infty$. Define $f = [1/\mu(E)] \cdot \chi_E$ when $g > 0$ and $f = -[1/\mu(E)] \cdot \chi_E$ when $g \leq 0$. Consider $\Phi_g(f)$.]

× **3.5.19.** Let $(X, \mathscr{A}, \mu)$ be any measure space and $g \in L_q$, $q > 1$. Let $p > 1$ and $1/p + 1/q = 1$. For each $f \in L_p$, let $\Phi_g(f) = \int fg \, d\mu$. Show that

$$\| g \|_q = \sup\{| \Phi_g(f) | : \| f \|_p \leq 1\}.$$

(All functions here are real functions.) [Hint: Use Hölder's inequality and then consider $f = g^{q-1}/\| g \|_q^{q-1}$ when $g > 0$ and $f = - g^{q-1}/\| g \|_q^{q-1}$.]

× **3.5.20.** Let $(X, \mathscr{A}, \mu)$ be any measure space and g be a real integrable function. Define Φ_g on L_∞ by $\Phi_g(f) = \int fg \, d\mu$. Show that

$$\| f \|_1 = \sup\{| \Phi_g(f) | : \| f \|_\infty \leq 1\}.$$

× **3.5.21.** Show that the L_p space, $1 \leq p < \infty$, of Lebesgue-measurable functions on R is a separable Banach space. [Hint: Use Problem 3.5.17 and recall that polynomials with rational coefficients uniformly approximate continuous functions on a finite interval.]

3.5.22. Show that the l_p space, $1 \leq p < \infty$, is a separable Banach space.

× **3.5.23.** Show that l_∞, as well as L_∞ (over the Lebesgue measure space), is not separable in its usual norm. [Hint: Consider the family $\{\chi_{[0,x]}: 0 < x < 1\}$ in L_∞. Note that for any f, g in this family, $\| f - g \|_\infty = 1$.]

× **3.5.24.** Show that the L_p space, $1 \le p \le \infty$, over an arbitrary measure space is not, in general, separable in its usual norm. [Hint: Let S be an uncountable set with the counting measure. Then for any s, t in S, $\| \chi_{\{s\}} - \chi_{\{t\}} \|_p = 2^{1/p}$.]

★ **3.5.25.** Let $1 \le p < \infty$ and f be a real function in $L_p[0, 1]$ over the Lebesgue measure space. For $a > 0$, define f_a on $[0, 1]$ by

$$f_a(x) = (1/2a) \cdot \int_{x-a}^{x+a} f(t)\, dt.$$

Prove that f_a is a well-defined continuous function such that

(i) $| f_a(x) | \le (2a)^{-1/p} \cdot \| f \|_p$, $x \in [0, 1]$, and

(ii) $\| f_a \|_p \le \| f \|_p$.

Finally show that a closed and bounded subset $S \subset L_p[0, 1]$ is compact in the L_p norm if and only if for each $\varepsilon > 0$ there exists a $\delta > 0$ such that for all $f \in S$ and $0 < a < \delta$, $\| f - f_a \|_p < \varepsilon$.

4

Differentiation

There are two theorems from the theory of Riemann integration which indicate the sense in which integration and differentiation are inverse processes. They read as follows:

(1) If f is Riemann integrable on $[a, b]$ and F is the function given by the equation

$$F(x) = \int_a^x f(t)\, dt \qquad \text{for } x \in [a, b]$$

then F is continuous on $[a, b]$; also if f is continuous at x_0 in $[a, b]$, then F is differentiable at x_0 and $F'(x_0) = f(x_0)$.

(2) If f is differentiable on $[a, b]$ and f' is Riemann integrable on $[a, b]$, then

$$\int_a^b f'(x)\, dx = f(b) - f(a).$$

In this chapter we shall first explore the analogs of these theorems in the theory of Lebesgue integration of real-valued functions on an interval $[a, b]$. In doing so, we shall study absolutely continuous functions and functions of bounded variation and prove Lebesgue's well-known result that every monotone increasing function is differentiable almost everywhere. This exploration will lead us to more abstract notions of indefinite integrals and absolute continuity, and in particular we shall prove the famous Radon–Nikodyn Theorem. This outstanding result has many uses in measure theory and probability theory. At the close of this chapter, we shall present two applications of this key theorem—namely, the Riesz Representation Theorem and a change of variable formula for integration.

4.1. Differentiation of Real-Valued Functions

In order to define the derivative of a real function f on R, we first define four quantities known as the *derivates* of f at a point a.

Definition 4.1. Let $a \in R$ and let δ be a real number such that $\delta > 0$. If f is defined in the interval $[a, a+\delta)$, the *upper right derivate* $D^+f(a)$ and the *lower right derivate* $D_+f(a)$ of f at a are the extended real numbers given by

$$D^+f(a) = \limsup_{h \to 0^+} \frac{f(a+h) - f(a)}{h}$$

and

$$D_+f(a) = \liminf_{h \to 0^+} \frac{f(a+h) - f(a)}{h}.$$

If f is defined in the interval $(a-\delta, a]$, the *upper left derivate* $D^-f(a)$ and the *lower left derivate* $D_-f(a)$ of f at a are the extended real numbers given by

$$D^-f(a) = \limsup_{h \to 0^+} \frac{f(a) - f(a-h)}{h}$$

and

$$D_-f(a) = \liminf_{h \to 0^+} \frac{f(a) - f(a-h)}{h}.$$

If $D^+f(a) = D_+f(a)$, then the common value is called the *right derivative* $f_+'(a)$ of f at a, and f is said to be *right differentiable* at a if this common value is finite. In a similar fashion, the *left derivative* $f_-'(a)$ is defined as well as *left differentiability*. If $f_+'(a)$ and $f_-'(a)$ exist and are equal, their common value $f'(a)$ is the *derivative* of f at a, and f is *differentiable* at a if this common value $f'(a)$ is finite. ∎

It is clear that $D_+f(a) \leq D^+f(a)$ and $D_-f(a) \leq D^-f(a)$.

Proposition 4.1. Let f be a nondecreasing function on a bounded or unbounded interval I. The four derivate functions of f are measurable functions on I (defined except perhaps at the endpoints of I). ∎

Proof. We will show that the function D^+f whose value at a in I (a not an endpoint of I) is $D^+f(a)$ is a measurable function. We may assume for convenience that I is open. The proofs that the other derivates are measurable are similar.

By definition

$$D^+f(a) = \limsup_{h \to 0^+} \frac{f(a+h) - f(a)}{h}.$$

For each positive integer n the function f_n given by

$$f_n(a) = \sup_{\substack{0<h<1/n \\ a+h\in I}} \frac{f(a+h) - f(a)}{h} \tag{4.1}$$

is a well-defined extended real-valued function on I. Also

$$D^+f(a) = \lim_{n\to\infty} f_n(a),$$

so that our proof will be complete if we can show that for all values of n, f_n is a measurable function.

Consider the function $\hat{f}_n$ defined on the interval I by

$$\hat{f}_n(a) = \sup_{\substack{0<h_i<1/n \\ a+h_i\in I \\ h_i \text{ rational}}} \frac{f(a+h_i) - f(a)}{h_i}.$$

Since $\hat{f}_n$ is thus the pointwise supremum of a sequence (g_i) of measurable functions, where g_i is given by

$$g_i(a) = \begin{cases} \dfrac{f(a+h_i) - f(a)}{h_i}, & \text{if } a + h_i \in I, \\ 0, & \text{otherwise,} \end{cases}$$

each $\hat{f}_n$ is measurable. The proof is therefore reduced to showing that $f_n = \hat{f}_n$.

Clearly, $\hat{f}_n \leq f_n$ for all n. Now let n be fixed and ε be an arbitrary positive number. Suppose first that $f_n(a) < \infty$. From equation (4.1), there exists an element h_0 [or $h_0(a)$] in $(0, 1/n)$ such that $a + h_0 \in I$ and

$$f_n(a) - \frac{\varepsilon}{2} < \frac{f(a+h_0) - f(a)}{h_0}.$$

Let r be a rational number in $(h_0, 1/n)$. Since f is nondecreasing,

$$f_n(a) - \frac{\varepsilon}{2} < \frac{f(a+h_0) - f(a)}{h_0} \leq \frac{f(a+r) - f(a)}{h_0}. \tag{4.2}$$

Multiplying the inequalities (4.2) by h_0/r, we obtain the inequality

$$f_n(a) + \left(\frac{h_0}{r} - 1\right) f_n(a) - \frac{\varepsilon h_0}{2r} < \frac{f(a+r) - f(a)}{r}. \tag{4.3}$$

Choose r very close to h_0 in $(h_0, 1/n)$; then (4.3) becomes

$$f_n(a) - \varepsilon < \frac{f(a+r) - f(a)}{r}$$

when r is chosen, so that $(h_0/r - 1)f_n(a) > -\varepsilon/2$. Thus, $f_n(a) \leq \hat{f}_n(a)$ when $\hat{f}_n(a) < \infty$. It is easy to see that $\hat{f}_n(a) = \infty$ whenever $f_n(a) = \infty$. Hence $f_n \leq \hat{f}_n$. ∎

Proposition 4.2. Let (a, b) be an open interval of R and f a real-valued function defined on (a, b). The set of all points in (a, b), where $f_-'(x)$ and $f_+'(x)$ both exist (although possibly infinite) and are not equal, is at most countable. ∎

Proof. Let $A = \{x \in (a, b): f_-'(x)$ and $f_+'(x)$ exist but $f_-'(x) < f_+'(x)\}$. For each x in A choose a rational number r_x such that $f_-'(x) < r_x < f_+'(x)$. Also choose rational numbers s_x and t_x such that $a < s_x < x$ and $x < t_x < b$ and such that

$$r_x < \frac{f(y) - f(x)}{y - x}, \qquad \text{for all } y \text{ in } (x, t_x) \tag{4.4}$$

and

$$\frac{f(y) - f(x)}{y - x} < r_x, \qquad \text{for all } y \text{ in } (s_x, x). \tag{4.5}$$

The inequalities (4.4) and (4.5) together give

$$f(y) - f(x) > r_x(y - x) \tag{4.6}$$

for all $y \neq x$ in (s_x, t_x). Define φ: $A \to Q \times Q \times Q$ by

$$\varphi(x) = (s_x, r_x, t_x).$$

Then φ is a one-to-one function since if $\varphi(x) = \varphi(z)$ for x and z in A, then the intervals (s_x, t_x) and (s_z, t_z) are identical and x and z are both in this interval. If $x \neq z$, then (4.6) gives us

$$f(z) - f(x) > r_x(z - x)$$

and

$$f(x) - f(z) > r_z(x - z),$$

whereas the addition of these inequalities establishes that

$$0 > 0.$$

Thus φ is a one-to-one function and A must be at most countable. The proof that the set

$$B = \{x \in (a, b): f_-'(x) \text{ and } f_+'(x) \text{ exist but } f_-'(x) > f_+'(x)\}$$

is at most countable is similar. ∎

It is easy to construct a function continuous on an interval $[a, b]$ but which fails to have a derivative at any given finite number of points in $[a, b]$. Perhaps contrary to what we would imagine, however, continuous functions can be found whose derivatives do not exist at points of a dense subset of the interval (see Problem 4.1.3). Furthermore, if a is a positive odd integer and $0 < b < 1$, the function f given by the series

$$f(x) = \sum_{n=0}^{\infty} b^n \cos(a^n \pi x)$$

is a continuous function; but if $ab > 1 + \frac{3}{2}\pi$ the function f is not differentiable at *any* point. This example is due to Weierstrass (see [22]). Another example of an everywhere continuous but nowhere differentiable function on R is given in problem 4.1.7. A careful examination of these functions reveals that they are continuous but not monotonic on *any* interval. That this should be the case is evident from the theorem we prove below that a real-valued monotonic function on an interval is differentiable almost everywhere. To prove this we use the theorem below of Vitali. In this theorem we need the following terminology.

Definition 4.2. Let $E \subset R$. A family $\mathscr{F}$ of intervals, each having positive length, is said to be a *Vitali covering* of E if for each $\varepsilon > 0$ and any $x \in E$ there is an interval $I \in \mathscr{F}$ such that $x \in I$ and $l(I) < \varepsilon$. ∎

To give an example, let $E = [a, b]$. The collection of $I_{n,i} = [r_n - 1/i, r_n + 1/i]$ for $n, i = 1, 2, 3, \ldots$, where r_n is an enumeration of the rationals in E, is a Vitali covering of E.

Theorem 4.1. *Vitali's Covering Theorem.*† Let E be an arbitrary subset of R and let $\mathscr{F}$ be a family of closed intervals of positive length forming a Vitali covering of E. Let μ be any measure on R defined on a σ-algebra containing all intervals in R such that $\mu^*(E) < \infty$, where μ^* is the outer measure induced by μ. Then there exists a sequence (I_n) of disjoint intervals in $\mathscr{F}$ such that

$$\lim_{m\to\infty} \mu^*\left(E - \bigcup_{n=1}^{m} I_n\right) = \mu^*\left(E - \bigcup_{n=1}^{\infty} I_n\right) = 0. \quad ∎$$

† This general form of Vitali's Theorem is due to Miguel de Guzman.

(When μ is the Lebesgue measure, this theorem holds even when the intervals in $\mathscr{F}$ are not closed since the endpoints of an interval have Lebesgue measure zero.)

Proof. The proof is given in the following three steps.

Step I. Let A be any subset of R. Suppose that for each $x \in A$, there corresponds a finite interval (of positive length) I_x containing x. Then there exists a sequence (x_i) in A such that $A \subset \cup I_{x_i}$.

Because of Remark 1.46, there is no loss of generality in assuming that each x in A is either the left or right endpoint of the corresponding I_x. It is sufficient to give the argument when each x in A is a left endpoint of the corresponding I_x.

For each integer n and positive integer k, let us write $A_k = \{x \in A: \text{diam } I_x > 1/k\}$ and $a_{nk} = \inf\{x: x \in A_k \cap [n/k, (n+1)/k]\}$, whenever $A_k \cap [n/k, (n+1)/k]$ is not empty. When a_{nk} is thus defined let $(x^j_{nk}) \subset A_k$ and $x^j_{nk} \to a_{nk+}$. Then we have

$$A_k \cap [n/k, (n+1)/k] \subset \bigcup_j I_{x^j_{nk}}.$$

By taking the union over all n and k for which $A_k \cap [n/k, (n+1)/k] \neq \varnothing$, the assertion of this step follows.

Step. II. From any finite collection of intervals in R, one can select two disjoint subcollections of disjoint intervals such that the union of the intervals in the subcollections contains the intervals in the given collection.

To prove this step, there is no loss of generality in assuming that the given collection is $\{I_1, I_2, \ldots, I_n\}$ such that for each i, $I_i \cap (\bigcup_{j\neq i} I_j)^c$ is nonempty and contains some point a_i. We can also assume (after renaming the intervals if necessary) that the a_i are arranged in increasing order. Then $\{I_1, I_3, I_5, \ldots\}$ and $\{I_2, I_4, I_6, \ldots\}$ are subcollections of disjoint intervals meeting the requirements of this step.

Step III. In this step, we complete the proof of the theorem. By Step I, there is a sequence (J_i) of intervals in $\mathscr{F}$ such that $E \subset \cup J_i$. Then using Problem 2.3.2, we have

$$\infty > \mu^*(E) = \mu^*\left(E \cap \bigcup_{i=1}^{\infty} J_i\right) = \lim_{n\to\infty} \mu^*\left(E \cap \bigcup_{i=1}^{n} J_i\right),$$

and we can choose N so that

$$\mu^*\left(E \cap \bigcup_{i=1}^{N} J_i\right) > \frac{3}{4} \cdot \mu^*(E).$$

By Step II, we can choose from the collection (J_i), $1 \leq i \leq N$, a subcollection of disjoint intervals (I_i), $1 \leq i \leq n_1$, such that

$$\mu^*\left(E \cap \bigcup_{i=1}^{n_1} I_i\right) > \frac{3}{8} \cdot \mu^*(E). \tag{4.7}$$

Now let B be a μ^*-measurable set such that $B \supset E$ and $\mu^*(B) = \mu^*(E)$. Then we have

$$\mu^*\left(B - \bigcup_{i=1}^{n_1} I_i\right) = \mu^*(B) - \mu^*\left(B \cap \bigcup_{i=1}^{n_1} I_i\right) \leq \mu^*(B) - \mu^*\left(E \cap \bigcup_{i=1}^{n_1} I_i\right),$$

and therefore,

$$\mu^*\left(E - \bigcup_{i=1}^{n_1} I_i\right) \leq \left(1 - \frac{3}{8}\right)\mu^*(E),$$

by (4.7) and above. Since $\bigcup_{i=1}^{n_1} I_i$ is closed, for each x in the set $E_1 = E - \bigcup_{i=1}^{n_1} I_i$ we can take an interval from $\mathscr{F}$ and disjoint from $\bigcup_{i=1}^{n_1} I_i$, and then repeat the above procedure with E_1 instead of E, obtaining disjoint intervals (I_i), $n_1 + 1 \leq i \leq n_2$, from $\mathscr{F}$ such that

$$\mu^*\left(E - \bigcup_{i=1}^{n_2} I_i\right) = \mu^*\left(E_1 - \bigcup_{i=n_1+1}^{n_2} I_i\right) < \frac{5}{8} \cdot \mu^*(E_1) < \left(\frac{5}{8}\right)^2 \cdot \mu^*(E).$$

By continuing this process, the theorem follows. ∎

The Vitali Covering Theorem is now used in the proof of the following theorem of Lebesgue.

Theorem 4.2. If f is a nondecreasing function on an interval I (bounded or unbounded), then f is differentiable almost everywhere. ∎

Proof. Recall that f is differentiable at a point a in the interior of I if $D^+f(a) = D^-f(a) = D_+f(a) = D_-f(a) \neq \pm\infty$. The idea of the proof is to show that the set of points for which the derivates are not all equal is a set of measure zero and then establish that the common value is also finite almost everywhere. The proof is given for the case when $I = (c, d)$, a bounded and open interval—the extensions being left to the reader.

We shall show that $E = \{x \in I\colon D^+f(x) > D_+f(x)\}$ is a null set. The proof that $\{x \in I\colon D_-f(x) < D^-f(x)\}$ is a null set is similar and omitted. By Proposition 4.2, these facts establish that $D^+f(x) = D_+f(x) = D_-f(x) = D^-f(x)$ almost everywhere.

For every pair of positive rationals u and v, let

$$E_{uv} = \{x \in E\colon D_+f(x) < u < v < D^+f(x)\}.$$

Clearly $E = \bigcup_{u<v} E_{uv}$ and this is a countable union with u and v from the positive rationals. It suffices to show that $m(E_{uv}) = 0$ for all rational u and v with $0 < u < v$.

Let us assume the contrary: There exist rationals u and v with $0 < u < v$ and $m(E_{uv}) = \alpha > 0$. Letting $\varepsilon > 0$ be arbitrary, choose an open set O containing E_{uv} such that $m(0) < \alpha + \varepsilon$. For each $x \in E_{uv}$ there exists an arbitrarily small interval $[x, x + h]$ contained in $O \cap I$ such that $f(x + h) - f(x) < uh$. The family of all such intervals forms a Vitali covering of E_{uv} and so there exists a finite pairwise disjoint subfamily $\{[x_i, x_i + h_i]: i = 1, 2, \ldots, N\}$ such that

$$m\left(E_{uv} - \bigcup_{i=1}^{N} [x_i, x_i + h_i]\right) < \varepsilon. \tag{4.8}$$

Clearly $m\left(E_{uv} - \bigcup_{i=1}^{N}(x_i, x_i + h_i)\right)$ is also less than ε, and since each $(x_i, x_i + h_i) \subset O$, with $A = \bigcup_{i=1}^{N}(x_i, x_i + h_i)$, we have

$$m(A) = \sum_{i=1}^{N} h_i \leq m(O) < \alpha + \varepsilon.$$

Thus the inequalities $f(x_i + h_i) - f(x_i) \leq uh_i$ imply

$$\sum_{i=1}^{N} f(x_i + h_i) - f(x_i) < u \sum_{i=1}^{N} h_i < u(\alpha + \varepsilon).$$

Likewise for each $y \in E_{uv} \cap A$, there is an arbitrarily small interval $[y, y + k] \subset A$ such that

$$f(y + k) - f(y) > vk. \tag{4.9}$$

The family of all such intervals forms a Vitali covering of $E_{uv} \cap A$ so that there exists a finite, pairwise disjoint subfamily $\{[y_i, y_i + k_i]: i = 1, \ldots, M\}$ of such intervals such that

$$m\left(E_{uv} \cap A - \bigcup_{i=1}^{M} [y_i, y_i + k_i]\right) < \varepsilon. \tag{4.10}$$

Hence, using inequalities (4.8) and (4.10),

$$\begin{aligned}
\alpha = m(E_{uv}) &\leq m(E_{uv} \cap A) + m(E_{uv} \cap A^c) \\
&\leq m\left((E_{uv} \cap A) - \bigcup_{i=1}^{M} [y_i, y_i + k_i]\right) + m\left(\bigcup_{i=1}^{M} [y_i, y_i + k_i]\right) + m(E_{uv} \cap A^c) \\
&< \varepsilon + \sum_{i=1}^{M} k_i + \varepsilon = 2\varepsilon + \sum_{i=1}^{M} k_i.
\end{aligned} \tag{4.11}$$

By inequalities (4.9) and (4.11),

$$v(\alpha - 2\varepsilon) < v\left(\sum_{i=1}^{M} k_i\right) < \sum_{i=1}^{M} f(y_i + k_i) - f(y_i).$$

Since

$$\bigcup_{i=1}^{M} [y_i, y_i + k_i] \subset \bigcup_{j=1}^{N} [x_j, x_j + h_j]$$

and f is nondecreasing, we have

$$v(\alpha - 2\varepsilon) < \sum_{i=1}^{M} f(y_i + k_i) - f(y_i) \leq \sum_{j=1}^{N} f(x_j + h_j) - f(x_j) < u(\alpha + \varepsilon).$$

Therefore $v\alpha < u\alpha +$ (an arbitrarily small positive number), so that $v \leq u$. This contradicts our original condition on v and u so that $m(E_{uv}) = 0$.

It remains to show that the set F of all x in $I = (c, d)$, where $f'(x) = \infty$, has measure zero. It is sufficient to show that for each $\varepsilon > 0$, $F_\varepsilon = F \cap (c + \varepsilon, d - \varepsilon)$ has measure zero. If β is an arbitrary positive number, then for each $x \in F_\varepsilon$ there is an arbitrarily small positive number h such that $[x, x + h] \subset (c + \varepsilon, d - \varepsilon)$ and

$$f(x + h) - f(x) > \beta h. \tag{4.12}$$

By Vitali's Theorem (Theorem 4.1), there exists a pairwise disjoint, countable family $\{[x_n, x_n + h_n]\}$ of such intervals such that $m(F_\varepsilon - \bigcup_n [x_n, x_n + h_n]) = 0$. Hence using inequality (4.12)

$$\beta m(F_\varepsilon) \leq \beta \sum_{n=1}^{\infty} h_n < \sum_{n=1}^{\infty} f(x_n + h_n) - f(x_n) \leq f(d - \varepsilon) - f(c + \varepsilon).$$

Since β is arbitrary, $m(F_\varepsilon)$ must be zero. ∎

(One could also show $m(F) = 0$ by using Proposition 4.4 of the next section since the integrability of the function $f'(x)$ implies it is finite almost everywhere.)

The next result characterizes all functions on an interval $[a, b]$ that can be expressed as the difference of two nondecreasing functions. They are the functions described in the next definition.

Definition 4.3. Let f be a real-valued function defined on an interval $I = [a, b]$. For each partition P: $a = x_0 < x_1 < x_2 < \cdots < x_n = b$ of $[a, b]$, let

$$V_a{}^b(f, P) = \sum_{i=1}^{n} |f(x_i) - f(x_{i-1})|.$$

The extended real number $V_a^b f$ given by

$$V_a^b f = \sup\{V_a^b(f, P)\colon P \text{ is a partition of } [a, b]\}$$

is called the *total variation* of f. If $V_a^b f < \infty$, then f is said to be a function of *bounded variation* on $[a, b]$. ∎

Remarks

4.1. The equality $V_a^c f + V_c^b f = V_a^b f$ holds for all $c \in [a, b]$ when $V_d^d f$ is defined to be zero for all d.

4.2. $v_f(x) \equiv V_a^x f$ is a nondecreasing function of x on $[a, b]$.

Proposition 4.3. f is a function of bounded variation on $[a, b]$ if and only if it can be written as the difference of two nondecreasing functions g and h. ∎

Proof. The sufficiency of this statement is easily verified. We here establish the converse.

Assume $V_a^b f < \infty$. Define real-valued functions g and h on $[a, b]$ by the rules

$$g(x) = V_a^x f \quad \text{and} \quad h(x) = V_a^x f - f(x),$$

where, as noted above, $V_a^a f$ is defined to be zero. By Remark 4.2, $g(x)$ is a nondecreasing function of x. Also $V_a^x f - f(x)$ is nondecreasing. To see this, suppose $x_1 < x_2$. Then

$$[V_a^{x_2} f - f(x_2)] - [V_a^{x_1} f - f(x_1)] = V_{x_1}^{x_2} f - [f(x_2) - f(x_1)] \geq 0.$$

This establishes the converse. ∎

Corollary 4.1. A real-valued function f on $[a, b]$ of bounded variation has a finite derivative $f'(x)$ a.e. ∎

Problems

4.1.1. Let f be the function given by

$$f(x) = \begin{cases} ax \sin^2 \dfrac{1}{x} + bx \cos^2 \dfrac{1}{x}, & \text{if } x > 0, \\ 0, & \text{if } x = 0, \\ cx \sin^2 \dfrac{1}{x} + dx \cos^2 \dfrac{1}{x}, & \text{if } x < 0, \end{cases}$$

where $a < b$ and $c < d$. Find the four derivates of f at 0.

× **4.1.2.** Let

$$f(x) = \begin{cases} 0, & \text{if } x = 0, \\ x^\alpha \sin 1/x^\beta, & \text{if } 0 < x \leq 1. \end{cases}$$

(a) If $0 < \beta < \alpha$ show that f is absolutely continuous.† {Hint: Observe that for every $\varepsilon > 0$, $f(1) - f(\varepsilon) = \int_\varepsilon^1 f'(x)\,dx$ and then show that $f'(x) \in L_1[0, 1]$.}

(b) If $0 < \alpha \leq \beta$ show that f is not of bounded variation.

4.1.3. Let $\{a_1, a_2, \ldots\}$ be an enumeration of the rationals in [0, 1]. Prove that the function f defined on [0, 1] by

$$f(x) = \sum_{n=1}^{\infty} \frac{|x - a_n|}{3^n}$$

is continuous on [0, 1], but not differentiable at each rational a_n by showing $f_+'(a_n) \neq f_-'(a_n)$.

4.1.4. Let f be a continuous real function on an interval I with a zero right derivative at each interior point of I. Prove f is constant on I. [Hint: Prove that if g is a continuous function on an interval $[a, b]$ with a right derivative at each point in (a, b) and $g(a) = g(b) = 0$, then there exist points c and d in (a, b) with $g_+'(c) \leq 0$ and $g_+'(d) \geq 0$. Apply this result to $g(x) = f(x) - f(a) - \{[f(b) - f(a)]/(b - a)\}(x - a)$, where $a < b$ in I.]

4.1.5. State and prove the Vitali Covering Theorem for the case when $m^*(E) = \infty$, m being the Lebesgue measure.

4.1.6. If f is continuous on $[a, b]$ and one of its derivates is nonnegative in the interval (a, b), show that $f(a) \leq f(b)$. [Hint: Say $D^+f \geq 0$. Suppose that $f(b) - f(a) < -\varepsilon(b - a)$ and let $\varphi(x) = f(x) - f(a) + \varepsilon(x - a)$. Let ξ be the largest value in (a, b) such that $\varphi(\xi) = 0$. Then $D^+\varphi(\xi) \leq 0$.]

4.1.7. B. L. Van der Waerden gave the following example of an everywhere nondifferentiable but continuous function. The reader should verify the succeeding statements. For each x let $f_n(x)$ denote the distance from x to the nearest number that can be written in the form $m/10^n$, where m is an integer. The function f given by $f = \sum_{n=1}^{\infty} f_n$ is then continuous since each

† Note that we define "absolute continuity" in the next section; at this point it suffices to show $f(1) - f(0) = \int_0^1 f'(x)\,dx$.

f_n is continuous and the series converges uniformly. To show f is not differentiable, write any x in (0, 1) as a decimal $x = .a_1a_2a_3 \cdots a_q \cdots$. If the qth term is 4 or 9, let $x_q = x - 10^{-q}$; otherwise let $x_q = x + 10^{-q}$. The sequence (x_q) thus obtained converges to x, but

$$f_n(x_q) - f_n(x) = \begin{cases} \pm (x_q - x), & \text{if } n < q, \\ 0, & \text{if } n \geq q, \end{cases}$$

so that $f(x_q) - f(x) = p(x_q - x)$, where p is an integer that is odd or even with $q - 1$.

4.1.8. This result is due to Fubini. Let $S = f_1 + f_2 + \cdots$ be a pointwise convergent series of monotonic functions—either all increasing or all decreasing—defined on $[a, b]$. Prove S is differentiable a.e. and $S'(x) = f_1'(x) + f_2'(x) + \cdots$ a.e. [Hint: Assuming each f_i is increasing we may also assume $f_i(a) = 0$ for all i. Set $S_n = \sum_{i=1}^{n} f_i$. Then $S_n \to S$, S is differentiable a.e., and $S_n'(x) \leq S_{n+1}'(x) \leq S'(x)$. Since the S_n' form an increasing sequence, S_n' converges, and to show $S_n' \to S'$ it suffices to show $S' - S_{n_k}' \to 0$ for a suitably chosen subsequence (n_k) of positive integers. Choose n_k so that $\sum_{k=1}^{\infty} S(b) - S_{n_k}(b) < \infty$. Then the series $\sum_{k=1}^{\infty} S - S_{n_k}$ converges everywhere and hence so does the series formed by $S' - S_{n_k}'$. Hence $S'(x) - S_{n_k}'(x) \to 0$.]

4.1.9. Let $E(\subset R)$ be the union of an arbitrary family of intervals, each being open, closed, half-open, or half-closed. Show that E is Lebesgue measurable. [Hint: Use a Vitali covering.]

4.1.10. Give an example of a continuous function that is not of bounded variation in any interval.

4.1.11. Show that if $f(x)$ is not of bounded variation on [0, 1], then there is $x \in [0, 1]$ such that f is not of bounded variation in any open interval containing x.

4.1.12. Prove the following result due to Wetzel. Suppose f: $I(\subset R) \to R$, where I is an open interval containing 0. Suppose that (i) $f(x + y) \geq f(x)f(y)$ and that (ii) $f(x)\, f(x) \geq 1 + x$ whenever x, $y \in I$ and $x + y \in I$. Then $f(x) = e^x$ for all $x \in I$. [Hint: Show $f(x) > 0$ for all $x \in I$ by means of $f(x) \geq [f(x/2^n)]^{2n}$. Show $f'(x) = f(x)$.]

× **4.1.13.** *A Continuous, Strictly Increasing Function with Derivative 0 a.e.* Let (I_n) be intervals $[0, 1]$, $[0, \frac{1}{2}]$, $[\frac{1}{2}, 1]$, $[0, \frac{1}{4}]$, ..., $[0, \frac{1}{8}]$, etc., and let f_n be the composition $f \circ g_n$, where $g_n : I_n \to [0, 1]$ is the natural homeomorphism and f is the Cantor function on [0, 1]. Let $f(x) = \sum_{n=1}^{\infty}(1/2^n)f_n(x)$. Prove that h is the desired function. [Use Problem 4.1.8.]

4.1.14. (i) Prove that if f is of bounded variation on $[a, b]$, then

$$v_f(c+) - v_f(c) = | f(c+) - f(c) | \text{ for each } c \in [a, b)$$

and

$$v_f(c) - v_f(c-) = | f(c) - f(c-) | \text{ for each } c \in (a, b].$$

Conclude that f is continuous on the right (left) at c if and only if v_f is continuous on the right (left) at c. (v_f is defined in Remark 4.2.)

(ii) Prove that if f is of bounded variation on $[a, b]$, then so is v_f and $V_a{}^b(v_f) = V_a{}^b(f)$.

4.1.15. Let f be a continuous function on $[a, b]$. (i) Show that the length of the curve $C = \{(x, y)\colon y = f(x) \text{ for } a \leq x \leq b\}$, which is given by

$$L_a{}^b f = \sup\left\{ \sum_{i=1}^{n} [(x_i - x_{i-1})^2 + (f(x_i) - f(x_{i-1}))^2]^{1/2} : \\ a = x_0 < x_1 < \cdots < x_n = b \right\},$$

is finite if and only if f is of bounded variation on $[a, b]$.

(ii) If the function f is of bounded variation on $[a, b]$, then prove $L_a{}^b(v_f) = L_a{}^b(f)$.

4.1.16. Suppose f and g are real functions on $[a, b]$ and g is nondecreasing. f is said to satisfy *a uniform Lipschitz condition with respect to g* on $[a, b]$ if there exists a nonnegative number K such that

$$| f(x_1) - f(x_2) | \leq K \,| g(x_1) - g(x_2) |, \qquad \text{for all } x_1, x_2 \in [a, b].$$

The least such number K is called the *Lipschitz constant* for f and g on $[a, b]$ and is denoted by $K_a{}^b(f, g)$. Prove that if f is of bounded variation on $[a, b]$, then f satisfies a uniform Lipschitz condition with respect to g if and only if v_f does also. Moreover, in this case $K_a{}^b(v_f, g) = K_a{}^b(f, g)$.

4.1.17. The following result is due to M. Ecker. Let σ be a permutation of the digits $0, 1, \ldots, 9$. For each $x \in [0, 1]$ let $.x_1x_2\ldots$ be the decimal expansion of x where expansions with all but finitely many digits equal to 9 are not allowed. Define $f\colon [0, 1) \to [0, 1]$ by $f(x) = .x_{\sigma 1}x_{\sigma 2}\ldots$.

(i) Prove f is continuous except possibly from the left at numbers like $.a_1a_2\ldots a_n000\ldots$, $a_n \neq 0$. At such a number f is continuous if and only if either σ fixes 0 and 9 and $\sigma(a_n - 1) = \sigma(a_n) - 1$ or else σ transposes 0 and 9 and $\sigma(a_n - 1) = \sigma(a_n) + 1$.

(ii) Prove f is differentiable everywhere on $[0, 1)$ only if σ is the identity permutation or $\sigma(i) = 9 - i$ for all i; otherwise f is nowhere differentiable.

(iii) Prove $\int_0^1 f(x)\,dx = 1/2$.

4.1.18. (i) Let $g: [a, b] \to R$ be continuous and assume that there is a countable set D of $[a, b)$ such that for each $x \in [a, b) - D$, there is a $\delta > 0$ such that $g(y) > g(x)$ for all $y \in (x, x + \delta)$. Prove g is strictly increasing on $[a, b]$. [Hint: Assume there exist c and d with $a \leq c < d \leq b$ and $g(c) > g(d)$. For each y in $(g(d), g(c)]$, let x_y be the largest x in $[c, d)$ such that $g(x) = y$. Show $g(t) < g(x_y)$ for all t in (x_y, d).]

(ii) Let $f: [a, b] \to R$ be continuous. Assumed D is a countable subset of $[a, b)$ such that f_+' exists and is finite at all points of $[a, b] - D$. Suppose f_+' is integrable over $[a, b]$. Prove $\int_b^a f_+'(x)\,dx = f(b) - f(a)$. [Hint: Let $f_1(x) = f_+'(x)$ on $[a, b] - D$ and 0 on D. Choose by (Problem 3.2.30) a lower semicontinuous function u such that $u > f_1$ on $[a, b]$ and $\int_b^a (u - f_1)\,dx < \varepsilon$. If $\eta > 0$, define F_η on $[a, b]$ by

$$F_\eta(x) = \int_a^x u(t)\,dt - [f(x) - f(a)] + \eta(x - a)$$

Show that there exists a $\delta > 0$ such that $F_\eta(y) - F_\eta(x) > 0$ if $y \in (x, x + \delta)$ and $x \in [a, b) - D$. Use part (i) and conclude $F_\eta(b) > 0$. Repeat reasoning for $-f$.]

4.2. Integration versus Differentiation I: Absolutely Continuous Functions

As stated in the introduction to this chapter, we wish to examine the analogs of statements (1) and (2) in the theory of Lebesgue integration. Our first proposition is a beginning of this exploration.

Proposition 4.4. Suppose f is a nondecreasing function on $[a, b]$. Then f' is integrable on $[a, b]$ and in fact

$$\int_a^b f'(x)\,dx \leq f(b) - f(a). \qquad \blacksquare$$

Proof. We can extend the domain of definition of f to R by defining $f(x) = f(a)$ for $x \leq a$ and $f(x) = f(b)$ for $x \geq b$. By Proposition 4.1 and Theorem 4.2 f is differentiable a.e. and f' is a measurable function. The functions

$$g_n(x) = n[f(x + 1/n) - f(x)], \qquad \text{for } n = 1, 2, \ldots,$$

are nonnegative and $f'(x) = \lim_n g_n(x)$ for almost all x. Using Fatou's Lemma,

$$\begin{aligned}\int_a^b f'(x)\,dx &= \int_a^b \liminf_n g_n(x) \leq \liminf_n \int_a^b g_n(x)\\ &= \liminf_n \left[n \int_a^b \left\{f\left(x + \frac{1}{n}\right) - f(x)\right\} dx\right]\\ &= \liminf_n n\left[\int_{a+1/n}^{b+1/n} f(x)\,dx - \int_a^b f(x)\,dx\right]\\ &= \liminf_n n\left[\int_b^{b+1/n} f(x)\,dx - \int_a^{a+1/n} f(x)\,dx\right]\\ &= \liminf_n n\left[\frac{1}{n} f(b) - \int_a^{a+1/n} f(x)\,dx\right]\\ &\leq f(b) - f(a).\end{aligned}$$

∎

Can the inequality of Proposition 4.4 be "improved" to an equality? As the following examples show, the answer is no—unless some additional restrictions are placed on f.

Examples

4.1. Let $k\colon [0, 1] \to [0, 1]$ be the Cantor ternary function. Since k is a constant on each interval contained in the complement of the Cantor set, $k'(x) = 0$, a.e. Thus

$$0 = \int_0^1 k'(x)\,dx \neq k(1) - k(0) = 1.$$

4.2. Problem 4.1.13 presents an example of a strictly increasing continuous function with $f' = 0$ a.e. The reader is encouraged to work that problem.

Our goal is to characterize the class of functions for which the inequality in Proposition 4.4 becomes an equality. More precisely we wish to find necessary and sufficient conditions on any function f defined on a closed interval $[a, b]$ to ensure that for all $x \in [a, b]$

$$f(x) - f(a) = \int_a^x f'(t)\,dt.$$

Let us first list some necessary properties possessed by any function

G defined on $[a, b]$ by

$$G(x) = \int_a^x g(t)\, dt, \tag{4.13}$$

where g is an integrable function on $[a, b]$. These are listed as Remarks 4.3–4.5 and Proposition 4.5.

Remarks

4.3. The function G defined in equation (4.13) is continuous on $[a, b]$. (See Problem 3.2.1.)

4.4. The function G defined in equation (4.13) is a function of bounded variation on $[a, b]$ and thus differentiable a.e. Indeed, G is the difference of two nondecreasing functions:

$$G(x) = \int_a^x [g(t) \vee 0]\, dt - \int_a^x [-g(t)] \vee 0\, dt.$$

4.5. The function G defined in equation (4.13) is the zero function if and only if $g(t) = 0$ a.e.

Proof. It is clear that if $g(t) = 0$ a.e., then $G(x) = 0$ for all x in $[a, b]$. Conversely, suppose $g(t) > 0$ on a set E, necessarily measurable, of positive measure. Then by Problem 2.3.9 there is a closed set $F \subset E$ of positive measure. The set $H = (a, b) - F$ is open and can be written as a disjoint union of open intervals $\bigcup_n (a_n, b_n)$. Therefore

$$0 = \int_a^b g(t)\, dt = \int_F g(t)\, dt + \int_H g(t)\, dt = \int_F g(t)\, dt + \sum_{n=1}^{\infty} \int_{a_n}^{b_n} g(t)\, dt.$$

However, for each n,

$$\int_{a_n}^{b_n} g(t)\, dt = \int_a^{b_n} g(t)\, dt - \int_a^{a_n} g(t)\, dt = G(b_n) - G(a_n) = 0.$$

Therefore $\int_F g(t)\, dt = 0$; but since $g > 0$ on F and $m(F) > 0$, $\int_E g > 0$. This is a contradiction and E must have measure zero. ∎

The fourth necessary property possessed by G of equation (4.13) is the analog of Theorem (1) listed in the introduction to this chapter and is important enough to warrant a separate proposition.

Proposition 4.5. If g is integrable over $[a, b]$, then

$$G(x) = \int_a^x g(t)\,dt$$

is differentiable a.e. and $G' = g$ a.e. ∎

Proof. By Remark 4.4 G is differentiable a.e. It remains to show that $G' = g$ a.e. For this proof we may assume that g is nonnegative. We divide the proof into two cases.

Case 1. Suppose that g is bounded; that is, $|g| \leq K$ on $[a, b]$. Let

$$f_n(x) = n[G(x + 1/n) - G(x)], \qquad \text{for } x \in [a, b], \tag{4.14}$$

where $G(y)$ is defined to be $G(b)$ for $y \geq b$. Defining $g(y)$ to be 0 for $y \geq b$, then

$$f_n(x) = n \int_x^{x+1/n} g(t)\,dt, \tag{4.15}$$

so that $|f_n(x)| \leq K$ for all x in $[a, b]$. Since $\lim_n f_n(x) = G'(x)$ a.e., the Dominated Convergence Theorem (Theorem 3.3) implies that G' is integrable and

$$\begin{aligned}\int_a^x G'(t)\,dt &= \lim_n \int_a^x f_n(t)\,dt = \lim_n n \int_a^x \left[G\left(t + \frac{1}{n}\right) - G(t)\right] dt \\ &= \lim_n n\left[\int_x^{x+1/n} G(t)\,dt - \int_a^{a+1/n} G(t)\,dt\right].\end{aligned} \tag{4.16}$$

Since G is continuous on $[x, x + 1/n]$, the mean value theorem for integrals implies that there exist δ and δ' such that $0 \leq \delta, \delta' \leq 1$ and

$$\int_x^{x+1/n} G(t)\,dt = \frac{1}{n}\,G\left(x + \frac{\delta}{n}\right)$$

and (4.17)

$$\int_a^{a+1/n} G(t)\,dt = \frac{1}{n}\,G\left(a + \frac{\delta'}{n}\right).$$

Putting equation (4.17) into equation (4.16) we obtain

$$\int_a^x G'(t)\,dt = G(x) - G(a) = \int_a^x g(t)\,dt.$$

Since $\int_a^x [G'(t) - g(t)]\, dt = 0$ for all x, we infer from Remark 4.5 that $G'(t) = g(t)$ a.e.

Case 2. If g is not bounded, let

$$g_n(x) = \begin{cases} g(x), & \text{if } g(x) \leq n, \\ n, & \text{if } g(x) > n. \end{cases}$$

By the Lebesgue Convergence Theorem,

$$\int_a^b g(t)\, dt = \lim_n \int_a^b g_n(t)\, dt.$$

By Case 1, since

$$G(x) = \int_a^x [g(t) - g_n(t)]\, dt + \int_a^x g_n(t)\, dt,$$

$$G'(x) = \frac{d}{dx} \int_a^x [g(t) - g_n(t)]\, dt + g_n(x), \text{ a.e.}$$

However, since $[g - g_n] \geq 0$, $\int_a^x [g - g_n]\, dt$ is an increasing function of x so that its derivative is nonnegative. Therefore $G'(x) \geq g_n(x)$ a.e. for all n. Therefore $g(x) = \lim_n g_n(x) \leq G'(x)$ a.e. Using Proposition 4.4 we obtain

$$G(b) - G(a) = \int_a^b g(t)\, dt \leq \int_a^b G'(t)\, dt \leq G(b) - G(a),$$

so that $g(t) = G'(t)$ a.e. ∎

Clearly, the fact that a function f is continuous, differentiable a.e., and of bounded variation on an interval $[a, b]$ is not sufficient to guarantee that

$$f(x) - f(a) = \int_a^x f'(t)\, dt, \qquad \text{for all } x \text{ in } [a, b]. \tag{4.18}$$

Indeed, the Cantor ternary function is continuous and nondecreasing and yet equation (4.18) is not satisfied. The Cantor ternary function (or the function in Example 4.2) is peculiar in that it is a nonconstant function with derivative equal to zero a.e. What can we conclude about such functions? The answer is given in the following proposition.

Proposition 4.6. Suppose $f'(x) = 0$ a.e. on $[a, b]$ and $f(x)$ is not a constant function on $[a, b]$. Then there exists some $\varepsilon > 0$ such that for any

$\delta > 0$ there is a finite disjoint collection of intervals

$$\{(x_1, y_1), (x_2, y_2), \ldots, (x_n, y_n)\}$$

with $\sum_{i=1}^{n} y_i - x_i < \delta$, but $\sum_{i=1}^{n} | f(y_i) - f(x_i) | > \varepsilon$. ∎

Proof. Since $f(x)$ is not a constant function, there exists a c in $(a, b]$ such that $f(a) \neq f(c)$. Let $E_c = [a, c) \cap \{x\colon f'(x) = 0\}$. For each x in E_c and any positive number γ, there are arbitrarily small intervals $[x, x + h] \subset [a, c]$ such that

$$| f(x + h) - f(x) | < \gamma h.$$

For a fixed γ, the set of all these intervals forms a Vitali covering of E_c. Hence by the Vitali theorem, for any $\delta > 0$ there is a finite disjoint collection

$$\{[x_1, x_1 + h_1), \ldots, [x_n, x_n + h_n)\}$$

such that $m\big(E_c - \bigcup_{i=1}^{n}[x_i, x_i + h_i)\big) = m\big([a, c] - \bigcup_{i=1}^{n}[x_i, x_i + h_i)\big) < \delta$. Rename the intervals so that

$$a = x_0 \leq x_1 < x_1 + h_1 < x_2 < x_2 + h_2 < \cdots < x_n + h_n \leq x_{n+1} = c.$$

Then with $h_0 = 0$ and $\varepsilon > 0$ chosen so that $2\varepsilon < | f(c) - f(a) |$,

$$\begin{aligned} 2\varepsilon < | f(c) - f(a) | &\leq \sum_{i=0}^{n} | f(x_{i+1}) - f(x_i + h_i) | \\ &\quad + \sum_{i=1}^{n} | f(x_i + h_i) - f(x_i) | \\ &\leq \sum_{i=0}^{n} | f(x_{i+1}) - f(x_i + h_i) | + \gamma \sum_{i=1}^{n} h_i \\ &\leq \sum_{i=0}^{n} | f(x_{i+1}) - f(x_i + h_i) | + \gamma(b - a). \end{aligned}$$

Since γ is arbitrary, pick γ small enough so that $\gamma(b - a) < \varepsilon$. Then

$$\varepsilon < \sum_{i=0}^{n} | f(x_{i+1}) - f(x_i + h_i) |,$$

but

$$\sum_{i=1}^{n} [x_{i+1} - (x_i + h_i)] = m\Big(E_c - \bigcup_{i=1}^{n} [x_i, x_i + h_i]\Big) < \delta.$$ ∎

The negation of the conclusion of the preceding proposition is the property enjoyed by an important class of functions called *absolutely continuous* functions.

Definition 4.4. A function f on $[a, b]$ is said to be *absolutely continuous* if for each $\varepsilon > 0$ there exists a $\delta > 0$ such that, whenever

$$\{(x_1, y_1), (x_2, y_2), \ldots, (x_n, y_n)\}$$

is a finite collection of nonoverlapping subintervals of $[a, b]$ with

$$\sum_{i=1}^{n} [y_i - x_i] < \delta,$$

then

$$\sum_{i=1}^{n} |f(y_i) - f(x_i)| < \varepsilon.$$ ∎

Remark 4.6. The word "nonoverlapping" is important in Definition 4.4, and removing it results in what may be termed the definition of a "strongly" absolutely continuous function. It can be shown that a necessary and sufficient condition that function f be "strongly" absolutely continuous on $[a, b]$ is that it satisfies the Lipschitz condition: There exists a constant K such that $|f(x) - f(y)| < K|x - y|$ for all x and y in $[a, b]$. (See Problem 4.2.10.) In contrast, although every function satisfying the Lipschitz condition on $[a, b]$ is absolutely continuous, that the converse is false is evidenced by the function $f(x) = x^{1/2}$ on $[0, 1]$.

Is the function G on $[a, b]$ given by the indefinite integral

$$G(x) = \int_a^x g(t)\,dt$$

of an integrable function g an absolutely continuous function? The reader can readily verify using Problem 3.2.1 that the answer, which we record in the following lemma, is in the affirmative.

Lemma 4.1. The indefinite integral G of an integrable function g on $[a, b]$ is absolutely continuous. ∎

We will now establish that the absolute continuity of a function G on an interval $[a, b]$ is a sufficient condition for G to be defined as an indefinite integral of an integrable function g. According to what we have already shown, G must then be necessarily continuous and of bounded variation on $[a, b]$ with $G' = g$ a.e. However, below we establish independently the fact that any absolutely continuous function is of bounded variation.

Lemma 4.2. If f is an absolutely continuous function on $[a, b]$, then f is a function of bounded variation on $[a, b]$. ∎

Proof. Let $\delta > 0$ correspond to the value $\varepsilon = 1$ in the definition of absolute continuity. Choose an integer n such that $n > (b - a)/\delta$. Subdivide $[a, b]$ into adjacent subintervals with the points

$$a = x_0 < x_1 < x_2 < \cdots < x_n = b,$$

where $x_{i+1} - x_i = (b - a)/n < \delta$, for $i = 1, 2, \ldots, n - 1$. Then for each i, $V_{x_i}^{x_{i+1}} < 1$ since any finite collection of nonoverlapping subintervals of $[x_i, x_{i+1}]$ will have total length less than δ. Hence,

$$V_a^b f = \sum_{i=0}^{n-1} V_{x_i}^{x_{i+1}} < n. \qquad ∎$$

Corollary 4.2. An absolutely continuous function on $[a, b]$ is differentiable a.e. on $[a, b]$. ∎

The converse of Lemma 4.2 is false, as the Cantor ternary function illustrates. (See Problem 4.2.4.) Problem 4.2.9 gives a sufficient condition for the converse to hold. In addition, the next result gives a necessary and sufficient condition for functions of bounded variation to be absolutely continuous.

Proposition 4.7. If f is of bounded variation on $[a, b]$, then f is absolutely continuous on $[a, b]$ if and only if the total variation function v, given by $v_f(x) = V_a^x f$, is absolutely continuous on $[a, b]$. ∎

Proof. Assume f is absolutely continuous. If $\varepsilon > 0$ is arbitrary, there exists a $\delta > 0$ such that, for any finite collection

$$\mathscr{C} = \{(x_1, y_1), (x_2, y_2), \ldots, (x_n, y_n)\}$$

of nonoverlapping subintervals of $[a, b]$ with $\sum_i (y_i - x_i) < \delta$, we have $\sum_i | f(y_i) - f(x_i) | < \varepsilon$. For each i, let

$$P_i\colon x_i = a_{i0} < a_{i1} < a_{i2} < \cdots < a_{im_i} = y_i$$

be a partition of $[x_i, y_i]$. Since

$$\sum_{i=1}^{n} \sum_{j=1}^{m_i} [a_{ij} - a_{i,j-1}] = \sum_{i=1}^{n} [y_i - x_i] < \delta,$$

then

$$\sum_{i=1}^{n} \sum_{j=1}^{m_i} | f(a_{ij}) - f(a_{i,j-1}) | < \varepsilon.$$

Fixing the collection $\mathscr{C}$ but varying the partitions P_i of each $[x_i, y_i]$, we obtain, upon taking the supremum over all such partitions P_i,

$$\sum_{i=1}^{n} V_{x_i}^{y_i} f < \varepsilon$$

or

$$\sum_{i=1}^{n} v_f(y_i) - v_f(x_i) < \varepsilon.$$

This proves v_f is absolutely continuous.

Conversely, since $| f(x_i) - f(x_{i-1}) | \leq v_f(x_i) - v_f(x_{i-1})$ for $a \leq x_{i-1} < x_i < b$, the absolute continuity of v_f implies that of f. ∎

If G is an indefinite integral as in equation (4.13), we know from Lemmas 4.1 and 4.2 that $V_a{}^bG < \infty$. In fact we have a formula for $V_a{}^bG$.

Proposition 4.8. If

$$G(x) = \int_a^x g(t)\, dt \text{ on } [a, b],$$

where g is an integrable function on $[a, b]$, then

$$V_a{}^bG = \int_a^b | g(t) |\, dt.$$ ∎

Proof. For any partition P: $a = x_0 \leq x_1 \leq x_2 \leq \cdots \leq x_n = b$, we have

$$\sum_{i=1}^{n} | G(x_i) - G(x_{i-1}) | = \sum_{i=1}^{n} \left| \int_{x_{i-1}}^{x_i} g(t)\, dt \right| \leq \int_a^b | g(t) |\, dt,$$

so that $V_a{}^bG \leq \int_a^b | g(t) |\, dt$.

To prove the opposite inequality, we need the observation that if s is a step function such as $\sum_{i=1}^n c_i \chi_{[x_{i-1}, x_i)}$, where $a = x_0 < x_1 < x_2 < \cdots$

$< x_n = b$ and each $c_i = -1, 0$, or 1, then

$$\int_a^b sg\,dt = \sum_{i=1}^n c_i \int_{x_{i-1}}^{x_i} g\,dt \leq \sum_{i=1}^n \left| \int_{x_{i-1}}^{x_i} g\,dt \right|$$

$$= \sum_{i=1}^n | G(x_i) - G(x_{i-1}) | \leq V_a{}^bG. \tag{4.19}$$

For each n, there is a step function σ_n such that $m\{x: | g(x) - \sigma_n(x) | \geq 1/2^n\} < 1/2^n$. Hence the sequence σ_n converges to g outside the null set

$$\bigcap_{n=1}^{\infty} \bigcup_{m=n}^{\infty} \left\{x: | g(x) - \sigma_m(x) | \geq \frac{1}{2^m}\right\};$$

that is, $\sigma_n \to g$ a.e. on $[a, b]$. Let s_n be the step function given by

$$s_n(x) = \begin{cases} 1, & \text{if } \sigma_n(x) > 0, \\ 0, & \text{if } \sigma_n(x) = 0, \\ -1, & \text{if } \sigma_n(x) < 0. \end{cases}$$

Then by equation (4.19), $\int_a^b s_n g\,dt \leq V_a{}^bG$. However, $s_n g$ converges to $| g |$ a.e. on $[a, b]$ and $| s_n g | \leq | g |$ so that the Dominated Convergence Theorem gives

$$\int_a^b | g |\,dt = \lim_n \int_a^b s_n g\,dt \leq V_a{}^bG. \qquad \blacksquare$$

By examining the contrapositive of Proposition 4.6 we have also proved the next result.

Lemma 4.3. If f is absolutely continuous on $[a, b]$ and $f' = 0$ a.e., then f is a constant function. $\blacksquare$

This leads us to our goal.

Theorem 4.3. If G is a real-valued function on $[a, b]$, the following statements are equivalent:

(i) G is an absolutely continuous function on $[a, b]$.
(ii) G is defined by

$$G(x) = \int_a^x g(t)\,dt + G(a),$$

where g is an integrable function on $[a, b]$.

(iii) G is differentiable a.e. on $[a, b]$ and G is defined by

$$G(x) = \int_a^x G'(t)\,dt + G(a).$$ ∎

Proof. Referring back to Proposition 4.5 and Lemma 4.1, we see that all that remains to be proved is that (i) implies (ii).

We know by Corollary 4.2 that G' exists a.e. and by Proposition 4.4 that G' is integrable. Define F on $[a, b]$ by

$$F(x) = \int_a^x G'(t)\,dt - G(x). \tag{4.20}$$

The function F is differentiable a.e. on $[a, b]$ and

$$F'(x) = G'(x) - G'(x) = 0, \text{ a.e.}$$

Since F is also absolutely continuous, by Lemma 4.3 F is a constant function. Thus by equation (4.20)

$$G(x) = \int_a^x G'(t)\,dt + G(a).$$ ∎

Our final result of this section generalizes Theorem 4.3 to functions defined on all of R. If f is defined on R, define $V_{-\infty}^{\infty} f$ to be $\lim_{a\to\infty} V_{-a}^{a} f$. If $V_{-\infty}^{\infty} f$ is finite, f is said to be of *bounded variation on R*. A function f with domain R is *absolutely continuous on R* if for each $\varepsilon > 0$ there is a $\delta > 0$ such that $\sum |f(y_i) - f(x_i)| < \varepsilon$ for every finite collection $\{(x_1, y_1), (x_2, y_2), \ldots, (x_n, y_n)\}$ of nonoverlapping intervals with $\sum^n (y_i - x_i) < \delta$. Clearly a function that is absolutely continuous on R is absolutely continuous on every closed interval of R, but the converse is not true, as the function $f(x) = x^2$ illustrates.

Theorem 4.4. A function F on R has the form

$$F(x) = \int_{-\infty}^x f(t)\,dt, \tag{4.21}$$

for some integrable function f on R, if and only if F is absolutely continuous on R, of bounded variation on R, and $\lim_{x\to-\infty} F(x) = 0$. ∎

Proof. Suppose F has the form of equation (4.21). That F is absolutely continuous on R follows from Problem 3.2.1. Also, since on $[-a, a]$ for $a > 0$,

$$F(x) - F(-a) = \int_{-\infty}^{x} f(t)\,dt - \int_{-\infty}^{-a} f(t)\,dt = \int_{-a}^{x} f(t)\,dt,$$

using Proposition 4.8 we have

$$V_{-\infty}^{\infty} F = \lim_{a\to\infty} V_{-a}^{a} F = \lim_{a\to\infty} \int_{-a}^{a} |f(t)|\,dt \leq \int_{-\infty}^{\infty} |f(t)|\,dt < \infty.$$

Using the Lebesgue Convergence Theorem with the functions $f_n = |f| \cdot \chi_{(-\infty,-n)}$, we see that the limit of $F(x)$ as $x \to -\infty$ is zero.

Conversely, if F is absolutely continuous on R, then for all $a > 0$,

$$F(x) = \int_{-a}^{x} F'(t)\,dt + F(-a), \qquad \text{for } x \in [-a, a].$$

Let $a \to \infty$. Then

$$F(x) = \lim_{a\to\infty} \int_{-a}^{x} F'(t)\,dt$$

for any real x since $F(-a) \to 0$. However, using the Monotone Convergence Theorem and Proposition 4.8

$$\begin{aligned}
\int_{-\infty}^{\infty} |F'(t)|\,dt &= \int_{-\infty}^{\infty} \lim_{n\to\infty} |F'(t)|\,\chi_{(-n,n)}\,dt \\
&= \lim_{n\to\infty} \int_{-\infty}^{\infty} |F'(t)|\,\chi_{(-n,n)}\,dt \\
&= \lim_{n\to\infty} \int_{-n}^{n} |F'(t)|\,dt = \lim_{n\to\infty} V_{-n}^{n} F = V_{-\infty}^{\infty} F < \infty.
\end{aligned}$$

Therefore $|F'(t)|$ and hence $F'(t)$ is integrable. Also since

$$F(x) = \lim_{n\to\infty} \int_{-n}^{x} F'(t)\,dt = \lim_{n\to\infty} \int_{-\infty}^{x} F'(t)\chi_{(-n,\infty)}\,dt,$$

the Lebesgue Convergence Theorem implies that

$$F(x) = \int_{-\infty}^{x} F'(t)\,dt.$$

∎

Problems

× **4.2.1.** Let f and g be absolutely continuous functions on $[a, b]$. Prove

(a) $f \pm g$ are absolutely continuous,

and

(b) $f \cdot g$ is absolutely continuous.

× **4.2.2.** Let f be defined on [0, 1]. Then show that:

(a) If f is differentiable on [0, 1] with a bounded derivative, then f is absolutely continuous.

(b) f need not be absolutely continuous on [0, 1] even if f is so on $[\varepsilon, 1]$ for every $\varepsilon > 0$ and continuous at 0.

(c) f is necessarily absolutely continuous on [0, 1] if f is so on $[\varepsilon, 1]$ for every $\varepsilon > 0$, continuous at 0, and of bounded variation on [0, 1]. {Hint: if f is of bounded variation on [0, 1], then

$$\int_0^1 f'(x)\,dx = \lim_{\varepsilon_n \to 0} \int_{\varepsilon_n}^1 f'(x)\,dx, \quad 0 < \varepsilon_n.\}$$

4.2.3. Show that

$$\int_a^b |f'(t)|\,dt = V_a^b f < \infty$$

if and only if f is absolutely continuous.

4.2.4. Prove directly from the definition of absolute continuity that the Cantor ternary function is not absolutely continuous.

× **4.2.5.** This problem gives integration by parts formulas.

(a) If f and g are integrable functions on $[a, b]$, let

$$F(x) = \int_a^x f(t)\,dt$$

and

$$G(x) = \int_a^x g(t)\,dt.$$

Prove

$$\int_a^b G(t)f(t)\,dt + \int_a^b g(t)F(t)\,dt = F(b)G(b) - F(a)G(a).$$

(b) If f and g are absolutely continuous functions on $[a, b]$, prove

$$\int_a^b f(t)g'(t)\,dt + \int_a^b f'(t)g(t)\,dt = f(b)g(b) - f(a)g(a).$$

(c) Let f and g be functions on R each of which is absolutely continuous on R, of bounded variation on R, and vanishes as $x \to -\infty$. Prove

$$\int_R f(t)g'(t)\,dt + \int_R f'(t)g(t)\,dt = \lim_{x\to\infty} f(x) \cdot \lim_{x\to\infty} g(x)$$

$$= \int_R f'(t)\,dt \cdot \int_R g'(t)\,dt.$$

× **4.2.6.** Show that any nondecreasing function f on $[a, b]$ can be written as $g + h$, where g and h are nondecreasing, g is absolutely continuous, and h is singular, i.e., $h' = 0$ a.e.

4.2.7. (a) If f is a function defined on the set $[a, b]$ and E is any subset of $[a, b]$ on which f' exists and $|f'| \leq K$, show $m^*f(E) \leq Km^*(E)$. (Hint: Let

$$E_n = \{x \in E: |y - x| < 1/n \Rightarrow |f(y) - f(x)| \leq (K + \varepsilon)|y - x|\}.$$

Then $m^*(E_n)$ increases to $m^*(E)$. Consider intervals I_{nj} of length $1/n$ such that $E_n \subset \bigcup_{j=1}^{\infty} I_{nj}$ and $\sum_{j=1}^{\infty} m(I_{nj}) < m^*(E) + \varepsilon$.)

(b) Let f be defined and measurable on $[a, b]$ and let E be any set on which f is differentiable. Then

$$m^*(f(E)) \leq \int_E |f'(x)|\,dx.$$

[Hint: Suppose first $|f'(x)| < N$ (an integer) on E. Let $E_k{}^n = \{x \in E: (k-1)/2^n \leq |f'(x)| < k/2^n\}$ for $k = 1, 2, \ldots, N \cdot 2^n$ and $n = 1, 2, \ldots$. Show for each n that

$$m^*(f(E)) \leq \sum_{k=1}^{N\cdot 2^n} \frac{k}{2^n} m(E_k{}^n) = \sum_{k=1}^{N\cdot 2^n} \frac{k-1}{2^n} m(E_k{}^n) + \frac{1}{2^n} \sum_{k=1}^{N\cdot 2^n} m(E_k{}^n).\Big]$$

The next two problems give another characterization of absolutely continuous functions.

× **4.2.8.** (a) An absolutely continuous function maps null sets into null sets. (b) Prove that an absolutely continuous function maps measurable sets into measurable sets.

4.2.9. Prove that if f is continuous and of bounded variation on $[a, b]$ and if f maps sets of measure zero into sets of measure zero, then f is absolutely continuous on $[a, b]$. [Hint: If $\{(a_k, b_k): k = 1, 2, \ldots, n\}$ is any collection of nonoverlapping subintervals of $[a, b]$, let $E_k = \{x \in (a_k, b_k):$

f is differentiable at $x\}$. Show $m(f(a_k, b_k)) = m(f(E_k))$ and using Problem 4.2.7 show

$$\sum^{n} | f(b_k) - f(a_k) | \leq \sum_{i=1}^{n} \int_{a_k}^{b_k} | f'(x) | \, dx.\Big]$$

4.2.10. Verify Remark 4.6; that is, show that a function is "strongly" absolutely continuous on $[a, b]$ if and only if it satisfies the Lipschitz condition on $[a, b]$. [In establishing the necessity of this result the following hint may be useful. If there does not exist an $\varepsilon > 0$ and a positive K such that $| f(x) - f(y) | \leq K | x - y |$ whenever $| x - y | < \varepsilon$, then, given any positive integer n, there exist x_n and y_n in $[a, b]$ such that $| x_n - y_n | < 1/n$ and $| f(y_n) - f(x_n) | > n | x_n - y_n |$. Use the strong absolute continuity of f on each $[x_n, y_n]$ to obtain a contradiction for some n.]

4.2.11. Show that $v_f'(x) = | f'(x) |$ a.e. if $v_f(x)$ is the total variation on $[a, x]$ of a function f of bounded variation on $[a, b]$. {Hint: Choose a partition P_n for $[a, b]$ such that $\sum | f(x_k) - f(x_{k-1}) | > v_f(b) - 1/2^n$. In each segment $x_{k-1} \leq x \leq x_k$ of P_n, let $f_n(x) = f(x) + c$ or $-f(x) + c$ according as $f(x_k) - f(x_{k-1}) \geq 0$ or ≤ 0, where c is chosen so that $f_n(a) = 0$ and the values of f_n at x_k agree. Then the function $v_f(x) - f_n(x)$ is increasing and $\sum_{n=1}^{\infty} [v_f(x) - f_n(x)] < \infty$. By Problem 4.1.8 of this chapter, $v_f'(x) - f_n'(x) \to 0$ a.e. as $n \to \infty$. Since $f_n'(x) = \pm f'(x)$ a.e., $v_f'(x) = | f'(x) |$ a.e.}

4.2.12. If f is of bounded variation on $[a, b]$, then show that its total variation $v_f(x)$ satisfies $v_f(b) \geq \int_a^b | f'(x) | \, dx$. (Compare with Problem 4.2.3.) Also show that $m^*(v_f(E)) \geq m(g(E))$, where

$$E = \{x \in [a, b] : v_f'(x) \text{ exists}\}$$

and $v_f = g + h$, where g is absolutely continuous and nondecreasing, and h is singular and nondecreasing. Finally, show that $m^*(v_f(E)) = \int_a^b v_f'(x) \, dx$.

4.2.13. Suppose that f is a continuous function defined on $[a, b]$ and that $N(y)$, called the *Banach indicatrix*, is the number of solutions x of the equation $y = f(x)$. Show that $v_f(x)$, the total variation of f, satisfies

$$v_f(b) = \int_{-\infty}^{\infty} N(y) \, dy.$$

{Hint: Let $P_1 \subset P_2 \subset \cdots$ be a sequence of partitions of $[a, b]$ such that $| P_n | \to 0$ as $n \to \infty$. If $N_n(y) = \sum_k \chi_{E_{n,k}}(y)$, where $E_{n,k} = f([x_{k-1}^{(n)}, x_k^{(n)}])$ and $[(x_{k-1}^{(n)}, x_k^{(n)}]$ is the kth segment in P_n, then show that $N_n(y) \to N(y)$ a.e. and apply the Monotone Convergence Theorem.}

4.2.14. Give an example of a real-valued, absolutely continuous function on [0, 1] that is monotone on no interval. {Hint: Using Cantor sets of positive Lebesgue measure, construct a Lebesgue-measurable set $A \subset [0, 1]$ such that $m(A \cap I) > 0$ and $m(A^c \cap [0, 1]) > 0$ for any interval I. Then consider $f(x) = \int_0^x [\chi_A(y) - \chi_{A^c}(y)]\, dy$.}

At this point the reader should be aware of the fact that there exists a real-valued, everywhere differentiable function f defined on R such that f is monotone on no subinterval of R and f' is bounded. For a proof of this, the reader is referred to the paper of Katznelson and Stromberg.†

4.2.15. Extend the result in Problem 4.2.12 as follows: If f and v_f are as in Problem 4.2.12 and A is a Lebesgue-measurable subset of $[a, b]$, then

$$m^*(v_f(A)) \geq \int_A |f'(x)|\, dx.$$

Here the equality holds for every $A \subset [a, b]$ if and only if f is absolutely continuous. [Hint: Suppose first that f is absolutely continuous; then v_f is also so. It is easy to show that for open sets G, $m^*(v_f(G)) = \int_G v_f'(x)\, dx$. By approximation through open sets prove this also for A, and then use Problems 4.2.6 and 4.2.11.] This result is due to D. E. Varberg.

★ **4.2.16.** If f is monotonically increasing or absolutely continuous on $[a, b]$ and A is any Lebesgue-measurable subset for which $m(f(A)) = 0$, then show that $f'(x) = 0$ a.e. on A. Is this result true for functions of bounded variation? (For related questions, the reader is referred to Varberg.‡)

4.2.17. If f is a real-valued continuous function on $[a, b]$ and f' exists for all but an at most countable set of points and if f' is Lebesgue integrable, then prove that f is absolutely continuous. (Hint: Let $E = \{x \in [a, b]: f'(x) \text{ exists}\}$. Then

$$\sum_{k=1}^{n} |f(b_k) - f(a_k)| \leq \sum_{k=1}^{n} mf([a_k, b_k])$$
$$= \sum_{k=1}^{n} mf([a_k, b_k] \cap E) \leq \sum_{k=1}^{n} \int_{[a_k, b_k]} |f'(x)|\, dx$$

by Problem 4.2.7(b).)

4.2.18. Let f be absolutely continuous on $[c, d]$ and g be absolutely continuous on $[a, b]$, where $[c, d] = g([a, b])$. Prove that $f \circ g$ is absolutely

† Y. Katznelson and K. Stromberg, *Amer. Math. Monthly* **81**(4), 349–354 (1974).

‡ D. E. Varberg, *Amer. Math. Monthly* **72**, 831–841 (1965).

continuous on $[a, b]$ if and only if $f \circ g$ is of bounded variation on $[a, b]$. Note that if $f(x) = x^{1/2}$ and $g(x) = x^2 \mid \sin(1/x) \mid$ for $x > 0$, $g(x) = 0$ for $x = 0$, then f and g are both absolutely continuous on [0, 1] even though $f \circ g$ is not.

4.2.19. *An example of a differentiable function f on* [0, 1] *such that f' is continuous only on* (0, 1].

(i) Show that the function given in Problem 4.1.2 is such a function.

(ii) Show that such a function f is absolutely continuous if and only if f' is Lebesgue integrable.

(iii) Prove that finding an example of such a function which is also absolutely continuous is equivalent to finding a function g on [0, 1] such that (a) g is continuous on (0, 1], (b) $\lim_{t \to 0^+} g(t)$ does not exist, and (c) $\lim_{x \to 0^+}(1/x) \int_0^x g(t)\, dt$ exists. Show that

$$g(t) = \begin{cases} 1, & \text{if } \dfrac{1}{2^n + 1} \leq t \leq \dfrac{1}{2^n}, \quad 0 \leq n < \infty, \\ 0, & \text{otherwise,} \end{cases}$$

is an example of such a function.

4.2.20. A real-valued function f on $[a, b]$ is *absolutely continuous with respect to the real function* g *on* $[a, b]$ if for each $\varepsilon > 0$ there exists a $\delta > 0$ such that if $\{(a_i, b_i)\}_{i=1}^n$ is a finite collection of nonoverlapping subintervals of $[a, b]$ and $\sum_{i+1}^n \mid g(b_i) - g(a_i) \mid < \delta$, then $\sum_{i=1}^n \mid f(b_i) - f(a_i) \mid < \varepsilon$. Prove the following assertions.

(i) If f is of bounded variation on $[a, b]$, then f is absolutely continuous with respect to v_f on $[a, b]$.

(ii) If f is of bounded variation on $[a, b]$, and g is nondecreasing on $[a, b]$, then f is absolutely continuous with respect to g if and only if v_f is absolutely continuous with respect to g on $[a, b]$.

4.2.21. *Absolute Continuity in Metric Spaces.* Let f: $M_1 \to M_2$ be a function where (M_1, d_1) and (M_2, d_2) are metric spaces. Define f to be *absolutely continuous* if for each $\varepsilon > 0$, there exists $\delta > 0$ such that if $\{a_1, a_2, \ldots, a_n\}$ is a finite set in M_1 such that $\sum_{i=1}^{n-1} d_1(a_i, a_{i+1}) < \delta$, then $\sum_{i=1}^{n-1} d_2(f(a_i), f(a_{i+1})) < \varepsilon$. Suppose that f is continuous. Do there exist equivalent metrics (inducing the same topology) to the given metrics in M_1 and M_2, respectively, such that f is absolutely continuous with respect to these new metrics? [Hint: Define $\varrho(x, y) = d_1(x, y) + d_2(f(x), f(y))$. Consider f: $(M_1, \varrho) \to (M_2, d_2)$.]

4.2.22. (L'Hospital Rule). Assume that f and g are absolutely continuous functions on R and that f' and g' exist except on a set Ω of measure zero. Prove the following assertion.

(i) If for $x \geq a$, $g'(x) > 0$, g is nondecreasing to ∞ as $x \to \infty$, and for some constant B

$$\frac{f'(x)}{g'(x)} < B,$$

then

$$\overline{\lim_{x\to\infty}}\, f(x)/g(x) \leq B.$$

(ii) Assume that as $x \uparrow \infty$, $x \downarrow -\infty$, $x \uparrow x_0$, or $x \downarrow x_0$, then for $x \in \Omega$, $|g(x)| \to \infty$ and either $g'(x) < 0$ or $g'(x) > 0$. Then, as $x \uparrow \infty$, $x \downarrow -\infty$, $x \uparrow x_0$, or $x \downarrow x_0$,

$$\underline{\lim}\,\frac{f'(x)}{g'(x)} \leq \underline{\lim}\,\frac{f(x)}{g(x)} \leq \overline{\lim}\,\frac{f(x)}{g(x)} \leq \overline{\lim}\,\frac{f'(x)}{g'(x)}.$$

[Hint: Prove the right inequality for the case where $x \uparrow \infty$ and $g'(x) > 0$ by using (i). By means of the transformations $x = -y$, $x = x_0 \pm 1/y$, $g = -g$, and $f = -f$ the other cases and the left inequality can be established.] For additional information the reader can consult the paper "Note on the Bernoulli–L'Hospital Rule" by A. M. Ostrowski in *Amer. Math. Monthly* **83**, 239–242 (1976).

(iii) Show that the result of (ii) may fail if g is not absolutely continuous [Hint: Let $g(x) = -1/x + k(x)$, where k is the Cantor ternary function defined for all nonnegative real numbers by the recurrent extension of k to each interval $(i, i+1]$, $i = 1, 2 \ldots$ given by

$$k(x) = k(i) + k(x - i) \qquad \text{for } 1 \leq i < x \leq i + 1.]$$

4.2.23. (i) Let $f(\geq 0)$ be a Lebesgue-integrable function on R. Show that there is a set A with Lebesgue measure zero such that, for $x \notin A$,

$$\lim_{\substack{h_1\to 0^+\\ h_2\to 0^+}} \frac{1}{h_1 + h_2} \int_{x-h_1}^{x+h_2} f(t)\, dt = f(x).$$

[Hint: Use Proposition 4.5.]

(ii) *The Lebesgue Density Theorem.* A point x is a point of density of a Lebesgue-measurable set A if

$$\lim_{\substack{m(I)\to 0\\ I\in\mathscr{I}_x}} \frac{m(I \cap A)}{m(I)} = 1,$$

where $\mathscr{I}_x = \{J\colon J \text{ is an open interval containing } x\}$. Prove that for every

Lebesgue-measurable set A, there is a measurable subset $B \subset A$ with $m(B) = 0$ such that each point of $A - B$ is a point of density of A.

★ **4.2.24.** *Periods of Measurable Functions.* A real-valued function f defined on R has period t if $f(x + t) = f(x)$ for all x. For a periodic function, *either* there exists the smallest positive period t_0 and all periods are of the form nt_0, where n is any integer, *or* the set of the periods is dense. The characteristic function of the rationals has any rational number as its period. Prove the following assertions:

(i) If t is a positive period of a measurable function f and $\int_0^t |f(x)|\, dx < \infty$, then the limit

$$L(f) = \lim_{x\to\infty} \frac{1}{2x} \int_{-x}^{x} f(s)\, ds$$

exists and equals $(1/t) \int_0^t f(s)\, ds$.

(ii) If a measurable function has a dense set of periods, then it is constant a.e.

{Hint: Let $g = f/(1 + |f|)$ and $G(x) = \int_0^x g(s)\, ds$. Then by (i),

$$L(g) = \frac{[G(x + t_n) - G(x)]}{t_n},$$

where (t_n) are periods of f, $t_n \to 0+$ as $n \to \infty$. Hence $g(x) = G'(x) = L(g)$ a.e..} This result is due to A. Lomnicki.

4.2.25. *The Cantor Function Revisited.* Consider the Lebesgue measure m on $[0, 1]$. For each $x \in (0, 1]$, consider the unique nonterminating binary representation $x = .x_1x_2x_3...$ where each $x_i = 0$ or 1. Define f_k: $(0, 1] \to \{0, 1\}$ by $f_k(x) = x_k$. Let $f_k(0) = 0$. Define the functions f, g, and h as follows:

$$f(x) = \sum_{k=1}^{\infty} (1/2^k) f_k(x), \qquad g(x) = \sum_{k=1}^{\infty} (1/2^{2k}) f_{2k}(x)$$

and

$$h(x) = \sum_{k=1}^{\infty} (1/2^{2k+1}) f_{2k+1}(x).$$

Let us also define $F(x) = m\{y \in [0, 1]: f(y) \leq x\}$, and similarly, the functions G and H corresponding to g and h, respectively. (The functions F, G, and H are called the distribution functions of f, g, and h, respectively.)

Show that the distribution function of $3g$, i.e., the function $G(x/3)$ is the Cantor function corresponding to the Cantor "middle-1/2" set [i.e., the set obtained by removing the intervals $(1/4, 3/4)$, $(1/4^2, 3/4^2)$, $(3/4 + 1/4^2, 3/4 + 3/4^2)$, etc.]. Thus, $G'(x) = 0$ a.e. Similarly, show also that $H'(x) = 0$ a.e., whereas the function F is given by

$$F(x) = \begin{cases} x, & \text{for } x \in [0, 1], \\ 0, & \text{for } x < 0, \\ 1, & \text{for } x > 1. \end{cases}$$

Note that for $k_1 < k_2 < \cdots < k_n$,

$$m\{x: f_{k_i}(x) = u_i \text{ when } i = 1, 2, \ldots, n\} = 1/2^n,$$

where $(u_1, u_2, \ldots, u_n)$ is any string of zeros and ones. Use this to show that g and h are independent, i.e.,

$$m\{x: g(x) \in B_1, h(x) \in B_2\} = m(g^{-1}(B_1)) \cdot m(h^{-1}(B_2)),$$

where B_1 and B_2 are any two Borel subsets of $[0, 1]$. Use this to show that $F(x) = \int_0^1 G(y - x)\, dH(y)$. [Hint for the first part: Let $g_0 = 3g$. Then $g_0(0) = 0$, $g_0(1) = 1$, and $0 \leq g_0 \leq 1$. Note that if $x_2 = 1$, then $g_0(x) \geq 3/4$; if $x_2 = 0$, then $g_0(x) \leq (3/4^2)[1 + 1/4 + 1/4^2 + \cdots] = 1/4$. Thus, $G_0(x)$ is constant on $(1/4, 3/4)$. Similarly, if $x_2 = 0$ and $x_4 = 1$, then $g_0(x) \geq 3/4^2$; and if $x_2 = 0 = x_4$, then $g_0(x) \leq (3/4^3)[1 + 1/4 + 1/4^2 + \cdots] = 1/4^2$. Thus, $G_0(x)$ is constant on $(1/4^2, 3/4^2)$, and so on.]

4.3. Integration versus Differentiation II: Absolutely Continuous Measures, Signed Measures, and the Radon-Nikodym Theorem

In the preceding section we characterized indefinite integrals of integrable functions f on $[a, b]$ as absolutely continuous functions. We shall now generalize and extend the notion of an indefinite integral to an arbitrary measure space $(X, \mathscr{B}, \mu)$. If f is an integrable function on X with respect to μ, for each E in $\mathscr{B}$ let us consider the real number $\nu(E)$ given by the "indefinite integral"

$$\nu(E) = \int_E f\, d\mu. \tag{4.22}$$

What are some properties of the real-valued set function ν defined on $\mathscr{B}$ by equation (4.22)? Clearly $\nu(\varnothing) = 0$ and $\nu(\bigcup_{i=1}^{\infty} E_i) = \sum_{i=1}^{\infty} \nu(E_i)$ for any sequence (E_i) of pairwise disjoint, measurable sets in $\mathscr{B}$. However, since ν is not necessarily nonnegative it falls short of being a measure. Nevertheless, it does fulfill the requirements of being a "signed measure."

Definition 4.5. Let $(X, \mathscr{B})$ be a measurable space. An extended real-valued function ν on $\mathscr{B}$ is called a *signed measure* provided the following conditions hold:

(i) $\nu(\varnothing) = 0$.

(ii) ν assumes at most one of the values $+\infty$ and $-\infty$.

(iii) For all pairwise disjoint sequences (E_i) of measurable sets in $\mathscr{B}$, $\nu(\bigcup_{i=1}^{\infty} E_i) = \sum_{i=1}^{\infty} \nu(E_i)$, where equality means that the series on the right converges absolutely if $\nu(\bigcup_{i=1}^{\infty} E_i)$ is finite and properly diverges otherwise. ∎

Condition (iii) immediately yields the following property of ν:

(iv) If $A_n \supset A_{n+1}$, $A_n \in \mathscr{B}$ for $1 \le n < \infty$, and $\nu(A_1) \ne \pm\infty$, then $\nu(\bigcap_{n=1}^{\infty} A_n) = \lim_{n\to\infty} \nu(A_n)$.

Observe that the signed measure defined by equation (4.22) has the following property: If $\mu(E) = 0$, then $\nu(E) = 0$. If ν is any signed measure on $(X, \mathscr{B})$ and μ is any measure on $(X, \mathscr{B})$ such that $\mu(E) = 0$ implies $\nu(E) = 0$ for all E in $\mathscr{B}$, then we say ν is "absolutely continuous" with respect to μ and write $\nu << \mu$. This concept will be precisely defined below.

Our goal in this section is to show that if ν is any signed measure on $(X, \mathscr{B})$ absolutely continuous with respect to a certain type of measure μ on $(X, \mathscr{B})$, then ν has an integral representation as in equation (4.22).

As a beginning let us look at a particular situation. Let f be a bounded distribution function—a bounded real-valued nondecreasing function on R continuous from the right—and let μ_f be the unique corresponding Borel measure such that, for all a and b in R,

$$\mu_f(a, b] = f(b) - f(a).$$

Since f is differentiable a.e. and f' is thus a nonnegative measurable function, we can define a measure ν on the class of Borel sets by

$$\nu(E) = \int_E f'(x)\, dx. \tag{4.23}$$

In particular, for the Borel set $(a, b]$,

$$\nu(a, b] = \int_a^b f'(x)\, dx \leq f(b) - f(a).$$

Thus ν is a Borel measure and the question of whether $\nu = \mu_f$ is equivalent to the question of whether

$$\int_a^b f'(x)\, dx = f(b) - f(a), \tag{4.24}$$

for all intervals $(a, b]$. By Theorems 4.3 and 4.4, a necessary and sufficient condition for equation (4.24) to be satisfied is that f be absolutely continuous on R. We have almost proved the necessity condition in the next proposition.

Proposition 4.9. Let μ_f be the Borel measure corresponding to a bounded distribution function f on R. The function f is absolutely continuous on R if and only if $\mu_f << m$, that is, whenever E is a Borel set with $m(E) = 0$ then $\mu_f(E) = 0$. ∎

Proof. If f is absolutely continuous, then $\mu_f = \nu$, where ν is defined in equation (4.23). Hence $\mu_f << m$.

Conversely, suppose $\mu_f << m$. The measure μ_f then satisfies the condition that given $\varepsilon > 0$, there exists a $\delta > 0$ such that for every Borel-measurable set with $m(E) < \delta$, $\mu_f(E) < \varepsilon$. Indeed, if not then there exists some $\varepsilon_0 > 0$ and a sequence E_n of measurable sets such that for each n

$$m(E_n) < 1/2^n, \qquad \text{but } \mu_f(E_n) \geq \varepsilon_0.$$

Let $E_0 = \bigcap_{k=1}^{\infty}(\bigcup_{n=k}^{\infty} E_n)$. Then

$$m(E_0) \leq \sum_{n=k}^{\infty} m(E_n) < \frac{1}{2^{k-1}}, \qquad \text{for all } k,$$

so that $m(E_0) = 0$. However since $\mu_f(\bigcup_{n=1}^{\infty} E_n) < \infty$, using Proposition 2.2 we obtain

$$\mu_f(E_0) = \lim_{k\to\infty} \mu_f\left(\bigcup_{n=k}^{\infty} E_n\right) \geq \lim_{k\to\infty} \mu_f(E_k) \geq \varepsilon_0 > 0,$$

which is a contradiction to the fact that $\mu_f << m$. Therefore, given any collection $(a_1, b_1), (a_2, b_2), \ldots, (a_n, b_n)$ of nonoverlapping intervals with

$$\sum_{i=1}^{n} m(a_i, b_i] = \sum_{i=1}^{n} b_i - a_i < \delta,$$

then

$$\mu_f\left(\bigcup_{i=1}^{n}(a_i, b_i]\right) = \sum_{i=1}^{n} \mu_f(a_i, b_i] = \sum_{i=1}^{n} |f(b_i) - f(a_i)| < \varepsilon.$$

Thus f is absolutely continuous. ∎

It is important to observe that f must be a bounded function in Proposition 4.9. For example if f is the function

$$f(x) = \begin{cases} 0, & \text{if } x < 0, \\ x^2, & \text{if } x \geq 0, \end{cases}$$

then

$$\mu_f(E) = \int_{E\cap[0,\infty)} 2x\, dx, \qquad \text{for all Borel sets } E,$$

so that $\mu_f << m$ but f is not absolutely continuous on R.

Let us note the following conclusions from the preceding discussion. If μ_f is the Borel measure corresponding to the bounded distribution function f, then $\mu_f << m$ if and only if f is absolutely continuous on R. However, a necessary and sufficient condition for

$$\int_{(a,b]} f'(x)\, dx = f(b) - f(a)$$

for all intervals $(a, b]$, is the absolute continuity of f on R. Thus we may conclude from the uniqueness of μ_f that, if $\mu_f << m$, then

$$\mu_f(E) = \int_E f'(x)\, dx \tag{4.25}$$

for all Borel sets E. Conversely, if equation (4.25) is satisfied for all Borel sets E, then $\mu_f << m$.

We can generalize this discussion in the following manner. Suppose f is a bounded variation function on R ($V_{-\infty}^{\infty} f < \infty$) that is continuous from the right. Then f can be written as the difference $f_1 - f_2$ of two bounded monotonic functions each continuous from the right ($f_1 = V_{-\infty}^{x} f = \lim_{a\to\infty} V_{-a}^{x} f$ and $f_2 = V_{-\infty}^{x} f - f$). Let μ_1 and μ_2 be the unique finite Borel measure corresponding to f_1 and f_2, respectively, such that

$$\mu_i(a, b] = f_i(b) - f_i(a), \qquad i = 1, 2,$$

for all intervals $(a, b]$. Let ν_1 and ν_2 be the measures given by

$$\nu_i(E) = \int_E f_i'(x)\, dx, \qquad i = 1, 2$$

for each Borel set E. Using Proposition 4.7, as above we can verify that f is absolutely continuous if and only if f_1 and f_2 are absolutely continuous, which is true if and only if $\nu_i = \mu_i$ for $i = 1$ and 2. Consequently, the absolute continuity of f implies that ν given by

$$\nu(E) = \nu_1(E) - \nu_2(E) = \int_E f'(x)\, dx$$

is a well-defined signed measure equal to $\mu_f \equiv \mu_1 - \mu_2$. Clearly $\nu << m$ so that $\mu_f << m$ when f is absolutely continuous.

Analogous to the argument given in Proposition 4.9, the reader can show that if $\mu_f << m$ then, necessarily f is absolutely continuous. We may conclude as before then that $\mu_f << m$ if and only if

$$\mu_f(E) = \int_E f'\, dx,$$

where f is the distribution corresponding to μ_f.

Having seen the implication of absolute continuity of a distribution function or a bounded variation function on the associated Borel measure, it does seem appropriate to say that μ_f is absolutely continuous with respect to m if $m(E) = 0$ implies $\mu_f(E) = 0$. Let us formalize this concept.

Definition 4.6. Let $(X, \mathscr{B}, \mu)$ be a measure space and ν a signed measure on $\mathscr{B}$. ν is *absolutely continuous* with respect to μ, written $\nu << \mu$, if whenever E is a set in $\mathscr{B}$ with $\mu(E) = 0$, then $\nu(E) = 0$. ▮

Remark 4.7. In the case that $|\nu(E)| < \infty$ whenever $\mu(E) < \infty$, $\nu << \mu$ if and only if for each $\varepsilon > 0$, there is a $\delta > 0$ such that whenever $\mu(E) < \delta$ then $|\nu(E)| < \varepsilon$. See Problem 4.3.1.

Later in this section we will further discuss the concept of absolute continuity of measures when we characterize signed measures absolutely continuous with respect to a measure. First we must "decompose" any signed measure as the difference of two measures.

Let ν be a signed measure on $(X, \mathscr{B})$. A set P in $\mathscr{B}$ is called *positive* with respect to ν if for every measurable set E in $\mathscr{B}$, $\nu(P \cap E) \geq 0$. A set N in $\mathscr{B}$ is called *negative* if for every measurable set E, $\nu(N \cap E) \leq 0$. A set that is both positive and negative with respect to ν is called *a null set*.

Remarks

4.8. A measurable set is a null set if and only if every measurable subset of it has ν-measure zero. However, a set of ν-measure zero is not necessarily a null set unless ν is a measure. For example, if ν is the signed measure on the Lebesgue measurable subsets of R given by

$$\nu(E) = \int_E f(x)\,dx,$$

where $f(x) = x$ for $-1 \leq x \leq 1$ and 0 otherwise, then $[-1, 1]$ is a set of ν-measure zero but is not a null set.

4.9. Every measurable subset of a positive (negative) set is itself positive (negative). A set of positive (negative) measure is not necessarily a positive (negative) set. A countable union of positive (negative) sets is positive (negative).

Theorem 4.5. If ν is a signed measure on $(X, \mathscr{B})$, then there exist two disjoint sets P and N in X such that $X = P \cup N$, P is positive, and N is negative. ∎

This theorem is known as the *Hahn Decomposition Theorem* and the sets P and N are said to form a *Hahn decomposition* of X with respect to ν.

To prove Theorem 4.5 we make use of the following lemma. We will assume in the lemma that $+\infty$ is the value omitted by ν.

Lemma 4.4. Every $E \in \mathscr{B}$ with $\nu(E) \neq -\infty$ contains a positive set P with $\nu(P) \geq \nu(E)$.

Proof. To every $\varepsilon > 0$, there corresponds a set E_ε in B such that $E_\varepsilon \subset E$, $\nu(E_\varepsilon) \geq \nu(E)$, and $\nu(A) > -\varepsilon$ whenever $A \subset E_\varepsilon$. Indeed, otherwise inductively there is a sequence $A_1 \subset E, \ldots, A_k \subset E - (A_1 \cup \cdots \cup A_{k-1}), \cdots$ such that $\nu(A_k) \leq -\varepsilon$. If $A = \bigcup_{k=1}^\infty A_k$, then $\nu(A) = -\infty$, so that

$$\nu(E) = \nu(A) + \nu(E - A) \tag{4.26}$$

would be $-\infty$, a contradiction.

Choose $\varepsilon_i \to 0$ and $E_{\varepsilon_1} \supset E_{\varepsilon_2} \supset \cdots$. Let $P = \bigcap_{i=1}^\infty E_i$. Then P is positive and $\nu(P) \geq \nu(E)$ by Definition 4.5(iv). ∎

Proof of Theorem 4.5. Since ν assumes at most one of the values of $+\infty$ or $-\infty$ we again assume that $+\infty$ is the value never attained. Let $\alpha = \sup\{\nu(A)\colon A \in \mathscr{B}\}$. Then there is a sequence P_n in $\mathscr{B}$ such that $\nu(P_n) \to \alpha$ and by Lemma 4.4 we may assume that the P_n are positive. Let $P = \bigcup_{n=1}^\infty P_n$. Then P is positive and $\nu(P) = \alpha$. However, $N = X - P$ is negative, for if $E \subset N$ and $\nu(E) > 0$ then $\nu(P \cup E) > \alpha$, which is impossible. ▮

The Hahn decomposition of a measurable space with respect to a signed measure ν need not be unique. However, it can easily be shown that if $X = P_1 \cup N_1$ and $X = P_2 \cup N_2$ are two Hahn decompositions of X, then

$$\nu(P_1 \triangle P_2) = 0 \qquad \text{and} \qquad \nu(N_1 \triangle N_2) = 0.$$

Therefore, if E is any measurable set in $\mathscr{B}$,

$$\nu(E \cap P_1) = \nu\big(E \cap (P_1 \cup P_2)\big) = \nu(E \cap P_2)$$

and

$$\nu(E \cap N_1) = \nu\big(E \cap (N_1 \cup N_2)\big) = \nu(E \cap N_2).$$

Definition 4.7. Two measures μ_1 and μ_2 on $(X, \mathscr{B})$ are said to be *mutually singular*, written $\mu_1 \perp \mu_2$, if there are disjoint measurable sets A and B in X with $X = A \cup B$ such that $\mu_1(A) = \mu_2(B) = 0$. ▮

The concepts of mutual singularity and absolute continuity are in a sense opposite concepts. If $\mu_1 \perp \mu_2$ then each nonzero value of μ_1 is taken on a set where μ_2 is zero. If in addition $\mu_1 << \mu_2$, then μ_1 must be the zero measure.

Proposition 4.10. Let ν be a signed measure on $(X, \mathscr{B})$. Then there is a unique pair of mutually singular measures ν^+ and ν^- such that $\nu = \nu^+ - \nu^-$. Moreover, if ν_1 and ν_2 are measures such that $\nu = \nu_1 - \nu_2$, then $\nu^+(E) \leq \nu_1(E)$ and $\nu^-(E) \leq \nu_2(E)$ for all E in $\mathscr{B}$. ▮

Proof. For E in $\mathscr{B}$ define ν^+ and ν^- by

$$\nu^+(E) = \nu(E \cap P) \qquad \text{and} \qquad \nu^-(E) = -\nu(E \cap N),$$

where P and N form a Hahn decomposition of X. The uniqueness of ν^+ and ν^- is left as Problem 4.3.2 as is the proof of the last statement (Problem 4.3.3). ▮

The measures ν^+ and ν^- defined in the proof of Proposition 4.10 are called the *positive variation* and *negative variation* of ν, respectively. The measure $|\nu|$ given by

$$|\nu|(E) = \nu^+(E) + \nu^-(E)$$

is called the *total variation* of ν. The representation of ν as $\nu^+ - \nu^-$ is called the *Jordan decomposition* of ν.

Example 4.3. Let $(X, \mathscr{B}, \mu)$ be a measure space and let ν be defined by

$$\nu(E) = \int_E f \, d\mu,$$

where f is an integrable function on X. Writing f as $f^+ - f^-$,

$$\nu(E) = \int_E f^+ \, d\mu - \int_E f^- \, d\mu.$$

Is $\nu^+(E) = \int_E f^+ \, d\mu$ and $\nu^-(E) = \int_E f^- \, d\mu$ for each E in $\mathscr{B}$? Let P be the measurable set $\{x \in X: f(x) > 0\}$. Clearly P is a positive set and P^c is negative. By definition,

$$\nu^+(E) = \nu(E \cap P) = \int_{E \cap P} f \, d\mu = \int_E f^+ \, d\mu$$

and

$$\nu^-(E) = -\nu(E \cap P^c) = -\int_{E \cap P^c} f \, d\mu = \int_E f^- \, d\mu.$$

Thus the Jordan decomposition of ν is generated by the decomposition of the measurable function f into its positive and negative parts.

The Jordan decomposition of a signed measure ν into the difference of two measures (nondecreasing set functions) is reminiscent of the decomposition of a function of bounded variation, in particular an absolutely continuous function, as the difference of two nondecreasing functions. In fact we have the following equality analogous to the definition of the total variation of a function.

Proposition 4.11. Let ν be a signed measure on $(X, \mathscr{B})$. Then

$$|\nu|(E) = \sup\left\{\sum_{k=1}^{n} |\nu(E_k)| : E = \bigcup_{k=1}^{n} E_k, \, E_k \in \mathscr{B}, \, E_k \cap E_j = \varnothing \text{ if } k \neq j\right\}. \blacksquare$$

The proof of this proposition is left as Problem 4.3.9.

We now state the main result of this section—a realization of the goal of this section.

Theorem 4.6. *Radon–Nikodym Theorem.* Let μ be a σ-finite measure and ν a signed measure on $(X, \mathscr{B})$. Assume ν is absolutely continuous with respect to μ. Then there is a μ-measurable function f_0 such that for each set E in $\mathscr{B}$ we have

$$\nu(E) = \int_E f_0 \, d\mu.$$

The function f_0 is unique in the sense that if g is any measurable function with this property, then $g = f_0$ a.e. $[\mu]$. ∎

In the proof of this outstanding result we need the following lemma.

Lemma 4.5. Let $(X, \mathscr{B}, \mu)$ be a measure space and let ν be a measure on $\mathscr{B}$. Assume both μ and ν are finite. Let $\mathscr{C}$ denote the collection of all μ-integrable nonnegative functions f such that

$$\nu(E) \geq \int_E f \, d\mu, \tag{4.27}$$

for all E in $\mathscr{B}$. Then there is a function f_0 in $\mathscr{C}$ such that

$$\int f_0 \, d\mu = \sup_{f \in \mathscr{C}} \int f \, d\mu. \quad ∎ \tag{4.28}$$

Proof. Let α be the finite real number which is the supremum in equation (4.28) and let (f_n) be a sequence of functions in $\mathscr{C}$ such that $\int f_n \, du \to \alpha$. Let (g_n) be the nondecreasing sequence of μ-integrable functions given by

$$g_n(x) = \max \{ f_1(x), \ldots, f_n(x) \}.$$

Then each g_n is in $\mathscr{C}$ since, for any set E in $\mathscr{B}$,

$$\int_E g_n \, d\mu = \sum_{i=1}^{n} \int_{E_i} g_n \, d\mu \leq \sum_{i=1}^{n} \nu(E_i) = \nu(E),$$

where E_i are disjoint measurable sets such that $E = \bigcup_{i=1}^{n} E_i$ and $g_n = f_i$ on E_i. Since for each n

$$\int g_n \, d\mu \leq \nu(X) < \infty,$$

the Monotone Convergence Theorem guarantees the existence of *a* μ-integrable function f_0 such that $g_n \uparrow f_0$ and

$$\int_E f_0 \, d\mu = \lim_n \int_E g_n \, d\mu \leq \nu(E)$$

for all E in $\mathscr{B}$. Since the g_n are in $\mathscr{C}$ we have the following inequalities:

$$\alpha = \lim_n \int f_n \, d\mu \leq \lim_n \int g_n \, d\mu = \int f_0 \, d\mu \leq \alpha,$$

which establishes the result. ∎

Proof of Theorem 4.6. We will divide the proof of the existence of the function f_0 into several parts, each part being a generalization of the preceding part.

(i) Assume first μ and ν are finite measures. By Lemma 4.5, there is a nonnegative μ-integrable function f_0 such that

$$\nu(E) \geq \int_E f_0 \, d\mu \tag{4.29}$$

for all E in $\mathscr{B}$ and $\int f_0 \, d\mu \geq \int f \, d\mu$ for any nonnegative function f such that

$$\nu(E) \geq \int_E f \, d\mu$$

for all E in $\mathscr{B}$. We wish to establish that inequality (4.29) is an equality.

Let ν_1 be the measure given by

$$\nu_1(E) = \nu(E) - \int_E f_0 \, d\mu.$$

Clearly $\nu_1(X) < \infty$, and since $\nu << \mu$, also $\nu_1 << \mu$. If ν_1 is not the zero measure, then $\nu_1(X) > 0$. Hence, for some constant $k > 0$,

$$\mu(X) - k\nu_1(X) < 0.$$

Let P and N be a Hahn decomposition for the signed measure $\mu - k\nu_1$. We assert that $\mu(N) > 0$. Indeed if $\mu(N) = 0$, then $\nu_1(N) = 0$ since $\nu_1 << \mu$. However, then

$$0 \leq (\mu - k\nu_1)(P \cap X) = \mu(P \cap X) - k\nu_1(P \cap X) = \mu(X) - k\nu_1(X) < 0.$$

Since this is clearly false, we must have $\mu(N) > 0$.

Define h on X by $h(x) = 1/k$ if $x \in N$ and $h(x) = 0$ if $x \notin N$. If $E \in \mathscr{B}$,

$$\int_E h \, d\mu = \frac{1}{k} \mu(N \cap E) \leq \nu_1(N \cap E) \leq \nu_1(E) = \nu(E) - \int_E f_0 \, d\mu.$$

Thus

$$\int_E (h + f_0)\, d\mu \leq \nu(E).$$

Since $h > 0$ on N, $h + f_0 > f_0$ on a set of positive μ-measure. Hence using inequality (4.29), we obtain

$$\int f_0\, d\mu < \int h + f_0\, d\mu,$$

a contradiction. Hence our original assumption that ν_1 is not the zero measure is false, so that (4.29) is an equality.

(ii) Assume μ is a finite measure and ν is a σ-finite measure. Let $X = \cup X_n$, X_n pairwise disjoint with finite ν-measure. By part (i) for each positive integer n, there exists a nonnegative measurable function f_n such that

$$\nu_n(E) = \int_E f_n\, d\mu,$$

for all E in $\mathscr{B}$, where ν_n is the finite measure $\nu_n(E) = \nu(E \cap X_n)$. Thus

$$\nu(E) = \sum_n \nu(E \cap X_n) = \int_E f_0\, d\mu,$$

where $f_0 = \Sigma f_n$.

(iii) Assume now that μ is a finite measure and ν is an arbitrary measure. Let $\mathscr{C}$ be the nonempty class of measurable subsets of X that can be written as a countable union of measurable sets with finite ν-measure. Let $\alpha = \sup \{\mu(A): A \in \mathscr{C}\}$; there exists a sequence (A_i) of sets from $\mathscr{C}$ such that $\mu(A_i) \to \alpha$. Let $A = \bigcup_{i=1}^{\infty} A_i$, a set in $\mathscr{C}$. Clearly $\alpha = \mu(A)$, and by (ii) there exists a nonnegative measurable function f_1 such that, for all sets E in $\mathscr{B}$,

$$\nu_A(E) = \int_E f_1\, d\mu, \tag{4.30}$$

where ν_A is the σ-finite measure given by $\nu_A(E) = \nu(A \cap E)$. Now if $\mu(E \cap A^c) = 0$, then $\nu(E \cap A^c) = 0$ since $\nu << \mu$. However, if $\mu(E \cap A^c) > 0$, then $\nu(E \cap A^c) = \infty$. Indeed, if $\nu(E \cap A^c) < \infty$, then $E \cap A^c \in \mathscr{C}$ and

$$\alpha = \mu(A) < \mu(A) + \mu(E \cap A^c) = \mu\big(A \cup (E \cap A^c)\big) \leq \alpha,$$

which is a contradiction. Therefore, no matter what the value of $\mu(E \cap A^c)$,

$$\nu(E \cap A^c) = \int_{E \cap A^c} \infty \, d\mu.$$

Defining f_0 on X as f_1 on A and ∞ on A^c, equation (4.30) gives

$$\nu(E) = \nu(A \cap E) + \nu(A^c \cap E) = \nu_A(E) + \nu(A^c \cap E) = \int_E f_0 \, d\mu$$

for all E in $\mathscr{B}$.

(iv) Assume now that μ is a σ-finite measure and ν is an arbitrary measure. Then $X = \cup X_n$, where the X_n are pairwise disjoint and $\mu(X_n) < \infty$ for each n. For each n there exists a nonnegative measurable function f_n such that for all sets E in $\mathscr{B}$ with $E \subset X_n$,

$$\nu(E) = \int_E f_n \, d\mu_n,$$

where μ_n is the finite measure given by $\mu_n(A) = \mu(A \cap X_n)$. Let $f_0 = \sum_{n=1}^{\infty} f_n$. Then f_0 is a nonnegative measurable function and

$$\nu(E) = \sum_n \nu(E \cap X_n) = \sum_n \int_{E \cap X} f_n \, d\mu_n = \sum_n \int_E f_n \, d\mu = \int_E f_0 \, d\mu$$

for all E in $\mathscr{B}$.

(v) Assume now finally that μ is a σ-finite measure and ν is an arbitrary signed measure. By Proposition 4.10 we can write ν as $\nu^+ - \nu^-$, where ν^+ and ν^- are measures—one being finite. By part (iv), there exist nonnegative functions f_1 and f_2 such that

$$\nu^+(E) = \int_E f_1 \, d\mu \quad \text{and} \quad \nu^-(E) = \int_E f_2 \, d\mu$$

for all E in $\mathscr{B}$. Letting $f_0 = f_1 - f_2$ (since one of f_1 or f_2 is integrable, one may be assumed to be real valued and hence $f_1 - f_2$ is well defined), we have

$$\nu(E) = \int_E f_0 \, d\mu$$

for all E in $\mathscr{B}$. The existence of f_0 is established.

The uniqueness of the function f_0 remains to be established. Suppose f_0 and g are both measurable functions such that, for all E in $\mathscr{B}$,

$$\nu(E) = \int_E f_0 \, d\mu \quad \text{and} \quad \nu(E) = \int_E g \, d\mu.$$

In particular, since μ is σ-finite and for each E in $\mathscr{B}$,

$$\int_E f_0 \, d\mu \leq \int_E g \, d\mu$$

we know from Proposition 3.8 that $f_0 \leq g$ a.e. $[\mu]$. Similarly $g \leq f_0$ a.e. $[\mu]$. ∎

Remark 4.10. The assumption that μ be σ-finite is vital and cannot be removed. As an example, let μ be a counting measure on the Lebesgue-measurable subsets of $[0, 1]$ and ν the Lebesgue measure on this same measure space. No function can be found satisfying Theorem 4.6.

Remark 4.11. It is important to point out that a theorem more general than Theorem 4.6 was first given by Bochner [8] for finitely additive bounded signed measures. He showed that if a finitely additive bounded signed measure ν on an algebra $\mathscr{A}$ of subsets of X is absolutely continuous with respect to a finite nonnegative finitely additive measure μ on $\mathscr{A}$, then given $\varepsilon > 0$, there is a simple function f such that, for any $A \in \mathscr{A}$,

$$\left| \int_A f \, d\mu - \nu(A) \right| \leq \varepsilon.$$

This means that the set Σ of all measures λ, where λ is of the form $\lambda(A) = \int_A f \, d\mu$ (f a simple function), is dense in the metric space Σ_0 of all bounded signed finitely additive measures that are absolutely continuous with respect to μ. [Here the metric is, of course, given by $d(\nu_1, \nu_2) = | \nu_1 - \nu_2 | (X)$.] We know from Chapter 3 that when μ is countably additive and $\mathscr{A}$ is a σ-algebra, then $L_1(\mu)$ is complete in the L_1-norm. Since the mapping $f \to \lambda_f$, $\lambda_f(A) = \int_A f \, d\mu$, from $L_1(\mu)$ into Σ is an isometry (i.e., distance preserving), the set Σ is also complete in Σ_0 and thus, closed in Σ_0. By Bochner's theorem, $\Sigma = \Sigma_0$ and this is the Radon–Nikodym theorem. For an elementary proof of Bochner's theorem, the reader may consult [8].

We have proved in particular that if ν is a measure absolutely continuous with respect to a σ-finite measure μ, then there is a nonnegative measurable function f_0 such that, for any measurable set,

$$\nu(E) = \int \chi_E \cdot f_0 \, d\mu.$$

If $\varphi = \sum a_i \chi_{E_i}$ is a simple function, the additivity of the integral gives

$$\int \varphi \, d\nu = \int \varphi \cdot f_0 \, d\mu.$$

Since any nonnegative measurable function f is the pointwise limit of a nondecreasing sequence (φ_n) of simple functions, the Monotone Convergence Theorem implies that

$$\int_E f \, d\nu = \int_E f \cdot f_0 \, d\mu$$

for all measurable E. Finally if f is integrable with respect to ν, it can be written as the difference of two nonnegative ν-integrable functions $f_1 - f_2$ so that $f_1 f_0$ and $f_2 f_0$ are each μ-integrable and

$$\int_E f \, d\nu = \int_E f_1 \, d\nu - \int_E f_2 \, d\nu = \int_E f_1 f_0 \, d\mu - \int_E f_2 f_0 \, d\mu = \int_E f f_0 \, d\mu$$

for all measurable sets E.

Thus we have proved the following corollary to the Radon–Nikodym Theorem, which is a strengthening of the result for the case when ν is a measure.

Corollary 4.3. Suppose μ and ν are measures on $(X, \mathscr{B})$ with μ σ-finite and $\nu << \mu$. Then there exists a unique nonnegative μ-measurable function f_0 such that (i) for each nonnegative measurable function f on X,

$$\int_E f \, d\nu = \int_E f f_0 \, d\mu$$

and (ii) for each ν-integrable function f, $f f_0$ is μ-integrable and in fact

$$\int_E f \, d\nu = \int_E f f_0 \, d\mu,$$

for all E in $\mathscr{B}$. ∎

The function f_0 whose existence and "uniqueness" is established in the Radon–Nikodym Theorem is appropriately called the *Radon–Nikodym derivative* of ν with respect to μ and is denoted by $d\nu/d\mu$. Observe that the Radon–Nikodym derivative of the Borel measure μ_f (generated by a bounded absolutely continuous distribution function f on R) with respect to Lebesgue measure m is the usual derivative f'.

It is interesting that the Radon–Nikodym derivatives satisfy rules that are analogous to some familiar rules for derivatives of real-valued functions of a real variable, primarily the following.

(i) If $\nu_1 << \mu$ and $\nu_2 << \mu$ with μ σ-finite and a and b reals,

$$\frac{d(a\nu_1 + b\nu_2)}{d\mu} = a\frac{d\nu_1}{d\mu} + b\frac{d\nu_2}{d\mu} \quad \text{a.e. } [\mu].$$

(ii) If $\nu << \mu$ and $\mu << \lambda$ (with μ and λ σ-finite measures), then $\nu << \lambda$ and

$$\frac{d\nu}{d\lambda} = \frac{d\nu}{d\mu} \cdot \frac{d\mu}{d\lambda} \quad \text{a.e. } [\lambda].$$

(iii) If $\nu << \mu$ and $\mu << \nu$ (ν and μ both σ-finite), then

$$\left(\frac{d\nu}{d\mu}\right)^{-1} = \frac{d\mu}{d\nu} \quad \text{a.e. } [\mu].$$

The proofs of these statements are trivial and left as exercises (Problem 4.3.6).

As a consequence of the Radon–Nikodym Theorem we can prove the following decomposition theorem for a σ-finite measure.

Theorem 4.7. *Lebesgue Decomposition Theorem.* Let $(X, \mathscr{B}, \mu)$ be a σ-finite measure space and ν a σ-finite measure defined on $\mathscr{B}$. Then there exist unique measures ν_1 and ν_2 with $\nu_1 \perp \mu$ and $\nu_2 << \mu$ such that $\nu = \nu_1 + \nu_2$. ∎

Proof. The measure $\lambda = \mu + \nu$ is also σ-finite and clearly $\mu << \lambda$ and $\nu << \lambda$. Hence the Radon–Nikodym Theorem asserts the existence of nonnegative measurable functions f_0 and g_0 such that, for all measurable sets E,

$$\mu(E) = \int_E f_0 \, d\lambda \quad \text{and} \quad \nu(E) = \int_E g_0 \, d\lambda.$$

Let $U = \{x\colon f_0(x) > 0\}$ and $V = \{x\colon f_0(x) = 0\}$. Then $X = U \cup V$, $U \cap V = \varnothing$, and $\mu(V) = 0$. Defining ν_1 by

$$\nu_1(E) = \nu(E \cap V),$$

$\nu_1(U) = 0$ and so $\nu_1 \perp \mu$. Defining ν_2 by

$$\nu_2(E) = \nu(E \cap U) = \int_{E \cap U} g_0 \, d\lambda,$$

we have $\nu = \nu_1 + \nu_2$. It remains to show that $\nu_2 << \mu$. If $\mu(E) = 0$, then

$$0 = \int_E f_0 \, d\lambda,$$

so that $f_0 = 0$ a.e. $[\lambda]$ on E. Since $f_0 > 0$ on U, $\lambda(U \cap E) = 0$. Hence,

$$\nu(U \cap E) = \int_{U \cap E} g_0 \, d\lambda = 0,$$

so that $\nu_2(E) = 0$. Uniqueness is left for the reader to prove. ∎

Example 4.4. Let f be a bounded distribution function and μ_f the corresponding Borel measure. As we have already seen in Proposition 4.9 $\mu_f << m$ if and only if f is absolutely continuous. In that case,

$$\mu_f(E) = \int_E f' \, dm$$

(f' being the derivative of f). We assert now that $\mu_f \perp m$ if and only if $f' = 0$ a.e.

To prove this assertion, suppose first that μ_f and m are not mutually singular. Since μ_f and m are σ-finite measures, Theorem 4.7 gives two measures ν_1 and ν_2 such that

$$\mu_f = \nu_1 + \nu_2, \; \nu_1 \perp m, \qquad \nu_2 << m.$$

Now ν_2 is a Borel measure so that to it corresponds a distribution function g_2 given by

$$g_2(x) = \begin{cases} -\nu_2(x, 0], & \text{if } x < 0, \\ 0, & \text{if } x = 0, \\ \nu_2(0, x], & \text{if } x > 0. \end{cases}$$

Since $\nu_2 << m$, g_2 is absolutely continuous and

$$\nu_2(E) = \int_E g_2'(t) \, dt$$

for all Borel sets E. Since ν_2 cannot be zero (otherwise $\mu_f \perp m$), g_2' is positive on a Borel set E such that $m(E) > 0$. If distribution function g_1 corresponds to ν_1, then defining g as $g_1 + g_2$, we have $g' = g_1' + g_2' \geq g_2'$, which is positive on E. Since g is the distribution function of μ_f up to a constant, $f' = g' > 0$. Hence $f' = 0$ implies $\mu_f \perp m$.

Conversely, suppose $f' > 0$ on a Borel set A of measure greater than zero. Since f' is Lebesgue measurable, we can define λ by

$$\lambda(E) = \int_E f'(t)\,dt.$$

Clearly $\lambda << m$ and $\lambda(A) > 0$. Also

$$\lambda(a, b] = \int_a^b f'(t)\,dt \leq f(b) - f(a) = \mu_f(a, b]$$

for all $a < b$ in R and thus for all Borel-measurable sets E, $\lambda(E) \leq \mu_f(E)$. Thus we have found a nonzero σ-finite measure λ such that $\lambda \leq \mu_f$ and $\lambda << m$.

This rules out the possibility that $\mu_f \perp m$. Indeed, if $\mu_f \perp m$, then since $\lambda \leq \mu_f$, $\lambda \perp m$; but $\lambda << m$, which implies that $\lambda = 0$. This is a contradiction.

Using this example, the reader should readily see the analogy between the Lebesgue Decomposition Thcorem and Problem 4.2.6.

Let us indicate that the Lebesgue Decomposition Theorem can be proved easily without using the Radon–Nikodym Theorem even when μ is not σ-finite. Note that if $\nu(X) < \infty$, $\mathscr{F} = \{E: \mu(E) = 0\}$, and $s = \sup\{\nu(E): E \in \mathscr{F}\}$, then there exists $B \in \mathscr{F}$ such that $\nu(B) = s$. Defining $\nu_1(A) = \nu(A - B)$ and $\nu_2(A) = \nu(B \cap A)$, it can be easily verified that ν_1 and ν_2 meet the requirements of the theorem. It is easy to extend this argument when ν is σ-finite.

In the rest of this section, we will show another application of the Radon–Nikodym theorem to characterize the continuous linear functionals on L_p spaces, $1 \leq p < \infty$. We know from Theorem 3.12 that the L_p spaces $(1 \leq p < \infty)$ are Banach spaces in the $\|\cdot\|_p$ norm. For a more complete and useful study of these Banach spaces, it is important to identify the continuous linear functionals on these spaces since applications of many results in functional analysis are only possible when this identification is done. The most important theorem for the L_p spaces in this context is the Riesz Representation Theorem, which is the subject of the rest of the section. To prepare for the presentation of this theorem, we will introduce, for convenience, the concept of a bounded linear operator.

Definition 4.8. Let X and Y be vector spaces over the same scalar field F. Then a mapping T from X into Y is called a *linear operator* if for all

$x_1, x_2 \in X$ and scalars α, β,

$$T(\alpha x_1 + \beta x_2) = \alpha T(x_1) + \beta T(x_2). \qquad \blacksquare$$

Examples

4.5. Let X be the real-valued continuous functions defined on [0, 1] under the 'sup' norm and Y be the reals. Let

$$T(f) = \int_0^1 f(x)\,dx, \qquad f \in X.$$

Then T is a *continuous linear operator* from X into Y.

4.6. Let X be the class of real-valued continuously differentiable functions on [0, 1] and Y be the class of real-valued continuous functions on [0, 1], both under the uniform norm. Let

$$T(f) = \frac{df}{dx}, \qquad f \in X.$$

Then T is a *linear* operator. But T is *not continuous*, since the sequence $x^n/n \to 0$ in X, but the sequence $T(x^n/n) = x^{n-1}$ does not converge to 0 in Y.

4.7. Let X be a n-dimensional normed linear space† over F and let Y be *any* normed linear space over F. Let

$$T\left(\sum_{i=1}^{n} \alpha_i x_i\right) = \sum_{i=1}^{n} \alpha_i y_i,$$

where α_i's are scalars, $\{x_1, x_2, \ldots, x_n\}$ is a basis of X, and $y_1, y_2, \ldots, y_n$ are arbitrarily chosen, but fixed elements of Y. Then T is a linear operator. But T is also *continuous*, since for $x = \sum_{i=1}^{n} \alpha_i x_i$,

$$\| T(x) \| \leq \sum_{i=1}^{n} \| y_i \| \cdot \sup_{1 \leq i \leq n} | \alpha_i | \leq \sum_{i=1}^{n} \| y_i \| \cdot K \cdot \| x \|,$$

where K is a constant such that

$$\sup_{1 \leq i \leq n} | \alpha_i | \leq K \left\| \sum_{i=1}^{n} \alpha_i x_i \right\|$$

for all n-tuples $(\alpha_1, \alpha_2, \ldots, \alpha_n)$. Such a K always exists because of a result

† Recall that we have introduced normed linear spaces in Chapter 3. (See Def. 3.12.)

(that will not be proven here), which states that in a finite-dimensional space any norm is equivalent to the "sup" norm. In particular, $K = 1$ if we take the "sup" norm as the norm in X. It is now clear that a linear operator from a finite-dimensional normed linear space into any normed linear space is continuous.

4.8. Let T be a mapping from l_1 into F defined by

$$T(x) = \sum_{i=1}^{\infty} x_i, \qquad x = (x_i)_{i=1}^{\infty} \in l_1.$$

Then T is *linear*, but *not continuous* if we define a new norm in l_1 by considering it as a subspace of l_∞. T is not continuous since $T(z_n) = 1$ for all n, where

$$z_n = \Big(\underbrace{\frac{1}{n}, \frac{1}{n}, \ldots, \frac{1}{n}}_{n \text{ terms}}, 0, 0, \ldots\Big)$$

and $\| z_n \|_\infty \to 0$ as $n \to \infty$.

Definition 4.9. A linear operator T from a normed linear space X into a normed linear space Y is called *bounded* if there is a positive constant M such that

$$\| T(x) \| \leq M \| x \|, \qquad \text{for all } x \in X. \qquad \blacksquare$$

Proposition 4.12. Let T be a linear operator from a normed linear space X into a normed linear space Y. Then the following are equivalent:

(a) T is continuous at a point.
(b) T is uniformly continuous on X.
(c) T is bounded. ■

Proof. (a) ⇒ (b). Suppose T is continuous at a point x_0. Then given $\varepsilon > 0$, there exists $\delta > 0$ such that $\| x - x_0 \| < \delta \Rightarrow \| Tx - Tx_0 \| < \varepsilon$. Now let y and z be elements in X with $\| y - z \| < \delta$. Then $\| (y - z + x_0) - x_0 \| < \delta$ and therefore $\| T(y - z + x_0) - T(x_0) \| < \varepsilon$, which means that $\| T(y) - T(z) \| < \varepsilon$, by the linearity of T. Hence (b) follows.

(b) ⇒ (c). Suppose T is uniformly continuous on X and not bounded. Hence for each positive integer n there exists $x_n \in X$ such that $\| T(x_n) \| > n \cdot \| x_n \|$. This means that $\| T(x_n/n \| x_n \|) \| > 1$. But this contradicts the continuity of T at the origin since $\| x_n/n \| x_n \| \| \to 0$ as $n \to \infty$.

(c) ⇒ (a) Boundedness of T trivially implies the continuity of T at the origin. ■

The bounded linear operators from a normed linear space X into a normed linear space Y, denoted by $L(X, Y)$, form a vector space where addition of vectors and scalar multiplication of vectors are defined by

$$(T_1 + T_2)(x) = T_1(x) + T_2(x), \qquad (\alpha T)(x) = \alpha \cdot T(x).$$

Let us define on this vector space

$$\| T \| = \sup_{x \neq 0} \frac{\| T(x) \|}{\| x \|}.$$

Equivalently,

$$\| T \| = \sup_{\|x\| \leq 1} \| T(x) \| = \sup_{\|x\| = 1} \| T(x) \| = \sup_{\|x\| < 1} \| T(x) \|.$$

This defines a norm on $L(X, Y)$ and $L(X, Y)$ becomes a normed linear space. The completeness of $L(X, Y)$ in this norm depends upon that of Y. More precisely, we have the following proposition.

Proposition 4.13. $L(X, Y)$ is a Banach space if Y is complete. ∎

Proof. Let (T_n) be a Cauchy sequence in $L(X, Y)$. Then for each $x \in X$, $\| T_n(x) - T_m(x) \| \leq \| T_n - T_m \| \, \| x \|$, so that $\lim_{n\to\infty} T_n(x)$ exists in Y, Y being complete. Let us define $T(x) = \lim_{n\to\infty} T_n(x)$. Then T is a linear operator from X into Y. We wish to show that T is bounded and $\| T_n - T \|$ converges to 0 as $n \to \infty$. To do this, let $\varepsilon > 0$. There exists N such that for $n, m \geq N$, we have $\| T_n - T_m \| < \varepsilon$ or $\| T_n \| \leq \| T_N \| + \varepsilon$. Therefore $\| T(x) \| = \lim_{n\to\infty} \| T_n(x) \| \leq (\| T_N \| + \varepsilon) \cdot \| x \|$. Hence T is bounded. Now, for $x \in X$ and $n \geq N$,

$$\begin{aligned} \| T_n(x) - T(x) \| &= \lim_{m\to\infty} \| T_n(x) - T_m(x) \| \\ &\leq \lim_{m\to\infty} \| T_n - T_m \| \cdot \| x \| < \varepsilon \cdot \| x \|. \end{aligned}$$

Therefore,

$$\| T_n - T \| = \sup_{\|x\| \leq 1} \| T_n(x) - T(x) \| < \varepsilon,$$

if $n \geq N$. The proposition follows. ∎

Definition 4.10. When Y is the scalar field F, which is a Banach space over itself under the absolute-value norm, the elements of $L(X, Y)$ are called the *bounded linear functionals* on X. The class $L(X, F)$ is denoted by X^*. It is called the dual of X. ∎

Example 4.9. Suppose X is a n-dimensional normed linear space under the "sup"-norm over the real numbers R. Let T be a bounded linear functional on X. Then $\| T \| = \sup_{\|x\|=1} | T(x) |$. To compute $\| T \|$, let $\{x_1, x_2, \ldots, x_n\}$ be a basis of X and $T(x_i) = r_i$, $1 \leq i \leq n$. For $x = \sum_{i=1}^n a_i x_i \in X$,

$$| T(x) | \leq \sup_{1 \leq i \leq n} | a_i | \cdot \sum_{i=1}^n | r_i |.$$

Again, if we define $b_i = 1$ if $r_i > 0$, and $b_i = -1$ if $r_i \leq 0$, then for $x = \sum_{i=1}^n b_i x_i$,

$$\| x \| = \sup_{1 \leq i \leq n} | b_i | = 1$$

and

$$| T(x) | = \sum_{i=1}^n b_i r_i = \sum_{i=1}^n | r_i |.$$

From this equality and the above inequality, it follows that

$$\| T \| = \sum_{i=1}^n | r_i |.$$

Now we present the Riesz Representation Theorem. This theorem is one of the many important contributions of F. Riesz in analysis and was proved by him in 1907 and 1909 for the Lebesgue measure space on [0, 1] when $1 < p < \infty$. The case $p = 1$ was treated later in 1919 by H. Steinhaus.

Theorem 4.8. (*The Riesz Representation Theorem*)†. Let $(X, \mathscr{A}, \mu)$ be a σ-finite measure space and $1 \leq p < \infty$. Let Φ be a bounded linear functional on the L_p space over the reals. Let $1/p + 1/q = 1$. Then there is a unique element g in L_q such that

$$\Phi(f) = \int fg \, d\mu, \qquad f \in L_p,$$

and

$$\| \Phi \| = \| g \|_q. \qquad \blacksquare$$

Proof. Suppose first that μ is finite. Let $\nu(E) = \Phi(\chi_E)$, for $E \in \mathscr{A}$.

† In this theorem, $\Phi(f) = \int f\bar{g} \, d\mu$ when the scalars are complex numbers. The reader should verify this.

Let $E_n \in \mathscr{A}$, $E_n \cap E_m = \varnothing$ $(n \neq m)$. Then whenever $\alpha_n \in F$ with $|\alpha_n| = 1$, $\sum_{n=1}^{k} \alpha_n \chi_{E_n} \to \sum_{n=1}^{\infty} \alpha_n \chi_{E_n}$ in L_p. Therefore, since Φ is continuous, we have

$$\begin{aligned}\sum_{n=1}^{\infty} |\nu(E_n)| &= \lim_{k\to\infty} \sum_{n=1}^{k} \Phi(\chi_{E_n} \cdot \operatorname{sgn} \Phi(\chi_{E_n}))\\ &= \Phi\Big(\sum_{n=1}^{\infty} \chi_{E_n} \cdot \operatorname{sgn} \Phi(\chi_{E_n})\Big)\\ &< \infty,\end{aligned}$$

and similarly,

$$\sum_{n=1}^{\infty} \nu(E_n) = \Phi\Big(\sum_{n=1}^{\infty} \chi_{E_n}\Big) = \Phi(\chi_{\cup E_n}) = \nu\Big(\bigcup_{n=1}^{\infty} E_n\Big).$$

Hence ν is a signed measure and so by the Radon–Nikodym Theorem there is an integrable function g such that $\nu(E) = \int_E g\, d\mu$, $E \in \mathscr{A}$. If f is a simple function, we have

$$\Big|\int fg\, d\mu\Big| = |\Phi(f)| \leq \|\Phi\| \cdot \|f\|_p.$$

By Problem 3.5.6, $g \in L_q$. Let χ be the linear functional defined on L_p by

$$\chi(h) = \int hg\, d\mu, \qquad h \in L_p.$$

Then χ is clearly bounded (by the Hölder Inequality) and $\chi - \Phi$ is a bounded linear functional vanishing on the class of all simple functions that are dense in L_p by Problem 3.5.2. Hence, $\chi - \Phi = 0$ and

$$\Phi(h) = \int hg\, d\mu, \qquad \text{for all } h \in L_p.$$

It follows easily that $\|\Phi\| = \|g\|_q$. (See Problems 3.5.18 and 3.5.19.) The uniqueness part is trivial.

Now to prove the theorem when μ is σ-finite, let $X = \bigcup_{n=1}^{\infty} X_n$, $X_n \subset X_{n+1}$, and $\mu(X_n) < \infty$ for each n. Then there exists a sequence of functions (g_n) such that $g_n \in L_q$, $g_n(x) = 0$ for $x \notin X_n$, and

$$\Phi(f) = \int fg_n\, d\mu,$$

where $f \in L_p$ and $f(x) = 0$ for $x \notin X_n$. Also $\|g_n\|_q \leq \|\Phi\|$. Because of the uniqueness of the g_n's (except for changes on sets of measure zero), we can assume

$$g_{n+1}(x) = g_n(x), \qquad \text{if} \quad x \in X_n.$$

Let us define

$$g(x) = g_n(x), \qquad \text{if} \quad x \in X_n.$$

Then, by the Monotone Convergence Theorem,

$$\int |g|^q \, d\mu = \lim_{n\to\infty} \int |g_n|^q \, d\mu \le \|\Phi\|^q,$$

and so $g \in L_q$. By the Lebesgue Convergence Theorem, if $f \in L_p$, then

$$\begin{aligned}\int f \cdot g \, d\mu &= \lim_{n\to\infty} \int_{X_n} f \cdot g \, d\mu \\ &= \lim_{n\to\infty} \int f \cdot \chi_{X_n} \cdot g_n \, d\mu \\ &= \lim_{n\to\infty} \Phi(f \cdot \chi_{X_n}) \\ &= \Phi(f).\end{aligned}$$

The rest is left to the reader. ∎

In the theorem above, σ-finiteness is necessary when $p = 1$. (Problem 4.3.20). But for $p > 1$, σ-finiteness is not necessary, as the following corollary shows.

Corollary 4.4. Let $(X, \mathscr{A}, \mu)$ be any measure space and Φ be a bounded linear functional on L_p, $1 < p < \infty$. Then there is a unique element $g \in L_q$, $1/p + 1/q = 1$ such that

$$\Phi(f) = \int f \cdot g \, d\mu, \qquad \text{for all } f \in L_p,$$

and $\|\Phi\| = \|g\|_q$. ∎

Proof. From Theorem 4.8 it follows that for $A \in \mathscr{A}$, A σ-finite, there exists a unique $g_A \in L_q$, vanishing outside A such that

$$\Phi(f) = \int f \cdot g_A \, d\mu$$

for every $f \in L_p$, vanishing outside A. Clearly, because of uniqueness $A \subset B$

implies $g_A = g_B$ a.e. on A. Let us define $\nu(E)$ for E σ-finite by

$$\nu(E) = \int |g_E|^q \, d\mu \qquad [= \| g_E \|_q^q \leq \| \Phi \|^q].$$

Then let (E_n) be an increasing sequence of σ-finite sets in $\mathscr{A}$ such that $\lim_{n\to\infty}\nu(E_n) = \sup\{\nu(E): E \in \mathscr{A}, E \ \sigma\text{-finite}\}$. Denote this supremum by s. Letting $C = \bigcup_{n=1}^{\infty} E_n$, then $\nu(C) = s$. Let us define

$$g(x) = \begin{cases} g_C(x), & x \in C, \\ 0, & x \notin C. \end{cases}$$

Then $g \in L_q$. If D is σ-finite, $D \in \mathscr{A}$ and $C \subset D$, then $g_D = g_C$ a.e. on C and $\int |g_D|^q \, d\mu = \nu(D) \leq s = \int |g_C|^q \, d\mu$. Hence $g_D = g$ a.e. in X. If $f \in L_p$, then $N = \{x: f(x) \neq 0\}$ is σ-finite; let $E = N \cup C \supset C$. Then $g_E = g$ a.e., as above. Hence $\Phi(f) = \int f \cdot g_E \, d\mu = \int f \cdot g \, d\mu$. The rest is left to the reader. ▮

Corollary 4.5. For $1 < p < \infty$, L_p^* is linearly isometric onto L_q, $1/p + 1/q = 1$. ▮

The proof follows from Corollary 5.6.

Corollary 4.6. For $1 \leq p < \infty$, l_p^* is linearly isometric to l_q, $1/p + 1/q = 1$. ▮

The proof follows from Theorem 4.8 since l_p, $1 \leq p < \infty$, is a special case of L_p in the σ-finite measure space of integers with each integer having measure 1.

Problems

× **4.3.1.** (a) Suppose ν is a signed measure and μ a measure on $(X, \mathscr{B})$ such that $|\nu(E)| < \infty$ whenever $\mu(E) < \infty$. Prove $\nu << \mu$ if and only if for each $\varepsilon > 0$, there is a $\delta > 0$ such that whenever $\mu(E) < \delta$ then $|\nu(E)| < \varepsilon$.

(b) Show that statement (a) is not necessarily valid if the condition that $\mu(E) < \infty$ implies $|\nu(E)| < \infty$ does not hold. [Hint: Let X = positive integers, $\mu(E) = \sum_{n\in E} 2^{-n}$ and $\nu(E) = \sum_{n\in E} 2^n$.]

× **4.3.2.** Show that the Jordan decomposition of a signed measure is unique. (Hint: Show that any such decomposition yields a Hahn decomposition.)

× **4.3.3.** (a) Let ν be a signed measure on $(X, \mathscr{B})$. Prove for all E in $\mathscr{B}$ that $\nu^+(E) = \sup \{\nu(F)\colon F \in \mathscr{B}, F \subset E\}$, and $\nu^-(E) = -\inf \{\nu(F)\colon F \in \mathscr{B}, F \subset E\}$, where $\nu = \nu^+ - \nu^-$ is the Jordan decomposition of ν.

(b) If $\nu = \nu_1 - \nu_2$, where ν_1 and ν_2 are measures, show that $\nu^+ << \nu_1$ and $\nu^- << \nu_2$.

× **4.3.4.** Let μ be a measure and ν a signed measure on the measurable space $(X, \mathscr{B})$. Let $\nu = \nu^+ - \nu^-$ be the Jordan decomposition of ν. Prove that the following are equivalent:

(i) $\nu << \mu$,

(ii) $\nu^+ << \mu$ and $\nu^- << \mu$,

(iii) $|\nu| << \mu$.

4.3.5. Show that, if E is any measurable set, then

$$-\nu^-(E) \leq \nu(E) \leq \nu^+(E) \quad \text{and} \quad |\nu(E)| \leq |\nu|(E).$$

× **4.3.6.** Prove the formulas for differentiation stated in the text before Theorem 4.7.

× **4.3.7.** Let $(X, \mathscr{B}, \mu)$ be an arbitrary measure space and let ν be the measure on $\mathscr{B}$ given by $\nu(A) = 0$ if $\mu(A) = 0$ and $\nu(A) = \infty$ if $\mu(A) > 0$. Prove ν is a measure, $\nu << \mu$, and find f_0 such that $\nu(E) = \int_E f_0 \, d\mu$ for all E in $\mathscr{B}$.

4.3.8. Let $\mathscr{B} = \{A \subset R\colon \text{either } A \text{ or } A^c \text{ is countable}\}$. Prove the following:

(i) $\mathscr{B}$ is a σ-algebra.

(ii) The set functions μ and ν on $\mathscr{B}$ given, respectively, by

$$\mu(A) = \text{cardinality of } A$$

and

$$\nu(A) = \begin{cases} 0, & \text{if } A \text{ is countable,} \\ \infty, & \text{if } A \text{ is not countable,} \end{cases}$$

are measures.

(iii) $\nu << \mu$ but the Radon–Nikodym Theorem is not satisfied.

4.3.9. Prove Proposition 4.11.

4.3.10. Let k be the Cantor ternary function on $[0, 1]$ and extend k to R by defining $k(x) = 1$ if $x \geq 1$ and $k(x) = 0$ if $x \leq 0$. Let μ_k be the corresponding Borel measure on the Borel sets of R. If m is the Lebesgue measure on the Borel sets, show $\mu_k \perp m$.

4.3.11. If μ is a signed measure and f is a measurable function such that f is $|\mu|$-integrable, define

$$\int f\,d\mu = \int f\,d\mu^+ - \int f\,d\mu^-.$$

If μ is a finite signed measure, show that

$$|\mu|(E) = \sup\left\{\left|\int_E f\,d\mu\right| : |f| \le 1\right\}.$$

4.3.12. Let μ and ν be two σ-finite measures on $(X, \mathscr{A})$. Prove that there is a unique decomposition $\mu = \mu_1 + \mu_2 + \mu_3$, where $\mu_1 << \mu$, μ_2 is *purely atomic* [i.e., $\mu_2(B) > 0$ implies that there exist $C \subset B$ such that C is an atom of μ_2], and $\mu_3 \perp \mu$ with $\mu_3(\{x\}) = 0$ for every x. [Hint: By Theorem 4.7, $\mu = \mu_0 + \mu_1$, $\mu_1 << \mu$, and $\mu_0 \perp \mu$. Write $\mu_2(E) = \mu_0(E \cap A)$ and $\mu_3(E) = \mu_0(E - A)$, where $A = \{x: \mu_0(\{x\}) \neq 0\}$.]

★ **4.3.13.** Let $(X, \mathscr{A})$ and $(Y, \mathscr{B})$ be two measurable spaces. Let μ_1, μ_2 be two nonzero σ-finite measures on $\mathscr{A}$ and ν_1, ν_2 two nonzero σ-finite measures on $\mathscr{B}$. Prove that

(i) $\mu_1 \times \nu_1 << \mu_2 \times \nu_2$ if and only if $\mu_1 << \mu_2$ and $\nu_1 << \nu_2$;

(ii) $\mu_1 \times \nu_1 \perp \mu_2 \times \nu_2$ if and only if $\mu_1 \perp \mu_2$ or $\nu_1 \perp \nu_2$.

★ **4.3.14.** *An Extension of the Lebesgue Decomposition Theorem* (R. A. Johnson). Let μ and ν be measures on $(X, \mathscr{A})$. ν is called *$\mathscr{A}$-singular* with respect to μ, denoted by $\nu S \mu$, if given $B \in \mathscr{A}$ there exists $A \in \mathscr{A}$ with $A \subset B$, $\nu(B) = \nu(A)$, and $\mu(A) = 0$. Prove the following assertions:

(i) Suppose $\nu S \mu$ and ν is σ-finite. Then $\mu S \nu$ and $\nu \perp \mu$.

(ii) If $\nu S \mu$ and $\lambda << \mu$, then $\nu S \lambda$.

(iii) If $\nu S \mu$ and $\nu << \mu$, then $\nu = 0$.

(iv) There exist $\nu_0 << \mu$, $\nu_1 S \mu$, and $\nu_0 S \nu_1$ such that $\nu = \nu_0 + \nu_1$, where ν_1 is always unique. If ν is σ-finite, then ν_0 is also unique.

[Hint: Let $\mathscr{A}_1 = \{E \in \mathscr{A}: \mu(E) = 0\}$. If $\nu_1(A) = \sup\{\nu(A \cap E): E \in \mathscr{A}_1\}$, then ν_1 is a measure on $\mathscr{A}$ and $\nu_1 S \mu$. Let $\mathscr{A}_2 = \{E \in \mathscr{A}: \nu_1(E) = 0\}$. If $\nu_0(A) = \sup\{\nu(A \cap E): E \in \mathscr{A}_2\}$, then ν_0 is a measure on $\mathscr{A}$ and $\nu_0 << \mu$. Show that $\nu = \nu_0 + \nu_1$.]

★ **4.3.15.** Prove the following result due to R. A. Johnson: Let $(X, \mathscr{A})$ be a measurable space. Then the following hold.

(i) If μ_1 and μ_2 are *purely atomic* (see Problem 4.3.12) measures on $\mathscr{A}$, then so is $\mu_1 + \mu_2$.

(ii) If $\mu = \mu_1 + \mu_2$ and $\mu_1 S \mu_2$, then μ_1 is purely atomic or nonatomic according to whether μ is so. (See Problem 4.3.14.)

(iii) If $\mu << \mu_1 + \mu_2$, $\mu_2 S \mu_1$, μ is purely atomic, and μ_2 is nonatomic, then $\mu_2 S \mu$. [Hint: Use part (iv) of Problem 4.3.14 to write $\mu_2 = \lambda_1 + \lambda_2$, $\lambda_1 << \mu$, $\lambda_2 S \mu$, and $\lambda_1 S \lambda_2$. Now show that $\lambda_1 = 0$.]

(iv) For any measure μ on $\mathscr{A}$, there exist measures μ_1 and μ_2 such that $\mu = \mu_1 + \mu_2$, where μ_1 is purely atomic and μ_2 is nonatomic. μ_1 and μ_2 may be chosen such that $\mu_1 S \mu_2$ and $\mu_2 S \mu_1$, and under these conditions the decomposition is unique. [Hint: Let $\mathscr{A}_0$ be the family of sets that are countable unions of atoms of μ. Let $\mu_1(E) = \sup\{\mu(E \cap A): A \in \mathscr{A}_0\}$ and $\mu_2(E) = \sup\{\mu(E \cap A): \mu_1(A) = 0\}$. Then $\mu = \mu_1 + \mu_2$ is the desired decomposition.]

4.3.16. *An Application of the Baire-Category Theorem* (Vitali–Hahn–Saks). Let (μ_n) be a sequence of finite measures in a measure space $(X, \mathscr{A}, \mu_0)$ such that for each measurable set E, $\lim_{n\to\infty}\mu_n(E) = \mu(E)$ exists. Suppose that for each n, μ_n is absolutely continuous with respect to the measure μ_0. Then (i) the sequence (μ_n) is uniformly absolutely continuous with respect to μ_0 (i.e., given $\varepsilon > 0$, there exists $\delta > 0$ such that $\mu_0(E) < \delta \Rightarrow \mu_n(E) < \varepsilon$ for each n), and (ii) if μ_0 is finite, then μ is a measure and absolutely continuous with respect to μ_0. (Hint: $\mathscr{A}$ is a complete pseudometric space with the metric d defined by $d(A, B) = \arctan \mu_0(A \triangle B)$. Then $\mathscr{A} = \bigcup_{k=1}^{\infty} \mathscr{A}_k$, where $\mathscr{A}_k = \{E \in \mathscr{A}: |\mu_n(E) - \mu_m(E)| \leq \varepsilon$ for $n, m \geq k\}$. By the Baire-Category Theorem, there exist $a > 0$, $A \in \mathscr{A}$ and a positive integer N such that $\mu_0(A \triangle E) < a$ implies that $|\mu_n(E) - \mu_m(E)| \leq \varepsilon$ for $n, m \geq N$.)

4.3.17. *The Setwise Limit of a Sequence of Finite Measures on a σ-Algebra Is a Measure* (Nikodym). Let (μ_n) be a sequence of finite measures in a measure space such that for each measurable set E, $\lim_{n\to\infty}\mu_n(E) = \mu(E)$ exists. Then μ is a measure. (Compare Problems 2.1.5 and 2.2.7.) [Hint: Write $\nu_n(E) = \mu_n(E)/\mu_n(X)$ and let $\nu = \sum(1/2^n)\nu_n$. Now use part (ii) of Problem 4.3.16.]

4.3.18. Let I be the interval $[0, 1]$, m the Lebesgue measure on I, and μ a finite measure on the Borel sets B of I. Suppose $0 < c < 1$ and that for every Borel set A, $\mu(A) = c$ whenever $m(A) = c$. Prove $\mu = m$ on B. [Hint: Prove $\mu << m$ and use Theorem 4.6. Prove $d\mu/dm = 1$ a.e. (m).]

4.3.19. Let μ_f be the Lebesgue–Stieltjes measure on R induced by a right continuous, monotonic increasing function f such that

$$0 = \lim_{x\to-\infty} f(x) < \lim_{x\to\infty} f(x) = a < \infty.$$

Prove the following assertions:

(i) If A is a Borel set such that for all $x \in A, f'(x)$ exists, then

$$f'(x) \leq \alpha \text{ for all } x \in A \text{ implies } \mu_f(A) \leq \alpha m(A)$$

and

$$f'(x) \geq \alpha \text{ for all } x \in A \text{ implies } \mu_f(A) \geq \alpha m(A).$$

(ii) $\mu_f \perp m$ if and only if $f'(x) = 0$ a.e. (m).

4.3.20. Let $X = \{x_1, x_2\}$ and μ be a measure on 2^X such that $\mu(\{x_1\}) = 1$ and $\mu(\{x_2\}) = \infty$. Show that $\dim L_1 = \dim L_1{}^* = 1$, whereas $\dim L_\infty = \dim L_\infty{}^* = 2$.

4.3.21. *Convolution of Integrable Functions* (*An Application of Theorem* 4.8). Let $f \in L_1$ and $g \in L_p$, $1 \leq p \leq \infty$ (with respect to the Lebesgue measure on R). Show that

(i) the convolution $f * g$ of f and g, defined by

$$f * g(x) = \int f(x - y)g(y)\, dy,$$

is finite a.e.;

(ii) $\| f * g \|_p \leq \| f \|_1 \cdot \| g \|_p$;

(iii) $f * g = g * f$, a.e.;

Also show that for $f_1, f_2 \in L_1$ and $g \in L_p$, $1 \leq p \leq \infty$,

(iv) $f_1 * (f_2 * g) = (f_1 * f_2) * g$, a.e.;

(v) $(f_1 + f_2) * g = f_1 * g + f_2 * g$, a.e.

[Hint: Let $1 \leq p \leq \infty$. Let $q \geq 1$ and $1/p + 1/q = 1$ and $h \in L_q$. Then $\int [\int f(x - y)g(y)h(x)\, dy]\, dx = \int h(x) [\int f(x - y)g(y)\, dy]\, dx = \int h(x) \times [\int f(y)g(x - y)\, dy]\, dx = \int f(y)[\int h(x)g(x - y)\, dx]\, dy \leq \| f \|_1 \cdot \| g \|_p \cdot \| h \|_q$. For $p = 1$, take $h = 1$. For $p > 1$ and $p = \infty$, take $h = \exp(-x^2)$. This establishes (i). Now consider on L_q the bounded linear functional $\Phi(h) = \int h(x) f * g(x)\, dx$. By Theorem 4.8, there is a function $u \in L_p$ such that $u = f * g$ a.e. and $\| u \|_p = \| \Phi \|$.]

4.3.22. Consider the linear space L_p, $0 < p < 1$, in the Lebesgue measure space on $[0, 1]$, with the pseudometric

$$d(f, g) = \int_0^1 | f(t) - g(t) |^p\, dt.$$

Then show the following:

(i) (L_p, d), $0 < p < 1$, is a pseudometric linear space [that is, a vector space with metric (pseudo) topology where the vector addition and scalar multiplication are continuous].

(ii) f can be written as $g + h$, $d(g, 0) = d(h, 0) = \frac{1}{2}d(f, 0)$. {Hint: Let $g_x = f \cdot \chi_{[0,x]}$, $h_x = f \cdot \chi_{[x,1]}$. Then $d(g_x, 0) + d(h_x, 0) = d(f, 0)$. But $d(g_x, 0)$ is a continuous function from [0, 1] onto $[0, d(f, 0)]$. Use the Intermediate Value theorem.}

(iii) The only continuous linear functional on L_p is the zero functional. [Hint: Let $\Phi \in L_p{}^*$ with $\Phi(f) = 1$. Then, by part (ii), there exists $g \in L_p$ with $\Phi(g) \geq \frac{1}{2}$, $d(g, 0) = \frac{1}{2}d(f, 0)$. Let $g_1 = 2g$. Then $\Phi(g_1) \geq 1$, $d(g_1, 0) = 2^{p-1}d(f, 0)$. Continue the process to get $g_1, g_2, \ldots$ with $\Phi(g_n) \geq 1$ and $d(g_n, 0) = 2^{n(p-1)}d(f, 0)$.]

• 4.4. Change of Variables in Integration

In this section, we shall give an application of the Radon–Nikodym Theorem by proving a general theorem that may by thought of as a change-of-variable formula. We shall first prove a change-of-variable formula for integration of real-valued functions of real variables by use of some results on differentiation of composite functions. Specifically, if g: $[a, b] \to [c, d]$ is a function differentiable a.e. on $[a, b]$ and f is a real-valued integrable function on $[c, d]$, we wish to analyze when the formula

$$\int_{g(\alpha)}^{g(\beta)} f(x)\, dx = \int_{\alpha}^{\beta} f(g(s))g'(s)\, ds \tag{4.31}$$

is valid for all α and β in $[a, b]$.

Let us first examine a few examples.

Examples

4.10. Let k be the Cantor ternary function on [0, 1] and let $f(x) = x$ on [0, 1]. Since

$$\frac{1}{2} = \int_{k(0)}^{k(1)} f(x)\, dx \neq \int_0^1 k(x)k'(x)\, dx = 0,$$

the change-of-variable formula (4.31) does not hold.

4.11. Let $g(x) = x \sin(1/x)$ if $x \neq 0$ and $g(0) = 0$ be defined on [0, 1] and let $f(x) = x$. Then g is not absolutely continuous (Problem 4.1.2) on [0, 1], but direct computation shows that the change-of-variable formula (4.31) holds.

4.12. Let $g(x) = x^6 \sin^3(1/x)$ if $x \neq 0$ and $g(0) = 0$. Let $f(x) =$

$x^{-2/3}/3$ if $x \neq 0$ and $f(0) = 0$. Then f is integrable on [0, 1], and using Problems 4.1.2 and 4.2.1, g is seen to be absolutely continuous on [0, 1]. However, the change-of-variable formula (4.31) does not hold.

These examples show in particular that the absolute continuity of g is in general neither sufficient nor necessary for equation (4.31) to hold. Theorem 4.9 below gives a necessary and sufficient condition for equation (4.31) to hold. To prove this result we need the following results (see also [51]).

Proposition 4.14. If g is differentiable on a set E with $g' = 0$ a.e. on E, then $m(g(E)) = 0$. Conversely, if g has a derivative (finite or infinite) on a set E with $m(g(E)) = 0$, then $g' = 0$ a.e. on E. ▌

Proof. Using Problem 4.2.7(a), if $E_K = \{x \in E\colon K - 1 < |g'(x)| \leq K\}$,

$$m^*(g(E)) \leq \sum_{K=0}^{\infty} m^*(g(E_K)) \leq \sum_{K=0}^{\infty} Km^*(E_K) = 0. \tag{4.32}$$

Conversely, let $B = \{x \in E\colon |g'(x)| > 0\}$. Then $B = \bigcup_{n=1}^{\infty} B_n$, where

$$B_n = \{x \in B\colon |g(t) - g(x)| \geq |t - x|/n, \text{ for } |t - x| < 1/n\}.$$

Fix n and let $A = I \cap B_n$, where I is any interval of length less than $1/n$. To show $m(B) = 0$, we will show $m(A) = 0$. Since $m(g(A)) = 0$, for any $\varepsilon > 0$ there exists a sequence (I_K) of intervals such that $g(A) \subset \bigcup_{K=1}^{\infty} I_K$ and $\sum_{K=1}^{\infty} m(I_K) < \varepsilon$. Let $A_K = g^{-1}(I_K) \cap A$. Since $A \subset \bigcup_{K=1}^{\infty} A_K$ and $A_K \subset I \cap B_n$, we have

$$m^*(A) \leq \sum_{K=1}^{\infty} m^*(A_K) \leq \sum_{k=1}^{\infty} \sup_{s,t \in A_K} |s - t| \leq \sum_{k=1} n \sup_{s,t \in A_K} |g(s) - g(t)|.$$

However, $\sup_{s,t \in A_K} |g(s) - g(t)| \leq m(I_K)$ since $g(A_K) \subset I_K$ so that $m^*(A) \leq \sum_{k=1}^{\infty} nm(I_K) \leq n\varepsilon$. Since ε is arbitrary, $m(A) = 0$. ▌

The function f of Example 4.2 on the interval [0, 1] is an example of a *strictly* increasing function with $f' = 0$ a.e. Letting $F = f^{-1}$, then $(F \circ f)(x) = x$ for all x in [0, 1] so that $(F \circ f)' = 1$. However, $f' = 0$ so that the "chain rule" $(F' \circ f)(x) f'(x) = (F \circ f)'(x)$ does not hold for any x. The next result gives a condition to ensure that the "chain rule" holds.

Proposition 4.15. Suppose F, g, and $F \circ g$ are differentiable almost everywhere on their domains $[c, d]$ and $[a, b]$, where $g([a, b]) \subset [c, d]$. If

$m(F(E)) = 0$ whenever $m(E) = 0$, then

$$(F \circ g)' = (F' \circ g)g' \tag{4.33}$$

holds almost everywhere on $[a, b]$. ∎

Proof. Let $Z = \{y \in [c, d]: F \text{ is not differentiable at } y\}$, $E = g^{-1}(Z)$ and $G = [a, b] - E$. For x in G, we can write

$$\Delta F = F(g(x + \Delta x)) - F(g(x)) = [F'(g(x)) + \varepsilon]\,\Delta g,$$

where $\Delta g = g(x + \Delta x) - g(x)$ and $\varepsilon \to 0$ as $\Delta g \to 0$. Dividing both sides of this equation by Δx and letting $\Delta x \to 0$, we obtain

$$(F \circ g)'(x) = F'(g(x))g'(x)$$

if g is differentiable at x, that is, almost everywhere on G.

Clearly $m(g(E)) = 0$. By our hypothesis, $mF(g(E)) = 0$. Using Proposition 4.14, we see that g' and $(F \circ g)'$ equal zero a.e. on E, so that equation (4.33) holds a.e. on $[a, b]$. ∎

Using Problem 4.2.8(a) we have immediately the following corollary.

Corollary 4.7. If g and $F \circ g$ are differentiable a.e. on $[a, b]$ and F is absolutely continuous on $[c, d]$, then equation (4.33) is valid almost everywhere on $[a, b]$. ∎

Theorem 4.9. Suppose that g is differentiable almost everywhere on $[a, b]$ and that f is integrable on $[c, d]$, where $g([a, b]) \subset [c, d]$. If $F(x) = \int_c^x f(t)\,dt$, then the absolute continuity of $F \circ g$ is a necessary and sufficient condition that

$$\int_{g(\alpha)}^{g(\beta)} f(x)\,dx = \int_\alpha^\beta f(g(s))g'(s)\,ds, \tag{4.34}$$

for all α and β in $[a, b]$. ∎

Proof. If the formula (4.34) holds, then for any x in $[a, b]$,

$$F \circ g(x) - F \circ g(a) = \int_{g(a)}^{g(x)} f(t)\,dt = \int_a^x f(g(s))g'(s)\,ds,$$

so that by Theorem 4.3, $F \circ g$ is absolutely continuous.

Conversely, if $F \circ g$ is absolutely continuous, then its derivative, which is $(F' \circ g)g'$ or $(f \circ g)g'$ a.e. by Proposition 4.15, is integrable and

$$\int_{g(\alpha)}^{g(\beta)} f(x)\,dx = F(g(\beta)) - F(g(\alpha)) = \int_{\alpha}^{\beta} f(g(s))g'(s)\,ds. \qquad \blacksquare$$

Problems 4.4.2–4.4.4 below give some important corollaries of Theorem 4.9.

We will now use the Radon–Nikodym Theorem to prove an analog of Theorem 4.9 in arbitrary measure spaces. Given two measurable spaces $(X, \mathscr{B})$ and $(Y, \mathscr{A})$, a *measurable transformation* is a function $\tau: X \to Y$ such that for each E in $\mathscr{A}$, $\tau^{-1}(E)$ is in $\mathscr{B}$. Thus a real-valued function f on $(X, \mathscr{B})$ is measurable if and only if f is a measurable transformation from $(X, \mathscr{B})$ to R with the class of Borel sets.

A function $\tau: X \to Y$ assigns in an obvious manner a function f on X to every function g on Y by the formula $f(x) = g(\tau(x))$. A measurable transformation $\tau: (X, \mathscr{B}) \to (Y, \mathscr{A})$ also assigns in an obvious manner a measure μ_τ on $\mathscr{A}$ to each measure μ on $\mathscr{B}$ by the formula $\mu_\tau(E) = \mu(\tau^{-1}(E))$.

Proposition 4.16. If $\tau: (X, \mathscr{B}) \to (Y, \mathscr{A})$ is a measurable transformation, μ is a measure on $\mathscr{B}$, and if g is a measurable function on Y, then

$$\int_Y g\,d\mu_\tau = \int_X g \circ \tau\,d\mu \tag{4.35}$$

in the sense that, if either integral exists, they both do and are equal. $\blacksquare$

Proof. Note first that $g \circ \tau$ is measurable, since for any real number α

$$\{x: g \circ \tau(x) > \alpha\} = \tau^{-1}\{y: g(y) > \alpha\}.$$

If $g = \chi_E$, where $E \in \mathscr{A}$, then $g \circ \tau = \chi_{\tau^{-1}(E)}$. Clearly the proposition is valid for such functions g and thereby for simple functions defined on Y and measurable with respect to $\mathscr{A}$. The assertion of the proposition now follows by writing any nonnegative measurable function as the limit of a nondecreasing sequence of simple functions, invoking the Monotone Convergence Theorem, and then writing any measurable function as the difference of two nonnegative measurable functions. $\blacksquare$

In the special case that $g = \chi_E \cdot h$ for $E \in \mathscr{A}$ and some measurable function h with respect to $\mathscr{A}$, equation (4.35) becomes

$$\int_E h\,d\mu_\tau = \int_{\tau^{-1}(E)} h \circ \tau\,d\mu.$$

This brings us to our goal—a formula for integration by substitution.

Theorem 4.10. Suppose τ: $(X, \mathscr{B}) \to (Y, \mathscr{A})$ is a surjective measurable transformation and μ and ν are measures on $(X, \mathscr{B})$ and $(Y, \mathscr{A})$, respectively, such that $\nu(\tau(E)) = 0$ whenever E is a measurable set in $\mathscr{B}$ with $\mu(E) = 0$. Suppose also that μ_τ is a σ-finite measure on Y. Then there is a nonnegative measurable function t_0 on X so that for a ν-integrable function f on Y, $(f \circ \tau)t_0$ is μ-integrable and

$$\int_E f(y)\, d\nu = \int_{\tau^{-1}(E)} (f \circ \tau) \cdot t_0\, d\mu, \tag{4.36}$$

for each measurable set E in $\mathscr{A}$. ∎

Proof. Suppose $E \in \mathscr{A}$ and $\mu_\tau(E) = \mu(\tau^{-1}(E)) = 0$. By hypothesis, $\nu(E) = \nu[\tau(\tau^{-1}(E))] = 0$ so that $\nu << \mu_\tau$. By Corollary 4.3 there exists a nonnegative measurable function f_0 on X such that for each ν-integrable function f on Y, ff_0 is μ_τ-integrable and

$$\int_Y f(y)\, d\nu = \int_Y f(y) f_0(y)\, d\mu_\tau. \tag{4.37}$$

By Proposition 4.16,

$$\int_Y f(y) f_0(y)\, d\mu_\tau = \int_X f(\tau(x)) f_0(\tau(x))\, d\mu. \tag{4.38}$$

Combining equations (4.37) and (4.38) we have, for each ν-integrable function f,

$$\int_Y f(y)\, d\nu = \int_X f(\tau(x)) t_0(x)\, d\mu,$$

where $t_0 = f_0 \circ \tau$. In the special case that f is replaced by $f \cdot \chi_E$ for some measurable set E in Y, we get equation (4.36). ∎

If τ is a monotonic, absolutely continuous function with domain $[a, b]$ and range $[\alpha, \beta]$, the function t_0 is equal to $|\tau'|$ a.e. on $[a, b]$. (See Problem 4.4.5.) Thus if f is Lebesgue integrable on $[\alpha, \beta]$, we have that $(f \circ \tau) |\tau'|$ is Lebesgue integrable on $[a, b]$ and

$$\int_\alpha^\beta f(y)\, dy = \int_a^b f \circ \tau(x) \,|\tau'(x)|\, dx.$$

In general the function t_0 plays the role of the absolute value of the Jacobian of a nonsingular transformation τ from R^n to R^n. The interested reader may consult Appendix A for a discussion of this situation.

Problems

4.4.1. Suppose F and g are real-valued functions on the closed intervals $[a, b]$ and $[c, d]$, respectively, with $g([a, b]) \subset [c, d]$. If F and g have finite derivatives a.e. on their respective domains and g' is zero at most on a null set, then prove that $F \circ g$ is differentiable a.e. and $(F \circ g)' = (F' \circ g)g'$ a.e.

In problems 4.4.2–4.4.4 function g is defined on $[a, b]$ and f is defined on $[c, d]$ with $g([a, b]) \subset [c, d]$.

4.4.2. If g is monotonic and absolutely continuous and f is integrable prove that $(f \circ g)g'$ is integrable and the change-of-variable-formula (4.31) holds.

4.4.3. If g is absolutely continuous and f is bounded and measurable, then prove that $(f \circ g)g'$ is integrable and the change-of-variable formula (4.33) holds.

4.4.4. If g is absolutely continuous, and f and $(f \circ g)g'$ are integrable, then prove that the change-of-variable formula (4.31) holds. (Hint: Use Problem 4.4.3 and the Dominated Convergence Theorem.)

4.4.5. Prove that if τ is a monotonic, absolutely continuous function with domain $[a, b]$ and range $[\alpha, \beta]$, then the function t_0 in Theorem 4.9 is equal to $| \tau' |$ a.e. on $[a, b]$. [Hint: Show $\tau(x) = \int_a^x t_0(s)\, ds + \tau(a)$ for each x when τ is increasing.]

5

Measure and Topology

As the chapter title indicates, our purpose here is to study measures on classes of subsets of certain topological spaces. Given a topological space X, measures with pertinent properties will be studied on σ-rings and σ-algebras generated by the compact subsets of X, the closed subsets of X, the compact G_δ subsets of X, or others. For the most part, X will be taken as a locally compact topological space.

The primary results of this chapter are the various formulations in Section 5.4 of what are called Riesz representation theorems. These theorems show the relationship between linear functionals on certain vector lattices and measures on topological spaces. The name of F. Riesz (1880–1956) is generally attached to these theorems since he first represented a continuous linear functional on $C[0, 1]$ by an integral.[†] Even though earlier representations had been given of such functionals, the representation given by Riesz avoided earlier defects. Later Riesz gave other proofs of his result; and later extensions to more general spaces than [0, 1] were given by such notables as Radon, who in 1913 considered continuous functions on compact sets in R^n; Banach, who in 1937 considered the space $C(S)$, where S is a compact metric space; Kakutani, who in 1941 extended the theorem to compact Hausdorff spaces; and Markov, who in 1938 extended the result to some spaces other than compact spaces. More recent contributors are Halmos [20, Chapter 10], Hewitt,[‡] and Edwards.[§]

† F. Riesz, Sur les opérations fonctionnelles linéaires, *C. R. Acad. Sci. Paris* **149**, 974–977 (1909).

‡ E. Hewitt, Linear functionals on spaces of continuous functions, *Fund. Math.* **37**, 161–189 (1950).

§ R. Edwards, A theory of Radon measures on locally compact spaces, *Acta Math.* **89**, 133–164 (1953).

Although there are various proofs of the results of Section 5.4, we give proofs based on the Daniell (1889-1946) approach to integration theory and specifically the Daniell–Stone Representation Theorem of Section 5.1. The Daniell approach, expounded first in (1918)† defines the integral as a linear functional on a certain class of functions and then derives the notions of measure and measurability of functions in terms of this linear functional. Various modifications of this point of view have taken place throughout the years, notable of which is the contribution of M. H. Stone.‡ The name of Theorem 5.1 acknowledges the contributions of Daniell and Stone to this point of view.

Section 5.1 is preparatory to Sections 5.2, 5.3, and 5.4. Sections 5.2 and 5.3 give a comprehensive survey of various classes of subsets of a topological space and measures on these classes. The fifth section is largely due to R. A. Johnson in [26]. The Fubini–Tonelli Theorem is the principal result here and gives a beautiful extension of the Fubini–Tonelli Theorem of Chapter 3.

5.1. The Daniell Integral

This section is basically preliminary for the sections to follow—in particular Section 5.4. Although it is possible to prove many of the results of Section 5.4 without making use of the Daniell approach to integration theory, the study of the Daniell integral is worth considering in its own right inasmuch as it gives an alternate approach to integration theory.

In Chapters 2 and 3 the concepts of measure, measurability of functions, and integrability of functions were studied extensively. In particular we know that if $(X, \mathscr{B}, \mu)$ is a measure space, then the set $\mathscr{L}^1(\mu)$ of *real-valued* integrable functions§ on X satisfies the following:

(i) If f, g are in $\mathscr{L}^1(\mu)$ and $\alpha, \beta \in R$, then $\alpha f + \beta g \in \mathscr{L}^1(\mu)$ and

$$\int_X (\alpha f + \beta g)\, d\mu = \alpha \int_X f\, d\mu + \beta \int_X g\, d\mu.$$

(ii) If f is in $\mathscr{L}^1(\mu)$, then $| f |$ is in $\mathscr{L}^1(\mu)$.

† P. J. Daniell, A general form of integral, *Ann. Math.* **19**, 279–294 (1918).

‡ M. H. Stone, Notes on integration I–IV, *Proc. Nat. Acad. Sci.* **34**, **35** (1948–1949).

§ In Chapter 3, $L_1(\mu)$ denoted the space of *all* extended-real measurable functions f such that $| f |$ is μ-integrable.

The consequence of (i) is that $\mathscr{L}^1(\mu)$ is a vector space and the integral is a linear functional on $\mathscr{L}^1(\mu)$. The second implication guarantees that $\mathscr{L}^1(\mu)$ is also a "lattice"; that is, if f and g are any two functions in $\mathscr{L}^1(\mu)$, then the "meet" $f \wedge g = \inf(f, g)$ and the "join" $f \vee g = \sup(f, g)$ are in $\mathscr{L}^1(\mu)$. This follows since

$$\sup(f, g) = \tfrac{1}{2}(f + g + | f - g |) \tag{5.1}$$

and

$$\inf(f, g) = -\sup(-f, -g) = \tfrac{1}{2}(f + g - | f - g |).$$

In short we can say that $\mathscr{L}^1(\mu)$ is an example of what is called a "vector lattice."

More precisely, we have the following definition.

Definition 5.1. Let X be any set. A *vector lattice* VL is a vector space of real-valued functions on X such that

(i) f in VL implies $| f |$ is in VL and
(ii) $\inf(f, 1)$ is in VL for all f in VL. (1 is the function χ_X.) ∎

Remarks

5.1. Condition (ii) of Definition 5.1 is generally not part of the definition and is known as Stone's condition. We include it in the definition since all the vector lattices to be considered will have this property.

5.2. If f is in VL, so are $f^+ = \sup(f, 0)$ and $f^- = \sup(-f, 0)$, so f can be written as the difference of nonnegative functions in VL.

As noted above, the integral is a linear functional on $\mathscr{L}^1(\mu)$. Moreover it is a "positive" linear functional in the sense that $\int f \, d\mu \geq 0$ whenever $f \geq 0$. In addition, whenever (f_n) is a nonincreasing sequence of functions in $\mathscr{L}^1(\mu)$ converging to the zero function, then $\lim_n \int f_n \, d\mu = 0$ by the Lebesgue Convergence Theorem. The integral on $\mathscr{L}^1(\mu)$ is an example of a Daniell integral, defined as follows.

Definition 5.2. If VL is a vector lattice, a *Daniell integral I* on VL is a positive linear functional on VL such that condition (D) is satisfied.

(D) If (f_n) is a nonincreasing sequence in VL converging to zero, then $I(f_n)$ converges to zero. ∎

Remark 5.3. If I is a positive linear functional on VL, then I satisfies (D) if and only if I satisfies (D′).

(D′) If (f_n) is a nondecreasing sequence in VL with $f = \sup f_n$ in VL, then $I(f) = \sup_n I(f_n)$.

Examples

5.1. If $\mathscr{A}$ is an algebra of sets, μ is a measure on $\mathscr{A}$, VL is the class of simple functions that vanish outside a set of finite measure, and I: VL $\to R$ is given by

$$I(f) = \sum \alpha_i \mu(A_i)$$

for $f \in$ VL, then I is a Daniell integral on vector lattice VL.

5.2. The set of continuous functions on R, each vanishing outside some finite interval, is a vector lattice. The Riemann integral $\int f(x)\,dx$ gives a Daniell integral on this vector lattice.

5.3. Let $C_+'[0, 1]$ be the set of continuous functions f on X which are (right) differentiable at 0 and have $f(0) = 0$. $C_+'[0, 1]$ is a vector lattice. $I(f) = f'(0)$ is a positive linear functional on $C_+'[0, 1]$ but not a Daniell integral since $f_n \downarrow 0$, where $f_n(x) = \inf(1/n, x)$, but $I(f_n) = 1$.

5.4. These examples will be useful in Section 5.2 and following sections. If X is a topological space, the space $C(X)$ of continuous real functions on X, the space $C_b(X)$ of bounded continuous functions on X, and the space $C_b(X)$ of continuous functions with compact support are examples of vector lattices. We will characterize Daniell integrals on these spaces in Section 5.4.

If VL is a vector lattice of real-valued functions on a set X, let VLU be the collection of all extended real-valued functions on X of the form $\sup f_n$, where (f_n) is a nondecreasing sequence of nonnegative functions in VL. Let I be a Daniell integral on VL. If (f_n) and $(\bar{f}_n)$ are two nondecreasing sequences of nonnegative functions in VL with $\sup_n f_n \leq \sup_n \bar{f}_n$, then $\sup_n I(f_n) \leq \sup_n I(\bar{f}_n)$. Indeed, using condition (D'), for any integer m we have

$$\sup_n I(\bar{f}_n) \geq \sup_n I(\bar{f}_n \wedge f_m) = I(f_m) \tag{5.2}$$

since $\bar{f}_n \wedge f_m \uparrow [(\sup \bar{f}_n) \wedge f_m] = f_m$. Inequality (5.2) means

$$\sup_n I(\bar{f}_n) \geq \sup_n I(f_n).$$

We can therefore safely define $I^*(f)$ for any f in VLU to be $\sup I(f_n)$, where (f_n) is any nondecreasing sequence in VL of nonnegative functions whose $\sup f_n$ is f. This extension of I on VL to I^* on VLU has some important properties.

Proposition 5.1. The extension of I on vector lattice VL to I^* defined on VLU satisfies the following properties:

(i) $0 \le I^*(f) \le \infty$ for all $f \in$ VLU.
(ii) If f and g are in VLU and $f \le g$, then $I^*(f) \le I^*(g)$.
(iii) If $f \in$ VLU and $0 \le c < \infty$, then $I^*(cf) = cI^*(f)$.
(iv) If $f, g \in$ VLU, then $f + g$, $f \wedge g$, and $f \vee g$ are in VLU and $I^*(f + g) = I^*(f \wedge g) + I^*(f \vee g) = I^*(f) + I^*(g)$.
(v) If (f_n) is a sequence in VLU and $f_n \uparrow f$, then $f \in$ VLU and $I^*(f_n) \uparrow I^*(f)$. ▮

The proofs are left to the reader.

Now let

$$\mathscr{G} = \{G \subset X: \chi_G \in \text{VLU}\}.$$

$\mathscr{G}$ will be called the collection of VL-open sets corresponding to vector lattice VL. For $G \in \mathscr{G}$ define $\mu(G)$ as $I^*(\chi_G)$. Then we have the following result.

Proposition 5.2. The VL-open sets and the extended nonnegative function μ on these sets satisfy the following properties:

(i) If G_1 and G_2 are VL-open sets, so are $G_1 \cup G_2$ and $G_1 \cap G_2$; and

$$\mu(G_1 \cup G_2) + \mu(G_1 \cap G_2) = \mu(G_1) + \mu(G_2).$$

(ii) If G_1 and G_2 are VL-open sets with $G_1 \subset G_2$, then $\mu(G_1) \le \mu(G_2)$.
(iii) If G_n for $n = 1, 2, \ldots$ are VL-open sets so is $\bigcup_{n=1}^{\infty} G_n$, and

$$\mu\left(\bigcup_{K=1}^{n} G_K\right) \uparrow \mu\left(\bigcup_{K=1}^{\infty} G_K\right). \qquad ▮$$

The proofs of these statements follow from (iv), (ii), and (v), respectively, of Proposition 5.1.

In general the collection of VL-open sets has the following "measurability" relation in regard to VL and VLU.

Proposition 5.3. If VL is a vector lattice, then

(i) for every real number $\alpha \geq 0$ and for every $f \in$ VLU, $A_\alpha \equiv \{x \in X: f(x) > \alpha\}$ is VL-open;

(ii) the smallest σ-algebra $\sigma(\mathscr{G})$ containing all VL-open sets coincides with the smallest σ-algebra σ(VL) of subsets of X for which all functions in VL are measurable. ∎

Proof. (i) If $f \in$ VLU, then there exists a nondecreasing sequence (f_n) of nonnegative functions in VL such that $f = \sup f_n$. Since inf $(f_n, \alpha) \in$ VL for each $\alpha \geq 0$,

$$g_n \equiv n[f_n - \inf(f_n, \alpha)]$$

is in VL and $g_n \geq 0$. If $x \in X - A_\alpha$, then $\lim_n g_n(x) = 0$; while if $x \in A_\alpha$, $\lim_n g_n(x) = \infty$. Since the sequence (g_n) is also nondecreasing, so is the sequence $(\inf(g_n, 1))_{n \in N}$—a nonnegative sequence in VL. Inasmuch as χ_{A_α} equals $\sup_n[\inf(g_n, 1)]$, χ_{A_α} is in VLU or A_α is VL-open.

(ii) If f is in VLU and α is any real number

$$\{x: f(x) > \alpha\} = \{x: \lim_n f_n(x) > \alpha\} = \bigcup_n \{x: f_n(x) > \alpha\},$$

where (f_n) is a nondecreasing sequence of nonnegative functions in VL with $f = \sup f_n$. Since $\{x: f_n(x) > \alpha\} \in \sigma(\text{VL})$ for each n, $\{x: f(x) > \alpha\} \in \sigma(\text{VL})$. This means $\mathscr{G} \subset \sigma(\text{VL})$ or $\sigma(\mathscr{G}) \subset \sigma(\text{VL})$. By part (i), every function in VLU is $\sigma(\mathscr{G})$-measurable. Since every function f in VL is the difference of two nonnegative functions in VL, each f in VL is $\sigma(\mathscr{G})$-measurable. This means we also have $\sigma(\text{VL}) \subset \sigma(\mathscr{G})$. ∎

Our objective now is to define a measure μ on the σ-algebra σ(VL) such that VL $\subset \mathscr{L}^1(\mu)$ and $I(f) = \int_X f\, d\mu$ for all f in VL. To this end we construct an outer measure on 2^X and show that its restriction to σ(VL) is a measure. If $A \subset X$, define

$$\mu^*(A) = \inf\{\mu(G): G \in \mathscr{G} \text{ and } A \subset G\}. \tag{5.3}$$

(We understand the infimum of the empty set is $+\infty$.) Clearly if $A \in \mathscr{G}$, then $\mu^*(A)$ equals $\mu(A)$, which was defined above to be $I^*(\chi_A)$. We may say that μ^* extends μ on $\mathscr{G}$ to all of 2^X.

Lemma 5.1. μ^* is an outer measure on 2^X. ∎

Proof. The only significant item to prove here is that

$$\mu^*\left(\bigcup_{n=1}^{\infty} A_n\right) \le \sum_{n=1}^{\infty} \mu^*(A_n)$$

for an arbitrary sequence (A_n) from 2^X. If $\sum_{n=1}^{\infty}\mu^*(A_n) = \infty$, there is nothing to prove. So we may assume this sum is finite. For each $\varepsilon > 0$ and each n there is a G_n in $\mathscr{G}$ such that

$$A_n \subset G_n \qquad \text{and} \qquad \mu(G_n) < \mu^*(A_n) + \varepsilon/2^n.$$

By Proposition 5.2, $\cup\, G_n \in \mathscr{G}$ and we may write by Proposition 5.1

$$\begin{aligned}\mu\left(\bigcup_{i=1}^{\infty} G_i\right) &= I^*\left(\chi_{\bigcup_{i=1}^{\infty} G}\right) \le I^*\left(\sum_{i=1}^{\infty} \chi_G\right) \\ &= I^*\left(\sup_{n\in N} \sum_{i=1}^{n} \chi_{G_i}\right) = \sup_{n\in N} \sum_{i=1}^{n} I^*(\chi_{G_i}) \\ &= \sum_{i=1}^{\infty} I^*(\chi_{G_i}) = \sum_{i=1}^{\infty} \mu(G_i).\end{aligned}$$

This means

$$\mu^*\left(\bigcup_{n=1}^{\infty} A_n\right) \le \mu\left(\bigcup_{n=1}^{\infty} G_n\right) \le \sum_{n=1}^{\infty} \mu(G_n) \le \sum_{n=1}^{\infty} \mu^*(A_n) + \varepsilon. \qquad \blacksquare$$

We know from Chapter 2 (Theorem 2.2) that the collection $\mathscr{B}$ of μ^* measurable sets—that is, the collection of all sets E in 2^X satisfying

$$\mu^*(A) \ge \mu^*(A \cap E) + \mu^*(A \cap E^c)$$

for all A in 2^X—is a σ-algebra. In addition we know that μ^* restricted to this σ-algebra $\mathscr{B}$ is a measure. We now wish to show that $\mathscr{G}$—and therefore $\sigma(\mathscr{G})$—is contained in the σ-algebra of μ^*-measurable sets.

Lemma 5.2. Each set G in $\mathscr{G}$ is μ^*-measurable. ∎

Proof. We must show that for each A in 2^X,

$$\mu^*(A) \ge \mu^*(A \cap G) + \mu^*(A \cap G^c).$$

By virtue of the definition [equation (5.3)] of μ^*, it suffices to show

$$\mu(E) \ge \mu(E \cap G) + \mu^*(E \cap G^c) \tag{5.4}$$

for each E in $\mathscr{G}$ with $\mu(E) < \infty$. Indeed, inequality (5.4) means that for $\mu^*(A) < \infty$,

$$\begin{aligned}\mu^*(A) &= \inf\{\mu(E): E \in \mathscr{G}, E \supset A\} \\ &\geq \inf\{\mu(E \cap G): E \in \mathscr{G}, E \supset A\} \\ &\quad + \inf\{\mu^*(E \cap G^c): E \in \mathscr{G}, E \supset A\} \\ &\geq \mu^*(A \cap G) + \mu^*(A \cap G^c).\end{aligned}$$

(Note that $E \cap G^c$ is not necessarily in $\mathscr{G}$.) Since E and $E \cap G$ are in $\mathscr{G}$, there exist nondecreasing sequences (f_n) and (g_n) of nonnegative functions in VL such that

$$\chi_E = \sup f_n \quad \text{and} \quad \chi_{E\cap G} = \sup g_n.$$

Inasmuch as $\chi_{E\cap G} \leq \chi_E$, the function h_n given by $h_n \equiv \chi_E - g_n$ is nonnegative and in VLU since h_n also satisfies $h_n = \sup_k\{f_k - g_n, 0\}$. Since also

$$h_n \geq \chi_E - \chi_{E\cap G} = \chi_{E\cap G^c},$$

the sets $A_\alpha{}^n = \{x \in X: h_n(x) > \alpha\}$ for $0 < \alpha < 1$ (sets in $\mathscr{G}$ by Proposition 5.3) contain $E \cap G^c$. Realizing that $\alpha\chi_{A_\alpha{}^n} \leq h_n$ or equivalently that $\chi_{A_\alpha{}^n} \leq (1/\alpha)h_n$ we have

$$\mu^*(E \cap G^c) \leq \mu(A_\alpha{}^n) = I^*(\chi_{A_\alpha{}^n}) \leq 1/\alpha I^*(h_n)$$

which means $\mu^*(E \cap G^c) \leq I^*(h_n)$ for each n since α is arbitrary in (0, 1). This implies

$$\begin{aligned}\mu^*(E \cap G^c) &\leq \underline{\lim}_n I^*(h_n) = \underline{\lim}_n I^*(\chi_E - g_n) \\ &= I^*(\chi_E) - I^*(\chi_{E\cap G}) \\ &= \mu(E) - \mu(E \cap G).\end{aligned}$$

Inequality (5.4) follows. ∎

We are now ready to prove the principal result of this section.

Theorem 5.1. *Daniell–Stone Representation Theorem.* Let VL be a vector lattice of functions on a set X and let I be a Daniell integral on VL. Then there exists a unique measure ν on the σ-algebra $\sigma(\text{VL})$ such that

$$\text{VL} \subset L_1(\nu), \tag{5.5}$$

$$I(f) = \int f\,d\nu \qquad \text{for all } f \text{ in VL}, \tag{5.6}$$

and

$$\nu(A) = \inf\{\mu(G): G \supset A \text{ and } G \text{ is VL-open}\}. \quad \blacksquare \quad (5.7)$$

Proof. Let ν be the restriction of μ^* defined by equation (5.3) to the σ-algebra $\sigma(\mathscr{G})$ or to what is the same, $\sigma(\text{VL})$. Obviously equation (5.7) is satisfied. Suppose G is VL-open. Then

$$I^*(\chi_G) = \mu(G) = \mu^*(G) = \nu(G) = \int_X \chi_G \, d\nu. \quad (5.8)$$

If $f \in \text{VL}$ and $f \geq 0$ then $f = \sup h_n$, where

$$h_n = \frac{1}{2^n} \sum_{K=1}^{n2^n} \chi_{\{K/2^n < f\}}.$$

Each $\{K/2^n < f\}$ is VL-open by Proposition 5.3 so that by Proposition 5.1, h_n is in VLU. Using equation 5.8 and Proposition 5.1, we have

$$I^*(h_n) = \frac{1}{2^n} \sum_{K=1}^{n2^n} I^*(\chi_{\{K/2^n < f\}}) = \frac{1}{2^n} \sum_{K=1}^{n2^n} \int_X \chi_{\{K/2^n < f\}} \, d\nu$$

$$= \frac{1}{2^n} \int_X \sum_{K=1}^{n2^n} \chi_{\{K/2^n < f\}} \, d\nu = \int_X h_n \, d\nu.$$

Since $I^*(h_n) \uparrow I^*(f)$ by Proposition 5.1, we have by the Monotone Convergence Theorem

$$I(f) = I^*(f) = \lim \int_X h_n \, d\nu = \int_X f \, d\nu.$$

Since $0 \leq I(f) < \infty$, f is ν-integrable or $f \in L_1(\nu)$. If f is an arbitrary function in VL (not necessarily nonnegative) the integrability of f and the equation $I(f) = \int_X f \, d\nu$ follows by writing f as the difference of the nonnegative functions f^+ and f^- in VL.

The uniqueness conclusion must yet be established. For any VL-open set G, there exists a sequence (f_n) from VL with $f_n \geq 0$ and $f_n \uparrow \chi_G$. Therefore if ν' is any measure such that $I(f) = \int_X f \, d\nu'$ for all f in VL then

$$\nu'(G) = \int \chi_G \, d\nu' = \lim_n \int f_n \, d\nu' = \lim_n I(f_n) = \mu^*(G).$$

Since $\nu'(G) = \mu^*(G)$ for each VL-open set G, equation (5.7) shows that $\nu = \nu'$. $\blacksquare$

The reader should have observed that the uniqueness conclusion of Theorem 5.1 is dependent on condition (5.7). In fact Problem 5.1.4 gives an example of a vector lattice VL and a Daniell integral I such that two measures exist on $\sigma(\text{VL})$ with conditions (5.5) and (5.6) satisfied. We have the following sufficiency, which will guarantee uniqueness in lieu of condition (5.7).

Proposition 5.4. Suppose VL is a vector lattice and I is a Daniell integral on VL. Suppose ν is a measure on $\sigma(\text{VL})$ such that

$$\text{VL} \subset L_1(\nu),$$

$$I(f) = \int_X f\, d\nu \qquad \text{for each } f \text{ in VL},$$

and

$$X = \bigcup_{i=1}^{\infty} X_i, \qquad \text{where } X_i \text{ are VL-open and } \nu(X_i) < \infty.$$

Then ν is the only measure on $\sigma(\text{VL})$ satisfying these properties. ∎

Proof. If $\bar{\nu}$ is any such measure, notice that $\bar{\nu}(G) = \nu(G)$ for any VL-open set G. Indeed, if (f_n) is a nonnegative sequence in VL with $f_n \uparrow \chi_G$, then by the Monotone Convergence Theorem

$$\bar{\nu}(G) = \int \chi_G\, d\bar{\nu} = \lim_n \int f_n\, d\bar{\nu} = \lim_n I(f_n) = \nu(G).$$

Now, for each n, let

$$D_n = \{A \in \sigma(\text{VL}) \colon \bar{\nu}(A \cap X_n) = \nu(A \cap X_n)\}.$$

For each n, D_n is a Dynkin system containing the VL-open sets so that it contains $\sigma(\text{VL})$ (see Chapter 1). Hence for each $A \in \sigma(\text{VL})$

$$\bar{\nu}(A) = \lim_n \bar{\nu}(A \cap X_n) = \lim_n \nu(A \cap X_n) = \nu(A). \qquad ∎$$

Problems

× **5.1.1.** Verify Remark 5.3.

5.1.2. Suppose VL is a vector lattice and I is a Daniell integral on VL. Suppose ν is a measure on $\sigma(\text{VL})$ such that $\text{VL} \subset L_1(\nu)$, $I(f) = \int_X f\, d\nu$ for each f in VL. Assume for some nondecreasing sequence (f_n) of nonnegative functions in VL, $\sup f_n(x) > 0$ for all x in X and $\sup I(f_n) < \infty$. Prove ν is σ-finite and ν is the only measure with these properties.

5.1.3. Let $\mathscr{R}$ be a ring of subsets of a set X.

(i) Prove that

$$\left\{ \sum_{i=1}^{n} \alpha_i \chi_{A_i} : \alpha_n \in R,\ A_i \in \mathscr{R},\ n \in N \right\}$$

is a vector lattice VL.

(ii) Prove that if $f \geq 0$ in VL, then

$$\{x \in X: f(x) > \alpha\} \in \mathscr{R} \text{ for all } \alpha > 0.$$

(iii) Prove that a set is VL-open if and only if it is the countable union of a sequence of sets in $\mathscr{R}$.

(iv) Let μ be a finitely additive measure on $\mathscr{R}$. For f in VL define

$$I(f) = \sum_{i=1}^{n} \alpha_i \mu(A_i)$$

where $f = \sum_{i=1}^{n} \alpha_i \chi_{A_i}$. Prove I is a Daniell integral if and only if μ is countably additive.

(v) Verify $\sigma(\text{VL}) = \sigma(\mathscr{R})$.

(vi) Use Theorem 5.1 to extend μ to a measure ν on $\sigma(\mathscr{R})$ in the case μ is countably additive. How does ν relate to the measure $\bar{\mu}$ constructed in Theorem 2.4?

5.1.4. Let $X = (-\infty, \infty) \cup \{\omega\}$ and let VL consist of all real-valued functions on X that are Lebesgue integrable on $(-\infty, \infty)$ and zero at ω. Let I be defined on VL by $I(f) = \int_R f(x)\, dx$. Prove the following:

(i) VL is a vector lattice.

(ii) $\sigma(\text{VL}) = \{B: B \cap (-\infty, \infty) \text{ is Lebesgue measurable}\}$.

(iii) I is a Daniell integral on VL.

(iv) There are two measures ν_1 and ν_2 on $\sigma(\text{VL})$ satisfying conditions (5.5) and (5.6) of Theorem 5.1.

5.1.5. Let VL $= C[0, 1]$, the continuous functions on $[0, 1]$, and let $I(f)$ be the Riemann integral of f on $[0, 1]$. Prove the following:

(i) I is a Daniell integral.

(ii) $\sigma(\text{VL})$ is the σ-algebra containing all open subsets of $[0, 1]$.

(iii) The measure of Theorem 5.1 is the Lebesgue measure on $[0, 1]$.

5.1.6. Let VL be a vector lattice of functions on a set X and let I be a Daniell integral on VL. Let ν be the unique measure on $\sigma(\text{VL})$ given by Theorem 5.1. Let $\mathscr{L}^p(\nu)$ be the real-valued functions from $L^p(\nu)$, where $1 \leq p < \infty$. Prove VL $\cap\, \mathscr{L}^p(\nu)$ is dense in $\mathscr{L}^p(\nu)$. [Hint: To show for each f in $\mathscr{L}^p(\nu)$ there exists u in VL $\cap\, \mathscr{L}^p(\nu)$ such that $\| f - u \|_p < \varepsilon$, show f can be assumed first to be nonnegative, then nonnegative and simple, and then of the form χ_A for $A \in \sigma(\text{VL})$. Then use condition (5.7).]

5.2. Topological Preliminaries. Borel and Baire Sets

Preparatory to our study of measures on certain topological spaces, we will examine in this section special classes of subsets of a topological space. These sets are called Borel and Baire sets. These sets are defined so that the concepts of measurability and continuity of functions coalesce.

First let us recall some topological concepts and derive some auxiliary results that will be used extensively in following sections. Recall that a topological space X is said to be T_4 if X is a Hausdorff space such that whenever A and B are disjoint closed subsets of X, then there are disjoint open sets $U \supset A$ and $V \supset B$. *Urysohn's Lemma* implies that if A and B are disjoint closed subsets of a T_4 space then there is a function f in $C(X)$ such that $0 \leq f \leq 1$ on X and $f \equiv 0$ on A while $f \equiv 1$ on B. The reader may readily verify that each compact Hausdorff space and each metric space is T_4. (See also Remark 1.37.)

Let us also recall that a *locally compact* topological space X is a topological space such that each point in X is contained in an open set whose closure is compact. Every compact space is locally compact while the Euclidean spaces R^n are examples of locally compact spaces that are not compact.

If X is a locally compact Hausdorff space we can form a compact Hausdorff space by adding to X a single point ω not in X. The result is a new compact space $X^* = X \cup \{\omega\}$ called the *one-point compactification*[†] of X provided we put the following topology on X^*: $A \subset X^*$ is open if A is either an open subset of X or A is the complement of a compact subset of X. If X is originally compact, then ω is an isolated point of X^*; otherwise X is dense in X^*. ω is called the *point at infinity* in X^*.

A subset of a topological space X is an F_σ *set* (respectively, G_δ set) if it is the union (intersection) of a sequence of closed (open) sets in X. A set A is a *σ-compact* set (respectively, *σ-bounded* set) if A is (is contained in) the union of a sequence of compact sets. Set A is bounded if A is contained in a compact set.

The *support* S_f of a real-valued function f on a topological space X is the closure of the set $\{x \in X: f(x) \neq 0\}$. The *zero set* of a real function

[†] See Remark 1.55.

is $\{x: f(x) = 0\}$. $C_c(X)$ denotes the set of all continuous real-valued functions on X whose support is compact. Clearly $C_c(X) \subset C_b(X)$, the space of bounded continuous functions on X.

The following lemmas are crucial to our study of measure and topology.

Lemma 5.3. Let K be a compact subset of a locally compact Hausdorff space X and let G be an open set with $K \subset G$. Then there is an open set O containing K such that $\bar{O}$ is compact and $\bar{O} \subset G$. In addition, for any such O, there is an $f \in C_c(X)$ with $0 \leq f \leq 1$, $f \equiv 1$ on K and $f \equiv 0$ on O^c. If K is also a G_δ set, then we can take $f < 1$ on K^c. ∎

Proof. By Remark 1.56, for each x in K there is an open set O_x such that $\bar{O}_x \subset G$ and $\bar{O}_x$ is compact. Let O be the union of a finite number of such O_x which will cover K. If O is any open set containing K with $\bar{O}$ compact, the sets K and $X^* - O$ are closed and disjoint in X^*, the one-point compactification of X. Since as a compact Hausdorff space, X^* is normal, Urysohn's Lemma gives the existence of a f^* in $C(X^*)$ such that $0 \leq f^* \leq 1$, $f^* \equiv 0$ on $X^* - O$ and $f^* \equiv 1$ on K. If f is the restriction of f^* to X then clearly f is the desired function of the lemma. The last statement is left for the reader to verify. ∎

Lemma 5.4. If K is compact, O is open, and $K \subset O$ in a locally compact Hausdorff space X, then there exist sets K_0 and O_0 in X such that K_0 is a compact G_δ, O_0 is a σ-compact open set, and

$$K \subset O_0 \subset K_0 \subset O.$$

In fact, O_0 is the countable union of compact G_δ sets. ∎

Proof. Since by Lemma 5.3 there exists a bounded open set U with $K \subset U \subset O$, we may assume O is bounded. Let f be a function of Lemma 5.3 such that $f \equiv 1$ on K, $0 \leq f \leq 1$, and $f \equiv 0$ on O^c. Let

$$K_0 = \{x: f(x) \geq 1/2\} = \bigcap_{n=1}^{\infty} \{x: f(x) > 1/2 - 1/n\}$$

and

$$O_0 = \{x: f(x) > 1/2\} = \bigcup_{n=1}^{\infty} \{x: f(x) \geq 1/2 + 1/n\}.$$

Then K_0 is a closed G_δ contained in the bounded set O so that K_0 is a compact G_δ. Clearly O_0 is σ-compact. ∎

Remark 5.4. It follows from Lemma 5.4 that the open Baire sets (see Definition 5.3) are a basis for the locally compact topology on X.

With these topological preliminaries we can now begin our study of special classes of sets in a topological space X. There are many classes of sets—both σ-rings and σ-algebras—that can be considered in a topological space. We have listed all these classes in Figure 4 together with relationships among these classes and under what conditions on topological space X these relationships are valid. In Figure 4 and the discussion to follow, if $\mathscr{C}$ is a class of sets, $\sigma_r(\mathscr{C})$ and $\sigma(\mathscr{C})$, respectively, represent the σ-ring generated by $\mathscr{C}$ and the σ-algebra generated by $\mathscr{C}$. Similarly if $\mathscr{L}$ is a class of real functions on X, $\sigma(\mathscr{L})$ is the smallest σ-algebra such that all functions in $\mathscr{L}$ are measurable. Our objective in this section is primarily to verify the relationships of Figure 4.

If X is a compact Hausdorff space, then since X is a compact G_δ set and every closed set is compact, we get equality in Figure 4 at all places

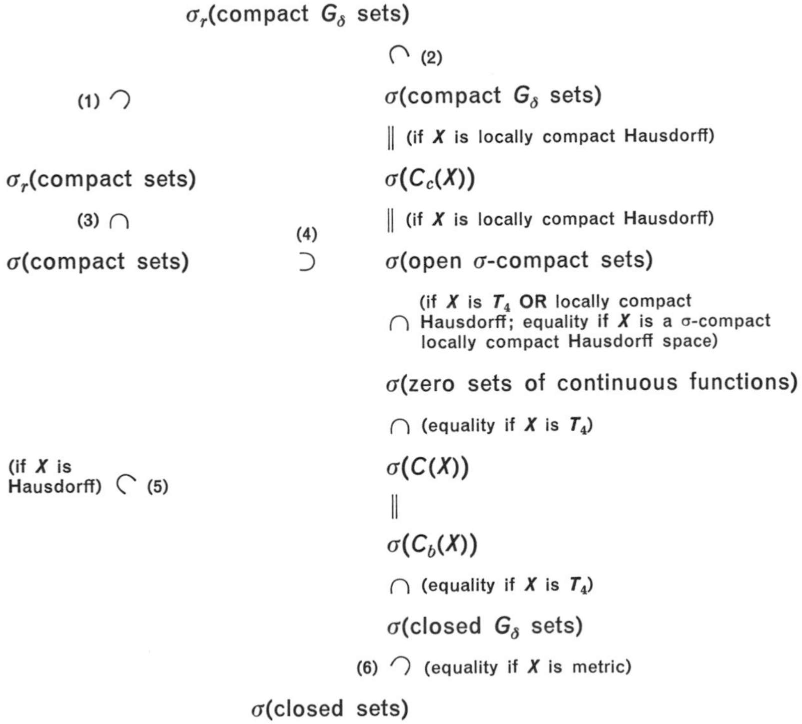

σ_r(compact G_δ sets)

$\cap$ (2)

(1) $\cap$ σ(compact G_δ sets)

‖ (if X is locally compact Hausdorff)

σ_r(compact sets) $\sigma(C_c(X))$

(3) $\cap$ ‖ (if X is locally compact Hausdorff)

(4)

σ(compact sets) $\supset$ σ(open σ-compact sets)

$\cap$ (if X is T_4 OR locally compact Hausdorff; equality if X is a σ-compact locally compact Hausdorff space)

σ(zero sets of continuous functions)

$\cap$ (equality if X is T_4)

(if X is Hausdorff) $\cap$ (5) $\sigma(C(X))$

‖

$\sigma(C_b(X))$

$\cap$ (equality if X is T_4)

σ(closed G_δ sets)

(6) $\cap$ (equality if X is metric)

σ(closed sets)

Fig. 4

except possibly at (1), (6), and (4). If X is a σ-compact locally compact Hausdorff space, we get equality at (2) and (3). In fact in this case we will show in Corollary 5.2 that we get equality along the right side of Figure 4 from σ_r (compact G_δ sets) to σ (closed G_δ sets.)

Definition 5.3. If X is a topological space, then

(i) the σ-algebra of *weakly Borel sets* $B_w(X)$ is the σ-algebra σ (closed sets),

(ii) the σ-algebra of *Borel sets* $B(X)$ is the σ-algebra σ (compact sets),

(iii) the σ-algebra of *weakly Baire sets* $\mathrm{Ba}_w(X)$ is the σ-algebra σ (closed G_δ),

(iv) the σ-algebra of *Baire sets* $\mathrm{Ba}(X)$ is the σ-algebra σ (compact G_δ).

The σ-rings σ_r (compact G_δ) and σ_r (compact) are called, respectively, the *σ-ring of Baire sets* and the *σ-ring of Borel sets.* ∎

Accordingly, we have the following scheme if X is a Hausdorff space:

$$\begin{array}{ccccc} \sigma_r\,(\text{compact } G_\delta) & \subset & \mathrm{Ba}(X) & \subset & \mathrm{Ba}_w(X) \\ \cap & & \cap & & \cap \\ \sigma_r\,(\text{compact}) & \subset & \mathrm{B}(X) & \subset & \mathrm{B}_w(X) \end{array}$$

It should be noted that not all authors use the same terminology. For some the Borel sets are what we have termed weakly Borel sets or the σ-ring σ_r (compact). For some the Baire sets are what we have termed weakly Baire sets or the σ-ring σ_r (compact G_δ).

It is understandable why we wish to consider such classes of sets. The objective is to study the interplay of measure and topology. It is natural therefore to consider σ-algebras generated by closed and compact sets. The σ-algebras $\mathrm{Ba}_w(X)$ and $\mathrm{Ba}(X)$ are so defined to ensure the measurability of continuous functions on X and continuous functions of compact support on X, respectively.

Examples

5.5. Let X be an uncountable set with the discrete topology (every subset is open). Then

$$\mathrm{Ba}(X) = \mathrm{B}(X) = \{A \in 2^X\colon A \text{ or } A^c \text{ is countable}\} \subsetneqq 2^X = \mathrm{Ba}_w(X) = B_w(X).$$

5.6. Let $X = R$ be the reals with the usual topology. Since every closed set is a G_δ (see proof of Corollary 5.1 below) we have

$$\text{Ba}(X) = \text{B}(X) \subset \text{Ba}_w(X) = \text{B}_w(X).$$

Moreover $\text{Ba}(X) = \text{Ba}_w(X)$ by Corollary 5.2 below.

5.7. Let $D = \{j/2^n\colon j \text{ an integer and } n \text{ a nonnegative integer}\}$. Let $E = R - D$ and for each integer j and nonnegative integer n let D_{nj} be the singleton set $\{[j/2^n, (j+1)/2^n]\}$. Let

$$X = E \cup \{D_{nj}\}.$$

Topologize X as follows: $N \subset X$ is a neighborhood of x in E if $x \in N$ and N contains all but finitely many of the D_{nj} with $j/2^n \leq x \leq (j+1)/2^n$; $N \subset X$ is a neighborhood of y from $\{D_{nj}\}$ if N is any subset of $X - E$ containing y. It is readily seen that X is then a locally compact Hausdorff space. Also

$$B(X) = \text{Ba}(X) = \{A \subset X\colon A \text{ or } A^c \text{ is countable}\},$$

and

$$B_w(X) = 2^X.$$

The following propositions serve to verify the relationships of Figure 4. Some of the inclusions of course are trivial.

Proposition 5.5. If X is any topological space, $\sigma(C_b(X)) = \sigma(C(X))$. ∎

Proof. Obviously $\sigma(C_b(X)) \subset \sigma(C(X))$. On the other hand, each f in $C(X)$ is measurable with respect to $\sigma(C_b(X))$ since it is the pointwise limit of the sequence (f_n) from $C_b(X)$, where

$$f_n \equiv \inf(\sup(f, -n), n). \qquad ∎$$

Proposition 5.6. If X is a T_4 topological space, $\text{Ba}_w(X) = \sigma(C(X))$. ∎

Proof. Obviously $\sigma(C(X)) \subset \text{Ba}_w(X)$, the σ-algebra generated by closed G_δ sets, since for any $\alpha \in R$

$$\{x \in X\colon f(x) \geq \alpha\} = \bigcap_{n=1}^{\infty} \{x\colon f(x) > \alpha - 1/n\},$$

so that $\{x \in X\colon f(x) \geq \alpha\}$ is a closed G_δ set. (Normality is not needed here.) The converse is accomplished by showing that if F is an open F_σ

set (that is, the complement of a closed G_δ set), then $F = \{x \in X: f(x) > 0\}$ for some nonnegative function f in $C_b(X)$. As an open F_σ set, F equals $\bigcup_{n=1}^\infty F_n$, where each F_n is closed. Since F_n and F^c are disjoint closed sets in X, by Urysohn's Lemma there is for each n an f_n in $C_b(X)$ with $0 \leq f_n \leq 1$, $f_n \equiv 1$ on F_n, and $f_n \equiv 0$ on F^c. Let f be defined as

$$f(x) = \sum_{n=1}^{\infty} \frac{f_n(x)}{n^2}$$

for each x in X. Since the convergence is uniform, $f \in C_b(X)$ and $F = \{x \in X: f(x) > 0\}$. ∎

Corollary 5.1. If (X, d) is a metric space, then

$$B_w(X) = \mathrm{Ba}_w(X) = \sigma(C(X)).$$ ∎

Proof. Since a metric space is T_4, the right equality follows from Proposition 5.6. Each closed subset F of X is a G_δ set as

$$F = \bigcap_{n=1}^{\infty} \{x \in X: \operatorname{dist}(x, F) < 1/n\},$$

whence follows the left equality.

Remark 5.5. Problem 5.2.5 gives an example of a compact *nonmetrizable* Hausdorff space such that each closed set is a G_δ set so that in this case the conclusion of Corollary 5.1 is again true.

In case X is a locally compact Hausdorff space, the following lemma is useful.

Lemma 5.5. Suppose X is a locally compact Hausdorff space.

(i) A subset G of X is $C_c(X)$-open (see Section 5.1) if and only if G is an open σ-compact set.

(ii) If X is σ-compact, then X can be written as the union of compact G_δ sets K_n with $K_n \subset K_{n+1}^\circ$, the interior of K_{n+1}, for each n. In particular, each compact subset K of X is contained in some K_n. ∎

Proof. (i) If G is $C_c(X)$-open, then $\chi_G = \lim f_n$, where (f_n) is a nondecreasing sequence of nonnegative functions in $C_c(X)$. Hence

$$G = \bigcup_{n=1}^{\infty} \{x \in X: f_n(x) > 0\} = \bigcup_{n,k=1}^{\infty} \{x \in X: f_n(x) \geq 1/k\},$$

so that G is open and a σ-compact set, since the set $\{x \in X: f_n(x) \geq 1/k\}$ is compact as a closed subset of the compact support of f_n. Conversely, if $G = \bigcup_{n=1}^{\infty} K_n$, where each K_n is compact, then by Lemma 5.3 for each n there exists an f_n in $C_c(X)$ with $0 \leq f_n \leq 1$, $f_n \equiv 1$ on K_n, and $S_{f_n} \subset G$. Letting $g_n = \sup\{f_1, f_2, \ldots, f_n\}$, $\chi_G = \lim g_n$ and (g_n) is a nondecreasing sequence in $C_c(X)$, whence G is $C_c(X)$-open.

(ii) By (i) X is $C_c(X)$-open. Hence there exists a nondecreasing sequence (f_n) in $C_c(X)$ such that $\chi_X = \lim f_n$. Let $K_n = \{x \in X: f_n(x) \geq 1/n\}$; K_n is a G_δ set and compact as a closed subset of the support of f_n. Also

$$K_n \subset \{x \in X: f_{n+1}(x) \geq 1/n\} \subset \{x \in X: f_{n+1}(x) > 1/n + 1\} \subset K_{n+1}$$

so that $K_n \subset K_{n+1}^{\circ}$. If K is compact in X, then $K \subset K_n$ for large n, as otherwise $\bigcup_{n=1}^{\infty} K_n{}^{\circ}$ is an open covering of K with no finite subcovering. ∎

Proposition 5.7. Let X be a locally compact Hausdorff space.

(i) The σ-algebra of Baire sets is identical to $\sigma(C_c(X))$.

(ii) The σ-algebra of Baire sets is identical to the σ-algebra generated by open σ-compact sets.

(iii) If X is σ-compact, the σ-algebra of Baire sets coincides with $\sigma(C(X))$. ∎

Proof. (i) The σ-algebra $\sigma(C_c(X))$ is the σ-algebra generated by sets of the form $\{x: f(x) \geq \alpha\}$ for $\alpha \in R$ and $f \in C_c(X)$. If $\alpha > 0$ these are compact G_δ sets and it follows by a simple argument that $\mathrm{Ba}(X)$ contains all sets of this form. Conversely, it follows from Lemma 5.3 that each compact G_δ set K equals $\{x \in X: f(x) \geq 1\}$ for some $f \in C_c(X)$.

(ii) By Proposition 5.3 the smallest σ-algebra containing $C_c(X)$-open sets equals $\sigma(C_c(X))$. The result now follows from (i) and Lemma 5.5 (i).

(iii) By Proposition 5.3, $\sigma(C_c(X))$ coincides with the σ-algebra generated by $C_c(X)$-open sets and $\sigma(C(X))$ coincides with the σ-algebra generated by the $C(X)$-open sets. It suffices to show that each $C(X)$-open set is $C_c(X)$-open. To this end let G be a $C(X)$-open set and let (f_n) be a nondecreasing sequence of nonnegative functions in $C(X)$ such that $\chi_G = \lim f_n$. Since X is an open σ-compact set, X is $C_c(X)$-open by Lemma 5.5 (i). This means there is a nondecreasing sequence (g_n) of nonnegative functions in $C_c(X)$ whose limit is χ_X. The result is that $\chi_G = \lim f_n g_n$ and $(f_n g_n)$ is a nonnegative sequence in $C_c(X)$ so that G is a $C_c(X)$-open set. ∎

Corollary 5.2. If X is a σ-compact locally compact Hausdorff space, then

$$\begin{aligned}\sigma_r(\text{compact } G_\delta \text{ sets}) &= \sigma(\text{open } \sigma\text{-compact sets}) = \sigma(C_c(X)) \\ &= \sigma(\text{compact } G_\delta) \\ &= \sigma(\text{zero sets of continuous functions}) \\ &= \sigma(C(X)) = \sigma(\text{closed } G_\delta \text{ sets}).\end{aligned}$$

In particular, $\text{Ba}(X) = \text{Ba}_w(X)$. ∎

Proof. Proposition 5.7 establishes the second and third equalities. Since by Lemma 5.4, X is the countable union of compact G_δ sets, whenever X is σ-compact the first equality is true. From Lemma 5.3, if G is a compact G_δ set, then there exists $f \in C_c(X)$ such that

$$G = \{x: (f - 1)(x) = 0\},$$

so that $\sigma(\text{compact } G_\delta) \subset \sigma(\text{zero sets})$. Trivially $\sigma(\text{zero sets}) \subset \sigma(C(X))$ for any topological space X. By Proposition 5.7 (iii), the fourth and fifth equalities hold. Since each σ-compact locally compact Hausdorff space is normal, the last equality holds by Proposition 5.6. ∎

It is natural to ask whether there are compact Baire sets that are not G_δ sets in the case that X is a locally compact Hausdorff space. If X is also second countable it is easy to show that each compact set is a G_δ so that in this case $\text{Ba}(X) = B(X)$. Regardless of whether X is second countable, we always have the following result.

Proposition 5.8. If X is a locally compact Hausdorff space, each compact Baire set C is a G_δ set. ∎

Proof. By Problem 5.2.6, there are compact G_δ sets C_n such that $C \in \sigma((C_n))$ and $C_n = \{x \in X: f_n(x) = 0\}$ for some f_n in $C(X)$ with $0 \leq f_n \leq 1$. Define the pseudometric [$d(x, y) = 0$ does not always imply $x = y$] on X by

$$d(x, y) = \sum_{n=1}^{\infty} \frac{1}{2^n} | f_n(x) - f_n(y) |.$$

For each x in X, let $[x] = \{y \in X: d(x, y) = 0\}$; that is, $[x]$ is the equivalence class of x with respect to the equivalence relation: $x \sim y$ if and only if $d(x, y) = 0$. Let $\hat{X}$ denote the set of all such equivalence classes and define metric δ on $\hat{X}$ by

$$\delta([x], [y]) = d(x, y).$$

If $n: X \to \hat{X}$ is the natural mapping given by $n(x) = [x]$, then n is continuous since if $E = \{[y]: \delta([y], [x]) < r\}$ for some $[x]$ in $\hat{X}$ and positive number r, then $n^{-1}(E) = \{y: d(y, x) < r\}$ and $d(y, x)$ is a continuous function of y.

A subset S of X equals $n^{-1}(\hat{S})$ for some $\hat{S} \subset \hat{X}$ if and only if S contains the set $[x]$ for any x in S. Since each C_n has this property, each C_n is the inverse image of some set in $\hat{X}$. Since the class of all inverse image sets is a σ-algebra, $C = n^{-1}(\hat{C})$ for some subset $\hat{C}$ of $\hat{X}$. Since $n: X \to \hat{X}$ is continuous, $n(C) = n(n^{-1}(\hat{C})) = \hat{C}$ is compact in $\hat{X}$. Since every compact subset of a metric space is a G_δ set, $\hat{C} = \bigcap_{n=1}^{\infty} \hat{O}_n$ for some open sets $\hat{O}_n$ in $\hat{X}$. This means

$$C = \bigcap_{n=1}^{\infty} O_n,$$

where O_n are the open sets $n^{-1}(\hat{O}_n)$ in X. ∎

The next proposition and the discourse following further illustrates the contrast between Borel and Baire sets. We will show that Baire sets "multiply" as described in Proposition 5.9, whereas Borel sets fail to "multiply." This information will motivate our study of product measures and Fubini's Theorem in Section 5.5.

Proposition 5.9. If X and Y are locally compact Hausdorff spaces, then

$$\sigma_r(\text{compact } G_\delta \text{ of } X) \times \sigma_r(\text{compact } G_\delta \text{ of } Y) = \sigma_r(\text{compact } G_\delta \text{ of } X \times Y). \quad (5.9)$$

[The left side of this equation is by definition the σ-ring generated by sets $A \times B$, where $A \in \sigma_r(\text{compact } G_\delta \text{ of } X)$ and $B \in \sigma_r(\text{compact } G_\delta \text{ of } Y)$.] ∎

Proof. From Problem 1.3.8 of Chapter 1 we know that the left side of equation (5.9) is identical to the σ-ring generated by the sets of the form $A \times B$, where A is a compact G_δ set in X, and B is a compact G_δ set in Y. Since $A \times B$ is a compact G_δ set in $X \times Y$ whenever A and B are compact G_δ sets in X and Y, respectively, clearly the left side of equation (5.9) is contained in the right side.

Conversely if C is a compact G_δ set in $X \times Y$, let O be an open set in $X \times Y$ containing C. Then from Remark 5.4, $C \subset E \subset O$, where E is a finite union of sets of the form $U \times V$, where U and V are open countable unions of compact G_δ sets in X and Y, respectively. Since $C = \bigcap_{i=1}^{\infty} O_i$ where each O_i is open, $C = \bigcap_{i=1}^{\infty} E_i$, where each E_i is a finite union of sets as $U \times V$. This means C is an element of the left side of equation (5.9). ∎

Corollary 5.3. If X and Y are σ-compact locally compact Hausdorff spaces, then

$$\text{Ba}(X) \times \text{Ba}(Y) = \text{Ba}(X \times Y)$$

and

$$\text{Ba}_w(X) \times \text{Ba}_w(Y) = \text{Ba}_w(X \times Y)$$ ∎

Proof. See Corollary 5.2. ∎

Example 5.8. This example shows that the condition that X and Y both be σ-compact is essential in Corollary 5.3. Let X be an uncountable set with the discrete topology and let Y be an infinite countable set with the discrete topology. Note that both X and Y are locally compact and Hausdorff; but whereas Y is σ-compact, X is not σ-compact. We have, since compact sets coalesce with finite sets,

$$\begin{aligned}
\text{Ba}(X) &= B(X) = \{A \subset X\text{: } A \text{ or } A^c \text{ is countable}\}, \\
\text{Ba}(Y) &= B(Y) = \{B \subset Y\text{: } B \text{ or } B^c \text{ is countable}\}, \\
\text{Ba}(X \times Y) &= B(X \times Y) = \{C \subset X \times Y\text{: } C \text{ or } C^c \text{ is countable}\}.
\end{aligned}$$

However, if A is an uncountable set in X with countable complement and $c \in Y$, then $A \times \{c\}$ is an element of $\text{Ba}(X) \times \text{Ba}(Y)$ but not an element of $\text{Ba}(X \times Y)$.

In contrast, the result of Proposition 5.9 is not true for Borel sets even though X and Y are σ-compact. If X and Y are σ-compact locally compact Hausdorff spaces, we know from Problem 1.3.8 in Chapter 1 that $B(X) \times B(Y)$ is generated by the sets of the form $A \times B$, where A is compact in X and B is compact in Y. Hence

$$B(X) \times B(Y) \subset B(X \times Y).$$

However, this inclusion is sometimes proper. Let us consider the following interesting examples.

Examples

5.9. Let X be the one-point compactification of a discrete space having cardinality greater than c. As every subset of X is either open or closed, $B(X) = 2^X$, the power set of X. As every subset of $X \times X$ is the difference of two compact subsets, $B(X \times X) = 2^{X \times X}$. Nevertheless, $B(X \times X) \neq B(X) \times B(X)$ as the diagonal $D = \{(x, y) \in X \times X\text{: } x = y\}$ is not in $B(X) \times B(X)$ (see Problem 5.2.8).

5.10. The classes of Borel and Baire sets can correspond in a σ-compact space X and yet be distinct in $X \times X$. If this is the case, we clearly have

$$B(X) \times B(Y) = \text{Ba}(X) \times \text{Ba}(Y) = \text{Ba}(X \times Y) \subsetneq B(X \times Y).$$

As an example, let X be a compact nonmetrizable space such that every closed subset is a G_δ set (see Problem 5.2.5, for instance). The diagonal D is compact and is thus a Borel set. If D were a Baire set, it would be a G_δ set by Proposition 5.8. Now the neighborhoods of D define a uniform structure which yields the given topology of X. (See Chapter 6 of [29].) If D were a G_δ set, it would have a fundamental sequence of neighborhoods so that X would be metrizable.

Do Borel sets ever multiply? A positive answer is given in the following proposition. It says in effect that if bounded subspaces of one of the factor spaces are second countable, then the Borel sets multiply.

Proposition 5.10. Suppose X and Y are locally compact Hausdorff spaces. Suppose that each bounded subspace of Y is second countable. (Since any locally compact Hausdorff space with a countable base is metrizable, it is equivalent to assume that each bounded subspace of Y is metrizable.) Then

$$\sigma_r(\text{compact sets in } X) \times \sigma_r(\text{compact sets in } Y) = \sigma_r(\text{compact sets in } X \times Y).$$

If X and Y are also σ-bounded, then

$$B(X) \times B(Y) = B(X \times Y). \qquad \blacksquare$$

Proof. For simplicity of notation we prove the result for the σ-bounded case. The other case is almost identical. Since the Borel sets are precisely the σ-algebra generated by the bounded open sets (each bounded open set is the difference of two compact sets and each compact set is the difference of two bounded open sets), it is sufficient to show that if W is a bounded open set in $X \times Y$, then $W \in B(X) \times B(Y)$. Let $F = \text{Pr}_Y(W)$, where Pr_Y is the projection of $X \times Y$ onto Y, a continuous open mapping. F is then a bounded open set in Y. By hypothesis, the subspace F has a countable base for the open sets, say $\mathscr{V}$. As F is a bounded open set, observe that each member of $\mathscr{V}$ is a bounded open set in Y. Since W is also an open subset of $X \times F$,

$$W = \bigcup \{U \times V\colon\ U \text{ is open in } X,\ V \in \mathscr{V},\ U \times V \subset W\}.$$

For each $V \in \mathscr{V}$, define

$$U_V = \bigcup\{U\colon\ U \text{ is open in } X,\ U \times V \subset W\}$$

an open and bounded subset of X. Obviously, $U_V \times V \in B(X) \times B(Y)$. Now

$$\begin{aligned} W &= \bigcup\{U \times V\colon\ U \text{ open in } X,\ V \in \mathscr{V},\ U \times V \subset W\} \\ &= \bigcup_{V \in \mathscr{V}} \bigcup\{U \times V\colon\ U \text{ open in } X,\ U \times V \subset W\} \\ &= \bigcup_{V \in \mathscr{V}} [\bigcup\{U\colon\ U \text{ is open in } X,\ U \times V \subset W\}] \times V \\ &= \bigcup_{V \in \mathscr{V}} U_V \times V. \end{aligned}$$

Since the last union is countable, $W \in B(X) \times B(Y)$. ∎

Corollary 5.4. If X and Y are locally compact σ-compact Hausdorff spaces and Y is metrizable, then

$$B(X \times Y) = B(X) \times B(Y).$$ ∎

Problems

× **5.2.1.** Prove that if K is a compact subset of a locally compact Hausdorff space and $\{O_\alpha\}$ is an open covering of K, then there are a finite number of nonnegative continuous functions $f_1, f_2, \ldots, f_n$ on X, each f_i vanishing outside a compact set contained in some O_{α_i} such that $f_1 + f_2 + \cdots + f_n \equiv 1$ on K. (Hint: For each $x \in K$, there is a nonnegative continuous function g that is positive at x and zero outside some O_α. Choose a finite number $g_1, g_2, \ldots, g_n$ of such g's so that $g_1 + g_2 + \cdots + g_n$ is positive on K. Let $h = 1/g$ and set $f_i = hg_i$.)

5.2.2. Prove that each element of σ_r(compact G_δ) and each element of σ_r(compact) is σ-bounded. Conclude that these are σ-algebras if and only if X is σ-bounded.

× **5.2.3.** If $\mathscr{A}$ is a σ-ring or σ-algebra of subsets of a set X, then a subset M of X is said to be *locally measurable* with respect to $\mathscr{A}$ if $M \cap A \in \mathscr{A}$ for each $A \in \mathscr{A}$. Prove the following:

(i) The class of locally measurable sets with respect to $\mathscr{A}$ is a σ-algebra containing $\mathscr{A}$.

(ii) Each weakly Borel set is locally measurable with respect to the σ-ring of Borel sets if X is a Hausdorff space.

(iii) Each weakly Baire set is locally measurable with respect to the σ-ring of Baire sets if X is Hausdorff.

5.2.4. Prove that if X is Hausdorff, the elements of σ_r(compact) are precisely the weakly Borel sets of X that are σ-bounded; the elements of σ_r(compact G_δ) are precisely the weakly Baire sets that are contained in a countable union of compact G_δ sets.

× **5.2.5.** Let $X = [-1, 1]$ with topology given by the subbase $\mathscr{S}$ given by

$$\mathscr{S} = \{A \subset [-1, 1]: A \text{ or } A^c \text{ has the form } [-b, b) \text{ for } 0 \leq b \leq 1\}.$$

Then:

(i) Prove that X is Hausdorff with this topology.

(ii) Show that X is compact by showing that every covering of X by subbasic members can be reduced to a covering of not more than two such members (see [29], page 139).

(iii) Show that X is not metrizable. (Hint: If X were compact and metrizable, it would have a countable base.)

(iv) Show that each closed subset of X is a G_δ set by showing that each open set is a countable union of basic sets.

× **5.2.6.** Let X be a locally compact Hausdorff space.

(i) Show that if C is a compact Baire set in X, then there is a sequence (C_n) of compact G_δ sets such that C is in the smallest σ-algebra $\sigma((C_n))$ containing each C_n.

(ii) Show that each compact G_δ set C equals $\{x: f(x) = 0\}$ for some continuous function f on X with $0 \leq f \leq 1$.

5.2.7. (a) Let $X^* = X \cup \{\omega\}$ be the one-point compactification of a locally compact Hausdorff space X. Prove

$$B(X^*) = B_w(X^*) = \{A \subset X^*: A \cap X \in B_w(X)\}.$$

(b) If in addition X is σ-compact, prove

$$\mathrm{Ba}(X^*) = \mathrm{Ba}_w(X^*) = \{A \subset X^*: A \cap X \in \mathrm{Ba}_w(X)\}.$$

× **5.2.8.** Show that if X is a second countable locally compact Hausdorff space, each compact set is a G_δ set.

5.2.9. Let X be the one-point compactification of a discrete space having cardinality greater than c. Prove that $D = \{(x, y) \in X \times X: x = y\}$ is not an element of $B(X) \times B(X)$. [Hint: If $D \in B(X) \times B(X)$, then there is a countable collection (R_n) of rectangles $R_n = A_n \times B_n$ such that $D \in \sigma((R_n))$. Let $\mathscr{E}$ be the collection of the sets A_n and B_n for $n = 1, 2, \ldots$.

Then $D \in \sigma(\mathscr{E}) \times \sigma(\mathscr{E})$ so that $D_y = \{x \in X: (x, y) \in D\}$ is an element of $\sigma(\mathscr{E})$ for each $y \in X$. However, $\sigma(\mathscr{E})$ has cardinality no greater than c. (See Proposition 1.10 in Chapter 1.)]

5.3. Measures on Topological Spaces; Regularity

On any topological space X we have defined six primary classes of sets in Definition 5.3—the σ-ring and σ-algebra of Borel sets, the σ-ring and σ-algebra of Baire sets, the σ-algebra of weakly Borel sets, and the σ-algebra of weakly Baire sets. In this section we wish to consider measures on these classes of sets.

Throughout this section X is assumed to be a locally compact Hausdorff space.

Definition 5.4. A (weakly) *Baire measure on* X is any measure on the σ-algebra of (weakly) Baire sets that is finite valued on compact Baire sets. A (weakly) *Borel measure on* X is any measure defined on the σ-algebra of (weakly) Borel sets that is finite valued on compact Borel sets. ∎

We will also concern ourselves with measures on σ_r (compact G_δ) and σ_r (compact) that are finite valued on compact sets belonging to these σ-rings. Such measures will be called *Baire measures on the σ-ring of Baire sets* and *Borel measures on the σ-ring of Borel sets*, respectively, in contrast to Baire and Borel measures on X.

Of primary interest are Borel and Baire measures whose values on Borel and Baire sets, respectively, can be approximated by the measures of the generating sets—the compact and compact G_δ sets, respectively. For this reason we introduce the concepts of inner and outer regularity.

Remark 5.6. In the following definition and in Examples 5.11–5.15 and Lemmas 5.6 and 5.7 below, the use of the term "measurable set" refers exclusively throughout to sets in whatever class is under consideration —the σ-ring of Baire sets, the σ-algebra of Baire sets, the σ-ring of Borel sets, or the σ-algebra of Borel sets.

Definition 5.5. Let μ be a Baire measure on X or the σ-ring of Baire sets, or let μ be a Borel measure on X or the σ-ring of Borel sets. A measurable set F is *inner regular* with respect to μ if

$$\mu(F) = \sup\{\mu(C): C \subset F,\ C \text{ a compact measurable set}\}. \tag{5.10}$$

F is *outer regular* with respect to μ if

$$\mu(F) = \inf\{\mu(O)\colon F \subset O,\ O \text{ an open measurable set}\}. \tag{5.11}$$

μ is *outer* (inner) *regular* if every measurable set is *outer* (inner) *regular* and μ is *regular* if it is both inner and outer regular. ▮

Our first goal in this section is to give conditions for which Borel or Baire measures are inner or outer regular. First some examples.

Examples

5.11. Not every Borel measure is regular. Let $\bar{X}$ be the set of all ordinals less than or equal to Ω, the first uncountable ordinal. Let $X = \bar{X} - \{\Omega\}$. On $\bar{X}$ put the order topology—the subbase consists of sets of the form $\{x \in \bar{X}\colon x < a\}$ or $\{x \in \bar{X}\colon a < x\}$ for some $a \in \bar{X}$. Since every cover of $\bar{X}$ by subbasic sets has a finite subcover, $\bar{X}$ is compact (see [29], Theorem 6, Chapter 5). Clearly $\bar{X}$ is Hausdorff. For each Borel set E in $\bar{X}$ define

$$\mu(E) = \begin{cases} 0, & \text{if } E \text{ contains no unbounded closed subset of } X, \\ 1, & \text{otherwise.} \end{cases}$$

μ is then a Borel measure that is not regular. (See Problems 5.3.9 and 5.3.13.)

5.12. Let X be as in Example 5.11 with the order topology. X is normal and locally compact, but not paracompact (see [29], page 172). Define μ on Baire sets $\mathrm{Ba}(X)$ as follows:

$$\mu(E) = \begin{cases} 1, & \text{if } E \text{ contains an uncountable closed set,} \\ 0, & \text{otherwise.} \end{cases}$$

μ is a finite nonregular Baire measure.

5.13. Let X be any space that is not σ-compact. Then the measure μ on $\mathrm{Ba}(X)$ given by

$$\mu(E) = \begin{cases} 0, & \text{if } E \text{ is a subset of a } \sigma\text{-compact set,} \\ \infty, & \text{otherwise} \end{cases}$$

is an infinite Baire measure that is not regular.

5.14. Let X be the locally compact space of Example 5.7. Then

$$\mathrm{Ba}(X) = \{A \subset X\colon A \text{ or } A^c \text{ is countable}\}.$$

Define μ on $\mathrm{Ba}(X)$ by

$$\mu(\{D_{nj}\}) = 1/2^n,$$

and

$$\mu(B) = \sum_{D_{nj} \in B} \mu(\{D_{nj}\}) \qquad \text{for any } B \in \mathrm{Ba}(X).$$

μ is a Baire measure that is inner regular but not outer regular. Since $B(X) = \mathrm{Ba}(X)$, μ is also a Borel measure on X that is inner regular but not outer regular.

5.15. Let X be an uncountable set with the discrete topology. Define μ on $\mathrm{Ba}(X)$ or $B(X)$

$$\mu(A) = \begin{cases} 0, & \text{if } A \text{ is countable,} \\ 1, & \text{if } A^c \text{ is countable.} \end{cases}$$

μ is a finite Baire and Borel measure that is outer regular but not inner regular.

It turns out that for Borel measures defined on the σ-ring of Borel sets or for Baire measures defined on the σ-ring of Baire sets outer and inner regularity are equivalent. In fact we show below that every such Baire measure is regular. Clearly these statements are not true for Baire measures on $\mathrm{Ba}(X)$ or Borel measures on $B(X)$, as Examples 5.11–5.15 above show.

First we need two lemmas whose proofs are straightforward and are left for the reader. The terminology is that of Remark 5.6 and Definition 5.5.

Lemma 5.6. If (E_n) is a sequence of outer (inner) regular sets, then $\bigcup_{n=1}^{\infty} E_n$ is outer (inner) regular. In addition if $\mu(E_n) < \infty$ for all n, then $\bigcap_{n=1}^{\infty} E_n$ is also outer (inner) regular. ∎

Lemma 5.7.

(i) If C and D are compact measurable sets and C is outer regular, then $C - D$ is outer regular.

(ii) If every bounded open measurable set is inner regular, then $C - D$ is inner regular for all compact measurable sets C and D. ∎

Proposition 5.11. The following statements are equivalent if μ is either a Borel measure on the σ-ring of Borel sets or a Baire measure on the σ-ring of Baire sets.

(i) μ is regular.
(ii) Every bounded open measurable set U is inner regular.
(iii) Every compact measurable set C is outer regular.
(iv) For each compact measurable set C there is a compact G_δ set G such that $C \subset G$ and $\mu(C) = \mu(G)$. ▮

Proof. Trivially (i) ⇒ (ii) and (i) ⇒ (iii). We prove first that (iii) implies (ii). Let U be a bounded open measurable set and let $\varepsilon > 0$ be arbitrary. Let C be a compact measurable set such that $U \subset C$. Since $C - U$ is measurable and compact, there exists by (iii) an open measurable set V such that $C - U \subset V$ and

$$\mu(V) \leq \mu(C - U) + \varepsilon.$$

Now $U = C - (C - U) \supset C - V$ and $C - V$ is a compact measurable set. Hence we can write

$$\begin{aligned}\mu(U) - \mu(C - V) &= \mu(U - (C - V)) = \mu(U \cap V) \\ &\leq \mu(V - (C - U)) = \mu(V) - \mu(C - U) \leq \varepsilon.\end{aligned}$$

This means U is inner regular.

Next we show that (ii) implies (iii). Let C be a compact measurable set and let $\varepsilon > 0$ be arbitrary. Using Lemma 5.4 there is an open bounded measurable set U such that $C \subset U$. Since $U - C$ is an open bounded measurable set, there exists a compact measurable set D such that

$$D \subset U - C \quad \text{and} \quad \mu(U - C) \leq \mu(D) + \varepsilon.$$

Now $C = U - (U - C) \subset U - D$ so that

$$\begin{aligned}\mu(U - D) - \mu(C) &= \mu((U - D) - C) = \mu((U - C) - D) \\ &= \mu(U - C) - \mu(D) < \varepsilon.\end{aligned}$$

Hence C is outer regular since $U - D$ is open.

Next we show that (ii) implies (i). By the above, for compact measurable sets C and D, the set $C - D$ is regular by Lemma 5.7. If $\mathscr{R}$ is the ring of all finite unions of the form $\bigcup_{i=1}^{n} C_i - D_i$, where C_i and D_i are compact measurable sets, then by Lemma 5.6 each element of $\mathscr{R}$ is regular. Now if for each compact measurable set C

$$\mathscr{M}_C = \{E\colon E \text{ is measurable and } E \cap C \text{ is regular}\},$$

then $\mathscr{M}_C$ is a monotone class by Lemma 5.6 since $\mu(E \cap C) < \infty$ for all measurable sets E. Since $\mathscr{R} \subset \mathscr{M}_C$, the smallest σ-ring containing $\mathscr{R}$ is contained in $\mathscr{M}_C$. This smallest σ-ring is the σ-ring of measurable sets. Now if E is any measurable set, there is a sequence of compact measurable sets (C_n) such that $E \subset \bigcup_{n=1}^{\infty} C_n$ (by Problem 1.3.2 in Chapter 1). Since $E = \bigcup_{n=1}^{\infty}(C_n \cap E)$ and since each $C_n \cap E$ is regular, E is regular by Lemma 5.6.

We now show that (iv) implies (iii). Let G be a compact G_δ set. Then $G = \cap O_i$ where the O_i are open sets. By Problem 5.3.4 we may assume the O_i are open measurable sets with $\mu(O_i) < \infty$. Since open measurable sets are outer regular, G is outer regular by Lemma 5.6. Assuming (iv), each compact measurable set is outer regular.

Finally we show (iii) implies (iv). Let C be a compact measurable set. For each n, there exists a sequence of open measurable sets U_n such that $C \subset U_n$ and $\mu(U_n - C) < 1/n$. For each n choose a compact G_δ set G_n such that $C \subset G_n \subset U_n$. Then $G = \cap G_n$ is the required G_δ set. ∎

Theorem 5.2. Every Baire measure on the σ-ring of Baire sets is regular. ∎

Proof. Every compact Baire set is a compact G_δ set by Proposition 5.8. The theorem now follows from Proposition 5.11.

When is a Baire measure on the σ-algebra $\text{Ba}(X)$ of Baire sets regular? A sufficient condition for outer regularity is given in the next result. Recall that a space X is *paracompact* if and only if each open cover $\mathscr{C}$ of X has an open locally finite refinement $\mathscr{C}'$—an open cover $\mathscr{C}'$ of X each member of which is a subset of a member of $\mathscr{C}$ and such that each point of X has a neighborhood intersecting only finitely many members of $\mathscr{C}'$. Each paracompact space is normal, while compact Hausdorff spaces and metric spaces are examples of paracompact spaces.

Theorem 5.3. Any Baire measure μ on a locally compact paracompact Hausdorff space X is outer regular.[†] ∎

Note that even metrizability is not enough to give us a corresponding result for inner regularity, as Example 5.15 shows.

Proof. As every member of $\text{Ba}(X)$ is itself σ-bounded or its complement is σ-bounded (see Problem 5.3.3), the proof is divided into these

[†] For this and other related results, the reader may also consult [30].

two cases. Note that if $\mu(E) = \infty$ for $E \in \text{Ba}(X)$ it is trivially true that E is outer regular. Accordingly, it may be assumed that $\mu(E) < \infty$. Suppose first, E in $\text{Ba}(X)$ is σ-bounded. Then $E \subset \bigcup G_i$, where G_i are disjoint sets in $\sigma_r(\text{compact } G_\delta)$ (see Lemma 5.4). Hence

$$E = E \cap \left(\bigcup_i G_i\right) = \bigcup_i (E \cap G_i),$$

where each $E \cap G_i$ is in the σ-ring of Baire sets (see Problem 5.2.3). Since every Baire measure on the σ-ring of Baire sets is outer regular, for each i there is an open Baire set such that $E \cap G_i \subset O_i$, and

$$\mu(O_i) \leq \mu(E \cap G_i) + \varepsilon/2^i.$$

Hence if $O = \cup O_i$, $E \subset O$, and

$$\mu(O) \leq \sum_{i=1}^{\infty} \mu(O_i) \leq \sum_{i=1}^{\infty} \mu(E \cap G_i) + \varepsilon/2^i = \mu(E) + \varepsilon.$$

Now suppose E^c is σ-bounded. Let $A = E^c$. Since X is paracompact and locally compact, $\bar{A}$ is σ-bounded. By Lemma 5.4, $\bar{A}$ is contained in a countable union U of open sets in $\sigma_r(\text{compact } G_\delta)$. Since a paracompact space is normal, Urysohn's Lemma can be used to obtain a continuous function f such that $f \equiv 1$ on $\bar{A}$ and $f \equiv 0$ on U^c. Let $B = \{x: f(x) = 1\}$. B is a closed G_δ set which contains $\bar{A}$ and since $B \subset U$, B is also σ-bounded. This means $B \in \sigma_r(\text{compact } G_\delta)$ and by the first part of this proof there is an open set O in $\text{Ba}(X)$ such that $B \cap E \subset O$ and

$$\mu(O) \leq \mu(B \cap E) + \varepsilon.$$

Now $O \cup B^c$ is an open set in $\text{Ba}(X)$ containing E and since $B^c = E - B$,

$$\mu(O \cup B^c) \leq \mu(O) + \mu(B^c) < \mu(B \cap E) + \varepsilon + \mu(E - B) \leq \mu(E) + \varepsilon. \quad ∎$$

We now turn our attention toward the extension of Baire measures to Borel measures. The hope is that the extension of a regular measure on a class of Baire sets to a larger class will also be regular. We will prove three extension theorems in the following discussion, namely, the extension of a regular Baire measure on the σ-ring of Baire sets to a unique regular Borel measure on the σ-ring of Borel sets, the extension of a regular Borel measure on the σ-ring of Borel sets to a unique inner regular weakly Borel measure, and the extension of an inner regular Baire measure to an outer regular weakly Borel measure.

Given a Borel measure μ on the σ-ring of Borel sets, it is a trivial matter to extend μ to a weakly Borel measure μ_w on $B_w(X)$ by simply defining $\mu_w(A) = \infty$ for each A in $B_w(X)$ but not in the σ-ring of Borel sets. However, the following proposition shows how to extend μ to $B_w(X)$ to assure the preservation of regularity.

• **Proposition 5.12.** If μ is any Borel measure on the σ-ring of Borel sets, then the formula for A in $B_w(X)$ given by

$$\mu_w(A) = \sup\{\mu(E)\colon\ E \subset A,\ E \in \sigma_r(\text{compact})\}$$

defines an extension of μ to a measure μ_w on $B_w(X)$. If μ is inner regular, then μ_w is the only extension of μ satisfying

$$\mu_w(A) = \sup\{\mu(E)\colon\ E \subset A,\ E \text{ compact}\}. \quad \blacksquare \quad (5.12)$$

Proof. The countable additivity of μ_w is essentially all there is to show in order that μ_w be a measure. It is important to observe that if $E \in \sigma_r$ (compact sets) and $A \,\varepsilon\, B_w(X)$, then $E \cap A \in \sigma_r$(compact sets) (see Problem 5.2.3). Let (A_i) be a disjoint countable collection of sets in $B_w(X)$. Then

$$\begin{aligned}\mu_w\left(\bigcup A_i\right) &= \sup\{\mu(E)\colon\ E \subset \bigcup A_i;\ E \in \sigma_r(\text{compact})\}\\ &= \sup\left\{\sum_{i=1}^{\infty} \mu(E \cap A_i)\colon\ E \subset \bigcup_i A_i;\ E \in \sigma_r(\text{compact})\right\}\\ &\leq \sum_{i=1}^{\infty} \mu_w(A_i).\end{aligned}$$

On the other hand

$$\begin{aligned}\sum_{i=1}^{\infty} \mu_w(A_i) &= \sum_{i=1}^{\infty} \sup\{\mu(E_i)\colon\ E_i \subset A_i;\ E_i \in \sigma_r(\text{compact})\}\\ &= \sup\left\{\sum_{i=1}^{\infty} \mu(E_i \cap A_i)\colon\ E_i \subset A_i;\ E_i \in \sigma_r(\text{compact})\right\}\\ &= \sup\left\{\mu\left[\left(\bigcup_{i=1}^{\infty} E_i\right) \cap \left(\bigcup_{j=1}^{\infty} A_j\right)\right]\colon\ E_i \subset A_i\right\}\\ &\leq \mu_w\left(\bigcup_{j=1}^{\infty} A_j\right).\end{aligned}$$

It is clear that μ_w satisfies equation (5.12) whenever μ is inner regular. The uniqueness of μ_w satisfying equation (5.12) follows from the fact that its values are determined by its values on compact sets. $\blacksquare$

Now let us extend a Baire measure (necessarily regular by Theorem 5.2) on the σ-ring of Baire sets to a regular Borel measure on the σ-ring of Borel sets and hence to an inner regular weakly Borel measure by Proposition 5.12.

• **Theorem 5.4.** Let μ be a Baire measure on σ_r(compact G_δ). Then there exists a unique regular Borel measure $\bar{\mu}$ on σ_r(compact) which extends μ. ▮

Proof. Define the set function μ^* on the class of open sets in σ_r (compact) by

$$\mu^*(O) = \sup\{\mu(K)\colon K \subset O,\ K \text{ a compact } G_\delta \text{ set}\}.$$

If $O \in \sigma_r$(compact G_δ), obviously $\mu^*(O) = \mu(O)$ since μ is regular. Using Problem 5.3.5, it is easy to verify that μ^* is countably subadditive. Let

$$\mathscr{M} = \{M \subset X\colon \text{for each } \varepsilon > 0, \text{ there is an open set } O_\varepsilon \text{ in } \sigma_r(\text{compact}) \text{ containing } M \text{ with } \mu^*(O_\varepsilon) < \varepsilon\}.$$

Observe that each subset of a set in $\mathscr{M}$ is also in $\mathscr{M}$, and $\mathscr{M}$ is closed under countable unions. Notice also that if $E \in \mathscr{M} \cap \sigma_r$(compact G_δ) and $\varepsilon > 0$ is arbitrary, then inasmuch as each compact G_δ set K contained in E is also in $\mathscr{M}$, for each such K there is an open set O_K in σ_r(compact) containing K with $\mu^*(O_K) < \varepsilon$. This means $\mu(E) = O$ as by the regularity of μ we have

$$\mu(E) = \sup\{\mu(K)\colon K \subset E,\ K \text{ compact } G_\delta\} \leq \varepsilon.$$

Let $\mathscr{R}$ be the class of sets

$$\mathscr{R} = \{E \triangle M\colon E\ \varepsilon\ \sigma_r(\text{compact } G_\delta) \text{ and } M \in \mathscr{M}\}.$$

Define on $\mathscr{R}$ the set function $\bar{\mu}$ given by

$$\bar{\mu}(E \triangle M) = \mu(E).$$

That $\bar{\mu}$ is well-defined follows from the fact that if $E \triangle M$ and $F \triangle N$ are equal elements of $\mathscr{R}$, then $E \triangle F = M \triangle N$ so that $E \triangle F\ \varepsilon\ \sigma_r$(compact G_δ) $\cap\ \mathscr{M}$. This means $\mu(E \triangle F) = 0$ and hence $\mu(E) = \mu(F)$. Observe that

(a) $\mathscr{R}$ is the smallest σ-ring containing σ_r(compact G_δ) and $\mathscr{M}$;

(b) $\mathscr{R}$ contains σ_r(compact); and

(c) the restriction of $\bar{\mu}$ to σ_r(compact) is a regular Borel measure on σ_r(compact) extending μ.

The verification of (a) is left for the reader, while the proof of (b) is accomplished by showing that each compact set is contained in $\mathscr{R}$. To this end let C be compact in X. By Lemma 5.4 we know there exists a compact G_δ set K_0 and an open σ-compact set O_0 in σ_r(compact G_δ) such that

$$C \subset O_0 \subset K_0.$$

Since $O_0 - C$ is an open set in σ_r(compact), by the definition of μ^* we can find a sequence of compact G_δ sets K_i such that $\mu^*(O_0 - C) = \lim_i \mu(K_i)$ with $K_i \subset O_0 - C$. Let $K = \cup K_i$. Then since

$$\mu^*((O_0 - C) - K) \leq \mu^*((O_0 - C) - K_i) \leq \mu^*(O_0 - C) - \mu(K_i) \to 0,$$

we have $\mu^*((O_0 - C) - K) = 0$ and $(O_0 - C) - K \in \mathscr{M}$. As $C = (O_0 - K) \triangle [(O_0 - C) - K]$ and $O_0 - K \in \sigma_r(\text{compact } G_\delta)$, $C \in \mathscr{R}$.

We complete the proof by showing that the restriction of $\bar{\mu}$ to σ_r(compact) is regular. (See Problem 5.3.7.) It suffices to show in light of Proposition 5.11 that each compact set is outer regular. If C is compact, we know from the above that

$$C = (O_0 - K) \triangle [(O_0 - C) - K],$$

where $O_0 - K \in \sigma_r(\text{compact } G_\delta)$. Hence for any $\varepsilon > 0$ there exists an open Baire set O with

$$O \supset O_0 - K \supset (O_0 - C) - K$$

and

$$\mu(O) \leq \mu(O_0 - K) + \varepsilon.$$

This means $O \supset C$ and $\mu(O) \leq \mu(O_0 - K) + \varepsilon = \bar{\mu}(C) + \varepsilon$. ∎

Although a Baire measure on the σ-ring of Baire sets can be extended uniquely to an inner regular weakly Borel measure by virtue of Proposition 5.12 and Theorem 5.4, such an extension is not always possible from $\text{Ba}(X)$ to $B_w(X)$ without additional assumptions. The remaining portion of this section discusses this situation and gives various additional assumptions making this extension possible.

Remember that X is always assumed to be a locally compact Hausdorff space in this section.

• **Theorem 5.5.** If μ is a regular Baire measure on Ba(X), then there exists an extension of μ to a unique weakly Borel measure $\bar{\mu}$ on $B_w(X)$ satisfying

(i) $\bar{\mu}(O) = \sup\{\mu(K): K \subset O, K \text{ compact } G_\delta\}$ for each open set O,

and (5.13)

(ii) $\bar{\mu}(E) = \inf\{\bar{\mu}(O): E \subset O, O \text{ an open set}\}$ for each weakly Borel set E.

If μ is finite, $\bar{\mu}$ is finite and regular. (See also Problem 5.3.8.) ∎

Proof. The proof is almost a replay of that of Theorem 5.4. The reader can supply omitted details. Define μ^* on the class of open sets by

$$\mu^*(O) = \sup\{\mu(K): K \subset O, K \text{ a compact } G_\delta \text{ set}\}, \tag{5.14}$$

μ^* is countably subadditive and agrees with μ on open sets in Ba(X). Let

$\mathscr{M} = \{M \subset X$: for each $\varepsilon > 0$, there is an open set O_ε containing M with $\mu^*(O_\varepsilon) < \varepsilon\}$.

Let $\mathscr{R}$ equal $\{E \triangle M: E \in \text{Ba}(X) \text{ and } M \in \mathscr{M}\}$. $\mathscr{R}$ is a σ-algebra containing Ba(X) and $\mathscr{M}$. Defining the set function μ_1 on $\mathscr{R}$ by

$$\mu_1(E \triangle M) = \mu(E),$$

it can be verified that μ_1 is a measure (Problem 5.3.7). Observe that if $\mu_1(E \triangle M) < \infty$, then by the outer regularity of μ and the definition of $\mathscr{M}$, there is an open set O containing $E \triangle M$ with $\mu^*(O) < \infty$. Observe in addition that each open set O with $\mu^*(O) < \infty$ is in $\mathscr{R}$. Indeed if $\mu^*(O)$ is finite, then for each positive integer n there is a compact G_δ set K_n with $K_n \subset O$ and

$$\mu^*(O - K_n) \leq \mu^*(O) - \mu^*(K_n) < 1/n.$$

Letting $K = \bigcup_{n=1}^\infty K_n$, $K \in \text{Ba}(X)$ and $O - K$ is in $\mathscr{M}$. Hence O is in $\mathscr{R}$ since

$$O = K \triangle (O - K).$$

Let $\mathscr{A}$ be the σ-algebra defined as

$$\mathscr{A} = \{A \subset X: A \cap B \in \mathscr{R} \text{ for each } B \in \mathscr{R} \text{ with } \mu_1(B) < \infty\}.$$

Then $\mathscr{A}$ contains each open set O' in X since if B is in $\mathscr{R}$ with $\mu_1(B) < \infty$, let O be the open set in $\mathscr{R}$ with $B \subset O$ and $\mu^*(O) < \infty$. Then

$$O' \cap B = (O' \cap O) \cap B$$

with $O' \cap O$ and B in $\mathscr{R}$. This means $\mathscr{A}$ contains $B_w(X)$.

Finally define $\bar{\mu}$ on $\mathscr{A}$ and hence on $B_w(X)$ by

$$\bar{\mu}(E) = \begin{cases} \mu_1(E), & \text{if } E \in \mathscr{R}, \\ \infty, & \text{otherwise.} \end{cases}$$

$\bar{\mu}$ satisfies the conclusion of the theorem. (i) is easily verified. To show (ii) suppose first that E is a weakly Borel set not in $\mathscr{R}$. Then $\bar{\mu}(E) = \infty$ and each open set O containing E has $\bar{\mu}(O) = \infty$. If not, then $\mu^*(O) < \infty$ and $O \in \mathscr{R}$, implying that E, equal to $E \cap O$, is in $\mathscr{R}$. Secondly, suppose E is a weakly Borel set in $\mathscr{R}$. Then $E = F \triangle M$, where $F \in \mathrm{Ba}(X)$ and $M \in \mathscr{M}$. From the outer regularity of μ and the definition of $\mathscr{M}$, respectively, we get open sets O_1 and O_2 with $F \subset O_1$ and $M \subset O_2$, and

$$\bar{\mu}(O_1 \cup O_2) \leq \bar{\mu}(O_1) + \bar{\mu}(O_2) \leq \mu(F) + \varepsilon. \qquad \blacksquare$$

• **Example 5.16.** Consider the space X of Examples 5.7 and 5.14. Let $E = R - D$ where $D = \{j/2^n \mid j \text{ an integer and } n \text{ a nonnegative integer}\}$. It is easy to verify that $\mu(E) = 0$, with μ being the measure on $\mathrm{Ba}(X)$ defined in Example 5.14. Recall that μ is inner regular but not outer regular. There is no outer regular extension $\bar{\mu}$ of μ to $B_w(X)$. For let $\bar{\mu}$ be any extension of μ to the class of weakly Borel sets. Let O be any open set containing E. For each n let

$$E_n = \{x \in E: x \in [j/2^m, (j+1)/2^m] \text{ and } D_{mj} \in O \text{ for all } m \geq n \text{ and for some } j \text{ depending on } m\}.$$

Since $E = \bigcup_{n=1}^{\infty} E_n$ (recall the topology defined on X), the outer Lebesgue measure λ^* of some E_{n_0} is not zero, that is, $\lambda^*(E_{n_0}) > 0$. This means that for each $m \geq n_0$

$$\lambda^*\Big(\bigcup_j [j/2^m, (j+1)/2^m]: D_{mj} \in O\Big) \geq \lambda^*(E_{n_0}) > 0,$$

whereby for each $m \geq n_0$, since $\lambda^*[j/2^m, (j+1)/2^m] = \mu(\{D_{mj}\})$,

$$\mu(O \cap X_m) \geq \lambda^*(E_{n_0}),$$

where $X_m = \{D_{mj}: j \text{ is an integer}\}$. (Notice that $O \cap X_m$ is a Baire set since it is a countable subset of X.) This means that

$$\bar{\mu}(O) \geq \sup\{\mu(K)\colon K \subset O,\ K \text{ compact } G_\delta \text{ sets}\} = \infty.$$

The preceding example shows that without regularity—even with inner regularity—the conclusion of Theorem 5.5 is invalid. However, we have the following corollaries of Theorem 5.5.

• **Corollary 5.5.** If X is a paracompact space and μ is an inner regular Baire measure on $\mathrm{Ba}(X)$, then there exists a unique extension $\bar{\mu}$ of μ to $B_w(X)$ satisfying equation (5.13). ∎

Proof. See Theorem 5.3 and use Theorem 5.5. ∎

• **Corollary 5.6.** If μ is a semifinite measure on $\mathrm{Ba}(X)$ such that for each E in $\mathrm{Ba}(X)$

$$\mu(E) = \inf\{\mu(G)\colon G \text{ open } \sigma\text{-compact set, } E \subset G\}, \tag{5.15}$$

then μ is regular and there exists a unique extension $\bar{\mu}$ of μ to $B_w(X)$ satisfying equations (5.13). If μ is finite, so is $\bar{\mu}$. ∎

Proof. The outer regularity of μ follows from the following inequality:

$$\begin{aligned}\mu(E) &\leq \inf\{\mu(O)\colon E \subset O,\ O \text{ an open set in } \mathrm{Ba}(X)\}\\ &\leq \inf\{\mu(G)\colon E \subset G,\ G \text{ open } \sigma\text{-compact}\}\\ &= \mu(E).\end{aligned} \tag{5.16}$$

We now show that μ is inner regular. Assume first $E \in \mathrm{Ba}(X)$ with $\mu(E) < \infty$. By equation (5.15), there exists for each $\varepsilon > 0$ an open σ-compact set $G = \bigcup_{i=1}^\infty K_i$ (where each K_i is a compact G_δ set) such that $E \subset G$ and $\mu(G - E) < \varepsilon$. Applying equation (5.15) again to $G - E$, there exists an open (σ-compact) set O containing $G - E$ such that $\mu(O) < 2\varepsilon$. Since $\mu(G) = \lim_n \mu(\bigcup_{i=1}^n K_i)$, there exists N such that $\mu(G - \bigcup_{i=1}^N K_i) < \varepsilon$. Let C be the compact set $(\bigcup_{i=1}^N K_i) \cap O^c$ contained in E. Now

$$\begin{aligned}\mu(E) &= \mu(C) + \mu(E - C)\\ &\leq \mu(C) + \mu\left(O \cup \left[G - \bigcup_{i=1}^N K_i\right]\right)\\ &\leq \mu(G) + \mu(O) + \mu\left(G - \bigcup_{i=1}^N K_i\right)\\ &\leq \mu(G) + 2\varepsilon + \varepsilon = \mu(G) + 3\varepsilon.\end{aligned}$$

Secondly, if $\mu(E) = \infty$, then as μ is semifinite there exists for each positive integer n a set F contained in E with finite measure greater than n. According to the preceding argument there exists a compact Baire set contained in F and hence in E with measure greater than n.

The conclusion of the corollary now follows from Theorem 5.5. ▮

The hypothesis of Corollary 5.5 includes the cases where X is σ-bounded or μ is finite or σ-finite. The only case left is that given in the next theorem.

• **Theorem 5.6.** Suppose μ is a nonfinite Baire measure on a non-σ-bounded space X. Assume also the hypothesis (5.15). Then μ has an extension to a unique weakly Borel measure $\bar{\mu}$ on $B_w(X)$ satisfying

$$\bar{\mu}(O) = \sup\{\mu(K): K \subset O,\ K \text{ a compact } G_\delta \text{ set}\} \tag{5.17}$$

for each open σ-bounded set O and

$$\bar{\mu}(E) = \inf\{\mu(G): E \subset G,\ G \text{ an open } \sigma\text{-bounded set}\} \tag{5.18}$$

for each E in $B_w(X)$. ▮

Proof. The proof is almost identical to that of Theorem 5.5 with the following exceptions. Replace equation (5.14) by

$$\mu^*(O) = \begin{cases} \sup\{\mu(K): K \subset O,\ K \text{ compact } G_\delta \text{ set}\}, & \text{if } O \text{ is } \sigma\text{-bounded}, \\ +\infty, & \text{otherwise} \end{cases} \tag{5.19}$$

for each open set O. Using the fact that for each open Baire set O either O or O^c is σ-bounded (Problem 5.3.3), it is easy to show that μ^* agrees with μ on open Baire sets (Problem 5.3.10). Define $\mathscr{M}$, $\mathscr{R}$, μ_1, $\mathscr{A}$, and $\bar{\mu}$ exactly as in the proof of Theorem 5.5 and obtain the desired conclusions (Problem 5.3.10.). ▮

Problems

5.3.1. Prove Lemma 5.6.

5.3.2. Prove Lemma 5.7.

× **5.3.3.** (i) Prove that every element of Ba(X) and $B(X)$ is either σ-bounded or its complement is σ-bounded.

(ii) Prove that a Borel measure on the σ-ring of Borel sets is σ-finite.

× **5.3.4.** Prove that any compact G_δ set G can be written as $\bigcap_{i=1}^{\infty} O_i$, where each O_i is an open set in σ_r(compact G_δ) and $\mu(O_i) < \infty$ for any Baire measure μ on σ_r(compact G_δ).

× **5.3.5.** Prove that if K is a compact G_δ set and $K \subset \bigcup_{i=1}^{\infty} O_i$ with each O_i open, then $K = \bigcup_{i=1}^{n} K_i$, where each K_i is a compact G_δ and $K_i \subset O_i$. [Hint: Use Problem 5.2.1.]

× **5.3.6.** Prove that every finite inner regular Baire measure is outer regular.

• **5.3.7.** Prove that $\bar{\mu}$ and μ_1 of Theorems 5.4 and 5.5, respectively, are measures on $\mathscr{R}$.

• **5.3.8.** Prove that the measure $\bar{\mu}$ of Theorem 5.5 also satisfies the following properties:

(i) For each weakly Borel set E of finite measure, $\bar{\mu}(E) = \sup\{\bar{\mu}(K): K \subset E$ and K is compact$\}$.

(ii) For each weakly Borel set A with $\bar{\mu}(A) < \infty$, there is a Baire set E and a weakly Borel set N with $\bar{\mu}(N) = 0$ and $A = E \triangle N$.

(iii) If μ is finite, $\bar{\mu}$ is finite. In this case $\bar{\mu}$ is regular.

5.3.9. (i) Prove that the set function μ defined in Example 5.11 is a measure.

(ii) Prove that μ is not regular. (Hint: Consider any interval containing Ω.) Compare Problem 5.3.13.

• **5.3.10.** Let μ be a Baire measure on Ba(X) satisfying equation (5.15).

(i) Prove that each σ-bounded set in Ba(X) is inner regular.

(ii) Prove that μ^* of equation (5.19) agrees with μ on open Baire sets if μ is nonsemifinite.

(iii) Complete the proof of Theorem 5.6.

⋆ **5.3.11.** *On the Measurability of Functions in Two Variables.* Let X and Y be two compact topological spaces and μ, ν be finite regular measures defined on the Borel sets of X and Y, respectively. By a *measurable modification* $\tilde{f}(x, y)$ of a function $f(x, y)$, we mean a $(\mu\times\nu)$-measurable function such that $\tilde{f}(x, \cdot) = f(x, \cdot)$ a.e. (ν) for every x. Prove the following result due to Mark Mahowald: If Y is metric, and if $f(x, y)$ has a measurable modification, and if $f(x, \cdot)$ is continuous for almost all x, then $f(x, y)$ is $(\mu\times\nu)$-measurable. [Hint: We may and do assume that $f(x, \cdot)$ is continuous for all x. Use Lusin's Theorem (Problem 3.1.13 in Chapter 3) to find a "large" compact set C in $X\times Y$ such that $\tilde{f}(x, y)$ is continuous on C and $\mu\times\nu(C \cap G) > 0$ whenever G is open and $C \cap G \neq \varnothing$. Let (U_n) be a countable basis for Y, and let $A_n = \{x: \nu((X\times U_n) \cap C)_x = 0\}$ and $B_n = (A_n\times Y) \cap (X\times U_n) \cap C$. Then $\cup B_n \equiv D$ has $(\mu\times\nu)$-measure zero. Let $E = C - D$. If $N_x = \{n: x \in A_n\}$, then $E_x =$

$\bigcap_{n\in N_x}(C_x - U_n)$, which is compact. Also $\nu(E_x \cap H) > 0$ whenever H is open and $H \cap E_x \neq \varnothing$. If $\nu(E_x) > 0$, $f(x, \cdot) = \tilde{f}(x, \cdot)$ *everywhere* (by continuity). It follows that $\tilde{f}(x, y) = f(x, y)$ a.e. $(\mu \times \nu)$ on C.]

5.3.12. Let X be a locally compact Hausdorff space and μ a weakly Borel measure such that for any weakly Borel B with $\mu(B) < \infty$,

$$\begin{aligned}\mu(B) &= \inf\{\mu(O)\colon O \text{ open} \supset B\}\\ &= \sup\{\mu(K)\colon K \text{ compact} \subset B\}.\end{aligned}$$

Then for $1 \leq p < \infty$, $C_c(X)$ is dense in $L_p(\mu)$. [Hint: Let $\mu(B) < \infty$ and $K \subset B \subset O$ such that there exists $f \in C_c(X)$ with $f(x) = 1$ for $x \in K$, $= 0$ for $x \notin O$ and $0 \leq f \leq 1$. If $\mu(O - K) < \varepsilon$, then $\| \chi_B - f \|_p \leq 2 \cdot \varepsilon^{1/p}$.]

★ **5.3.13.** *The Structure of Measures and Measurable Functions on the Borel Sets of* $[0, \Omega]$. Let $\mathscr{B}$ be the Borel sets of $[0, \Omega]$, Ω the first uncountable ordinal, with order topology. Let μ be a set function as in Example 5.11 on $\mathscr{B}$ such that $\mu(E) = 1$ or 0 according as E does or does not contain an unbounded, closed subset of $[0, \Omega)$. Then μ is a measure that is *not* regular. Prove the following assertions due to M. B. Rao and K. P. S. B. Rao:

(i) $\mathscr{B} = \{A \subset [0, \Omega]\colon A$ or A^c contains an unbounded closed subset of $[0, \Omega)\}$. [Hint: Let the set on the right be $\mathscr{F}$. Show that $\mathscr{F}$ is a σ-algebra and $\mathscr{F} \supset \mathscr{B}$. For the converse, let A be an unbounded closed subset of $[0, \Omega)$. Let $B = A^c - \Omega$. Assume $0 \in A$. Let α' stand for the first succeeding ordinal of α in A. Define a set-valued function g on A such that $g(\alpha) = \{\beta \in B\colon \alpha < \beta < \alpha'\}$. Then $g(A) = B$. Enumerate the elements of $g(\alpha)$ [whenever $\alpha \in A$ and $g(\alpha) \neq \varnothing$] and write $A_n - \bigcup_{\alpha\in A}$ {the nth element in the enumeration of $g(\alpha)$}. Then $g(A) = \bigcup_{n=1}^{\infty} A_n$; also A_n is Borel since

$$A_n = \bigcup_{\alpha\in A} \bigcap_{i=1}^{\infty} A_{i\alpha}^{(n)} = \bigcap_{i=1}^{\infty} \bigcup_{\alpha\in A} A_{i\alpha}^{(n)},$$

where $A_{i\alpha}^{(n)}$ are open sets $\subset (\alpha, \alpha')$ such that $\bigcap_{i=1}^{\infty} A_{i\alpha}^{(n)} =$ {the nth element in the enumeration of $g(\alpha)$}].

(ii) If λ is any Borel measure on $[0, \Omega]$ such that $\lambda(\{x\}) = 0$ for every singleton $\{x\}$, then $\lambda = c \cdot \mu$ for some $c \geq 0$. {Hint: By the Lebesgue Decomposition Theorem, $\lambda = \lambda_1 + \lambda_2$, $\lambda_1 << \mu$, and $\lambda_2 \perp \mu$. Then $\lambda_1 = c\mu$ for some $c \geq 0$; also there is $Y \subset [0, \Omega]$ such that $\lambda_2(Y) = \lambda_2([0, \Omega])$ and $\mu(Y) = 0$ and $Y \cap \mathscr{B} =$ the power set of Y by (i). By the Ulam Theorem, $\lambda_2 = 0$.}

(iii) There is *no nonatomic* Borel measure on $[0, \Omega]$.

(iv) Any Borel measure λ on $[0, \Omega]$ can be expressed as $c \cdot \mu + \nu$, where $c \geq 0$ and $\nu(E^c) = 0$ for some countable set $E \subset [0, \Omega]$.

(v) Every regular Borel measure on $[0, \Omega]$ is of the form ν in (iv).

(vi) A real-valued function on $[0, \Omega]$ is Borel measurable if and only if it is constant on an unbounded closed subset of $[0, \Omega)$.

5.3.14. Let X be a separable metric space. Prove $B_w(X)$ has a countable subfamily $\mathscr{C}$ such that $B_w(X)$ is the smallest σ-algebra of subsets of X containing $\mathscr{C}$. [In other words, $B_w(X)$ is separable.]

For the following problems, assume μ is a finite measure on the weakly Borel sets $B_w(X)$ of a metric space X.

5.3.15. Define μ to be *weakly regular* if for each set A in $B_w(X)$

$$\begin{aligned}\mu(A) &= \sup\{\mu(C): C \subseteq A, C \text{ closed}\}\\ &= \inf\{\mu(U): A \subseteq U, U \text{ open}\}.\end{aligned}$$

Prove any finite measure on $B_w(X)$ is weakly regular. [Hint: Prove that the class of all "weakly regular sets" is a σ-algebra.]

5.3.16. Let X be separable. Prove that there exists a unique closed set S_μ such that $\mu(S_\mu) = \mu(X)$, and if C is any closed set with $\mu(C) = \mu(X)$, then $S_\mu \subset C$. Show $S_\mu = \{x \in X: \mu(U) > 0$ for each open set U containing $x\}$. S_μ is called *the support* of μ. [Hint: Set $S_\mu = X - U_\mu$, where $U_\mu = \{U \subset X: U \text{ open}, \mu(U) = 0\}$.]

5.3.17. Suppose for each $\varepsilon > 0$ there is a compact set $K_\varepsilon \subset X$ such that $\mu(X - K_\varepsilon) < \varepsilon$. ($\mu$ is then called *tight*.) Prove that the support of μ (see Problem 5.3.16) is a separable set and that μ is inner regular. [Hint: For each n, let K_n be a compact set such that $\mu(X - K_n) < 1/n$. Let S be the separable set $\cup K_n$. Note $\mu(X - S) = 0$.]

5.3.18. Prove every finite weakly Borel measure μ on a complete separable metric space X is tight and therefore inner regular. [Hint: Let $\varepsilon > 0$. For each n, $X = \bigcup_{j=1} \bar{S}_{n,j}$, where $S_{n,j}$ is a sphere of radius $1/n$. Therefore $\mu(X - \bigcup_{j=1}^{k_n} \bar{S}_{n,j}) < \varepsilon/2^n$ for some integer k_n. Let X_n be the closed set $\bigcup_{j=1}^{n_k} \bar{S}_{n,j}$. Let $K_\varepsilon = \bigcap_{n=1}^\infty X_n$. Show that $K_\varepsilon \subset \bigcup_{j=1}^{n_k} \bar{S}_{n,j}$ for each n and that K_ε is compact. (Show that if $x_1, x_2, \ldots, x_n \ldots$ is a sequence in K_ε, there is a sequence $K_1 \supseteq K_2 \supseteq \ldots$ of closed sets defined by $K_i = K_\varepsilon \cap \bar{S}_{ij_i}$ where $\bar{S}_{ij_i}$ contains infinitely many sequence points. Show $\cap K_i$ contains a single point.) Show that $\mu(X - K_\varepsilon) \le \varepsilon$.]

5.3.19. (i) Prove that if f: $X \to R$ is measurable [with respect to $B_w(X)$], then for $\varepsilon > 0$ there exists a closed set C_ε such that $\mu(X - C_\varepsilon) \le \varepsilon$ and f restricted to C_ε, $f|_{C_\varepsilon}$, is continuous. (See also Problem 3.1.11.)

(ii) Prove that if μ is a tight measure, C_ε can be taken to be compact.

(iii) Assume μ is a tight measure. Prove that if f is measurable and $f^{-1}(A) \in B_w(X)$ for $A \subset R$, then there exist Borel sets A_1 and $A_2 \subset R$ such that $A_1 \subset A \subset A_2$ and $\mu f^{-1}(A_2 - A_1) = 0$. [Hint: Let $E = f^{-1}(A)$. Find a nondecreasing sequence of compact sets (K_n) such that $K_n \subset E$, $f_{|K_n}$ is continuous, and $\mu(E - K_n) \to 0$ as $n \to \infty$. Let $A_1 = \bigcup f(K_n)$, a Borel set in R. Then $A_1 \subseteq A$ and $\mu(E - f^{-1}(A)) = 0$. To find A_2 repeat the above argument obtaining $A_2{}^c \subset A^c$ and $\mu(f^{-1}(A^c) - f(A_2{}^c)) = 0$.]

5.4. Riesz Representation Theorems

In this section we will use the Daniell–Stone Representation Theorem to obtain results that establish relationships between linear functionals on certain vector lattices and measures on topological spaces. We will list several results of this nature which are generally called Riesz representation theorems. Proposition 5.13 and Theorems 5.7, 5.8, and 5.10 constitute various formulations of a result generally called the Riesz Representation Theorem.

Our first result is an almost immediate consequence of the Daniell–Stone Theorem.

Proposition 5.13. Let X be any topological space and let I be a Daniell integral on the vector lattice $C_b(X)$. There exists a unique measure μ on $\sigma(C_b(X))$ such that

$$C_b(X) \subset L_1(\mu) \tag{5.20}$$

and

$$I(f) = \int_X f\,d\mu, \qquad \text{for all } f \text{ in } C_b(X). \tag{5.21}$$

μ is necessarily finite. If X is T_4, μ is a weakly Baire measure. ∎

Proof. The Daniell–Stone Theorem gives the existence of a measure μ on $\sigma(C_b(X))$ satisfying conditions (5.20) and (5.21). Since $\chi_X \in C_b(X)$,

$$\mu(X) = \int \chi_X \, d\mu = I(\chi_X) < \infty.$$

If ν is any other measure on $\sigma(C_b(X))$ satisfying conditions (5.20) and (5.21), then $\nu = \mu$ by Proposition 5.4. The last assertion follows from Propositions 5.5 and 5.6. ∎

Proposition 5.13 actually gives a one-to-one correspondence between Daniell integrals on $C_b(X)$ and finite measures on $\sigma(C_b(X))$. Indeed, if μ

is any finite measure on $\sigma(C_b(X))$, $I(f) = \int_X f\,d\mu$ defines a Daniell integral on $C_b(X)$.

Let us now consider the vector lattice $C_c(X)$, where X is a locally compact Hausdorff space. If X is, moreover, σ-compact, then we know

$$\text{Ba}(X) = \sigma(C_c(X)) = \sigma(C_b(X)) = \text{Ba}_w(X).$$

Whether X is σ-compact or not, it is interesting that every positive linear functional on $C_c(X)$ is already a Daniell integral.

Proposition 5.14. If X is a locally compact Hausdorff space, then every positive linear functional on $C_c(X)$ is a Daniell integral. ▮

Proof. We must verify condition (D) of Definition 5.2: If (f_n) is a nonincreasing sequence in $C_c(X)$ converging to zero, then $I(f_n)$ converges to zero. To this end, let S_n be the support of the function f_n. Note that each S_n is compact and $S_{n+1} \subset S_n \subset S_1$ for each n. By Dini's Theorem (Problem 1.5.18 in Chapter 1) the sequence (f_n) converges uniformly to zero on S_1. Hence for any $\varepsilon > 0$ there exists an N such that for all $n \geq N$, $| f_n(x) | < \varepsilon$ for all x in S_1 and hence for all x in X, since $S_n \subset S_1$. Corresponding to S_1 there exists a function g in $C_c(X)$ with $0 \leq g \leq 1$ and $g \equiv 1$ on S_1 by Lemma 5.3. Therefore for $n \geq N$, $| f_n | < \varepsilon g$ and $| I(f_n) | \leq \varepsilon I(g)$. Since ε is arbitrary, $I(f_n) \to 0$. ▮

The next result is one of the key theorems of this section. It gives a one-to-one correspondence between the class of Baire measures on a σ-compact Hausdorff space X and positive linear functionals on $C_c(X)$.

Theorem 5.7. Let X be a locally compact Hausdorff space.

(i) For every Baire measure μ, the function

$$g \to \int_X g\,d\mu$$

is a Daniell integral on $C_c(X)$.

(ii) Corresponding to every positive linear functional I on $C_c(X)$ there is a Baire measure on $\text{Ba}(X)$ such that

$$I(f) = \int_X f\,d\mu$$

for all $f \in C_c(X)$. If X is σ-compact, μ is unique.

(iii) Corresponding to every positive linear functional I on $C_c(X)$ there is a unique Baire measure μ on $\mathrm{Ba}(X)$ such that

$$I(f) = \int_X f\,d\mu, \qquad \text{for all } f \text{ in } C_c(X), \tag{5.22}$$

and

$$\mu(A) = \inf\{\mu(G)\colon G \supset A,\ G \text{ open } \sigma\text{-compact}\} \tag{5.23}$$

for each $A \in \mathrm{Ba}(X)$. μ is also outer regular. ∎

Proof. (i) If $g \in C_c(X)$, then g is measurable with respect to Ba(X) as $\mathrm{Ba}(X) = \sigma(C_c(X))$, the smallest σ-algebra such that all such g are measurable (see Figure 4 of Section 5.2). Also g is integrable with respect to μ. This follows since by Lemma 5.4 there is a compact G_δ set A and $S_g \subset A$. Letting $\| g \|_\infty = \sup\{| g(x) |: x \in X\}$ we have $| g | \leq \| g \|_\infty \chi_A$. This means that g is integrable, since

$$\int | g |\, d\mu \leq \| g \|_\infty \int \chi_A\, d\mu = \| g \|_\infty\, \mu(A) < \infty.$$

Clearly, $g \to \int_X g\, d\mu$ is a positive linear functional on $C_c(X)$ and hence by Proposition 5.14 is a Daniell integral.

(ii) Since a positive linear functional on $C_c(X)$ is a Daniell integral, by the Daniell–Stone Theorem 5.1 there exists a measure μ on $\mathrm{Ba}(X)$ such that

$$C_c(X) \subset L_1(\mu)$$

and

$$I(f) = \int_X f\, d\mu, \qquad \text{for all } f \in C_c(X).$$

If K is any compact Baire set, then $\chi_K \leq f$ where f in $C_c(X)$ is a function given by Lemma 5.3 such that $f \equiv 1$ on K. Hence

$$\mu(K) = \int \chi_K\, d\mu \leq \int_X f\, d\mu = I(f) < \infty.$$

This means that μ is a Baire measure.

The uniqueness of μ remains to be proved in the case where X is σ-compact. By Lemma 5.5, X can be written as $\bigcup K_n$ with each K_n a compact Baire set and $K_n \subset K_{n+1}^\circ$. By Lemma 5.3, for each n there exists $f_n \in C_c(X)$ with $0 \leq f_n \leq 1$, $K_n \subset S_{f_n} \subset K_{n+1}^\circ$, and $\chi_{K_n} \leq f_n$. Clearly, $\chi_X = \sup f_n$.

Letting G_n be the $C_c(X)$-open set $\{x \in X: f_n(x) > 1/n\}$ (see Lemma 5.5), we have $X = \bigcup_{n=1}^{\infty} G_n$. Since $G_n \subset S_{f_n} \subset K_{n+1}$, the measure of G_n is finite for any Baire measure on Ba(X). By Proposition 5.4, μ is unique.

(iii) According to Proposition 5.14 and Theorem 5.1, there exists a unique measure μ on $\sigma(C_c(X))$ or on what is the same as Ba(X), such that

$$I(f) = \int_X f\,d\mu, \qquad \text{for all } f \in C_c(X)$$

and

$$\mu(A) = \inf\{\mu(G): G \supset A \quad \text{and} \quad G,\ C_c(X)\text{-open}\}$$

for each A in Ba(X). Equations (5.22) and (5.23) follow from these relations and Lemma 5.5. That μ is outer regular follows from Lemma 5.5 and the following inequalities:

$$\begin{aligned}\mu(A) &\leq \inf\{\mu(O): A \subset O \text{ and } O \text{ an open Baire set}\}\\ &\leq \inf\{\mu(G): A \subset G \text{ and } G \text{ an open } \sigma\text{-compact set}\}\\ &= \mu(A).\end{aligned}$$

▮

Using the extension theorems of Section 5.3, we can also correspond to any positive linear functional I on $C_c(X)$ a weakly Borel measure. More precisely, we have the following result.

Theorem 5.8. Corresponding to each positive linear functional I on $C_c(X)$ is a unique weakly Borel measure $\bar{\mu}$ on $B_w(X)$ satisfying

$$I(f) = \int f\,d\bar{\mu}, \qquad \text{for all } f \text{ in } C_c(X), \tag{5.24}$$

$$\bar{\mu}(O) = \sup\{\bar{\mu}(K): K \text{ compact } G_\delta \text{ set}, K \subset O\} \quad \text{for each open } \sigma\text{-bounded set } O, \tag{5.25}$$

and

$$\bar{\mu}(E) = \inf\{\bar{\mu}(O): E \subset O, O \text{ open } \sigma\text{-bounded set}\} \quad \text{for each weakly Borel set } E. \tag{5.26}$$

▮

Proof. According to Theorem 5.7 (iii), there exists a unique Baire measure μ such that equations (5.22) and (5.23) hold. If X is σ-compact, Corollary 5.6 guarantees a weakly Borel measure $\bar{\mu}$ satisfying equations (5.25) and (5.26) since in this case each open set is σ-bounded. If X is not σ-compact, then equation (5.23) implies that $\mu(X) = \infty$ or μ is not finite. In this case Theorem 5.6 gives a measure $\bar{\mu}$ satisfying equations (5.25) and

(5.26). In any case the unique measure $\bar{\mu}$ extending μ satisfies equation (5.24). That this is so follows from the fact that if f is any nonnegative function in $C_c(X)$, then

$$\int f\,d\bar{\mu} = \lim_n \int \Phi_n\,d\bar{\mu} = \lim_n \int \Phi_n\,d\mu = \int f\,d\mu$$

by the Monotone Convergence Theorem, where Φ_n is the simple function

$$\Phi_n = 2^{-n} \sum_{k=0}^{2^{2n}} k\chi_{E_{n,k}},$$

where $E_{n,k} = \{x\colon k2^{-n} \leq f(x) < (k+1)2^{-n}\}$ for nonnegative integers n and k.

The uniqueness part can also be established directly. Indeed, by equations (5.25) and (5.26) it is sufficient to show that any two measures μ_1 and μ_2 satisfying equations (5.24), (5.25), and (5.26) agree on compact G_δ sets. Let K be a compact G_δ set. By equation (5.26) for each $\varepsilon > 0$ there exists an open σ-bounded set O such that $K \subset O$ and

$$\mu_1(O) < \mu_1(K) + \varepsilon.$$

By Lemma 5.3 there exists f in $C_c(X)$ such that $f \equiv 1$ on K and $S_f \subset O$. Hence

$$\begin{aligned} \mu_2(K) &= \int \chi_K\,d\mu_2 \leq \int f\,d\mu_2 = I(f) = \int f\,d\mu_1 \\ &\leq \int \chi_O\,d\mu_1 = \mu_1(O) < \mu_1(K) + \varepsilon. \end{aligned}$$

Hence $\mu_2(K) \leq \mu_1(K)$. Similarly $\mu_1(K) \leq \mu_2(K)$. ∎

Examples

5.17. Let X be an uncountable set with the discrete topology. Define ν on $\text{Ba}(X)$ as the zero measure and I on $C_c(X)$ by $I(f) = \int f\,d\nu$. The unique measures guaranteed by Theorems 5.7 (iii) and 5.8 are the measures μ and $\bar{\mu}$ on $\text{Ba}(X)$ and $B_w(X)$, respectively, given by

$$\mu(A) = \begin{cases} 0, & \text{if } A \text{ is countable,} \\ \infty, & \text{if } A^c \text{ is countable,} \end{cases}$$

and

$$\bar{\mu}(A) = \begin{cases} 0, & \text{if } A \text{ is countable,} \\ \infty, & \text{if } A \text{ is not countable.} \end{cases}$$

5.18. Let X be any set with the discrete topology and let $x_0 \in X$. Define I on $C_c(X)$ by $I(f) = f(x_0)$. I is a Daniell integral by Proposition 5.14. The measures μ and $\bar{\mu}$ on $\mathrm{Ba}(X)$ and $B_w(X)$, respectively, given by Theorems 5.7 (iii) and (5.8) are the measures defined by

$$\mu(A) = \bar{\mu}(A) = \begin{cases} 0, & \text{if } A \text{ is countable not containing } x_0, \\ 1, & \text{if } A \text{ is countable containing } x_0, \\ \infty, & \text{if } A \text{ is not countable.} \end{cases}$$

Example 5.17 shows that in Theorem 5.8 the measure $\bar{\mu}$ corresponding to a *positive* (*even bounded*) linear functional need not be finite unless the space X is σ-compact. Despite the beauty of Theorems 5.7 and 5.8, this seems to be a slight drawback for these theorems. Fortunately, our next two theorems will remedy this. The proofs do not depend upon the Daniell theory.

Theorem 5.9. Let I be a positive linear functional on $C_c(X)$, where X is a locally compact Hausdorff space. There exists a *unique* weakly Borel measure μ satisfying the following conditions:

(i) $\mu(V) = \sup\{I(g): 0 \le g \le 1, g \in C_c(X) \text{ and } S_g \subset V\}$ for each open V.

(ii) $\mu(B) = \inf\{\mu(V): B \subset V \text{ open}\}$ for each weakly Borel set B.

(iii) If (A is open) *or* (A is weakly Borel and σ-finite), then $\mu(A) = \sup\{\mu(K): K \text{ compact} \subset A\}$. ∎

Proof. We give the proof in three steps.

Step I. We define, for any $B \subset X$,

$$\mu^*(B) = \inf\{\mu(V): B \subset V \text{ open}\}, \tag{5.27}$$

where $\mu(V)$ is defined as in the statement (i) of the theorem. We claim that μ^* is an outer measure on 2^X. Obviously, $\mu^*(\varnothing) = 0$; and whenever $A \subset B$, then $\mu^*(A) \le \mu^*(B)$. For open V_1 and V_2, we notice that

$$\mu(V_1 \cup V_2) \le \mu(V_1) + \mu(V_2). \tag{5.28}$$

Indeed, to verify inequality (5.28), let $h \in C_c(X)$ with $S_h \subset V_1 \cup V_2$ and $0 \le h \le 1$. Clearly we can find $g_1, g_2 \in C_c(X)$ with $0 \le g_1, g_2 \le 1$ and $h \le g_1 + g_2$ such that $S_{g_1} \subset V_1$ and $S_{g_2} \subset V_2$. Hence $I(h) \le I(g_1 + g_2) \le I(g_1) + I(g_2) \le \mu(V_1) + \mu(V_2)$, which implies inequality (5.28). Now let $B_n \subset X$ and $B = \bigcup_{n=1}^{\infty} B_n$. Given $\varepsilon > 0$, there exists open $V_n \supset B_n$ such that

$$\mu(V_n) \leq \mu^*(B_n) + \varepsilon/2^n.$$

Then $B \subset V = \bigcup_{n=1}^{\infty} V_n$ and $\mu^*(B) \leq \mu(V)$. If $0 \leq f \leq 1$, $f \in C_c(X)$ and $S_f \subset V$, then $S_f \subset \bigcup_{n=1}^{N} V_n$ (by compactness of the support of f); therefore,

$$I(f) \leq \mu\left(\bigcup_{n=1}^{N} V_n\right) \leq \sum_{n=1}^{N} \mu(V_n), \qquad \text{by inequality (5.28),}$$

$$\leq \sum_{n=1}^{\infty} \mu^*(B_n) + \varepsilon,$$

implying $\mu^*(B) \leq \mu(V) \leq \sum_{n=1}^{\infty} \mu^*(B_n) + \varepsilon$.

Step II. Let $\mathscr{A} = \{A \subset X: \mu^*(A) < \infty$ and $\mu^*(A) = \sup\{\mu^*(K)$: K compact $\subset A\}\}$. Then $\mathscr{A}$ is a ring of subsets of X containing open sets (with finite measure μ^*) and all compact sets. Furthermore, if (A_n) is a disjoint sequence in $\mathscr{A}$, then $\mu^*(\bigcup_{n=1}^{\infty} A_n) = \sum_{n=1}^{\infty} \mu^*(A_n)$; and if $\mu^*(\bigcup_{n=1}^{\infty} A_n) < \infty$, then $\bigcup_{n=1}^{\infty} A_n \in \mathscr{A}$.

The reader can verify the above assertions easily. Let us show only

$$\mu(V) = \sup\{\mu^*(K)\text{: } K \text{ compact} \subset V\} \tag{5.29}$$

for *all* open V, and

$$\mu^*\left(\bigcup_{n=1}^{\infty} A_n\right) = \sum_{n=1}^{\infty} \mu^*(A_n) \tag{5.30}$$

if (A_n) are in $\mathscr{A}$ and are pairwise disjoint.

To prove equation (5.29), let $0 < a < \mu(V)$. By the definition of $\mu(V)$, there exists $f \in C_c(X)$ with $0 \leq f \leq 1$ and $S_f = K \subset V$ such that $a < I(f) < \mu(V)$. By equation (5.27), $\mu^*(K) \geq I(f)$, and thus equation (5.29) is proved.

To prove equation (5.30), first let K_1, K_2 be two disjoint compact sets with W open such that $K_1 \cup K_2 \subset W$, and

$$\mu(W) \leq \mu^*(K_1 \cup K_2) + \varepsilon.$$

Let V_1, V_2 be open, $K_1 \subset V_1$, $K_2 \subset V_2$, and $V_1 \cap V_2 = \varnothing$. Let $g_1, g_2 \in C_c(X)$ with $0 \leq g_1$, $g_2 \leq 1$, $S_{g_1} \subset V_1 \cap W$, and $S_{g_2} \subset V_2 \cap W$ such that

$$\mu(W \cap V_i) \leq I(g_i) + \varepsilon, \qquad i = 1, 2.$$

Then

$$\begin{aligned}\mu^*(K_1) + \mu^*(K_2) &\leq \mu(W \cap V_1) + \mu(W \cap V_2)\\ &\leq I(g_1) + I(g_2) + 2\varepsilon\\ &= I(g_1 + g_2) + 2\varepsilon\\ &\leq \mu(W) + 2\varepsilon \leq \mu^*(K_1 \cup K_2) + 3\varepsilon\end{aligned}$$

showing μ^* is finitely additive for disjoint compact sets. Now let $(A_n) \in \mathscr{A}$ with the A_n's pairwise disjoint. Then there are compact $K_n \subset A_n$ such that

$$\mu^*(A_n) \leq \mu^*(K_n) + \varepsilon/2^n$$

and

$$\begin{aligned}\sum_{n=1}^{k} \mu^*(A_n) &\leq \sum_{n=1}^{k} \mu^*(K_n) + \varepsilon\\ &= \mu^*\left(\bigcup_{i=1}^{n} K_i\right) + \varepsilon\\ &\leq \mu^*\left(\bigcup_{i=1}^{\infty} A_i\right) + \varepsilon, \qquad \text{for all } k.\end{aligned}$$

This proves equation (5.30).

Step III. The outer measure μ^* is a measure on the σ-algebra $\mathscr{F} = \{A \subset X: A \cap K \in \mathscr{A} \text{ for all compact } K\}$ containing all the weakly Borel sets; and $A \in \mathscr{F}$ with $\mu^*(A) < \infty$ implies $A \in \mathscr{A}$.

To prove these assertions, let $(A_n) \in \mathscr{F}$. Then $A_n \cap K \in \mathscr{A}$ for compact K. Since $\mathscr{A}$ is a ring, $\bigcup_{n=1}^{\infty}(A_n \cap K)$ can be written as $\bigcup_{n=1}^{\infty} B_n$, where (B_n) is a disjoint sequence of sets in $\mathscr{A}$. By step (II), $\bigcup_{n=1}^{\infty} B_n (\subset K) \in \mathscr{A}$. This proves that $\mathscr{F}$ is closed under countable unions. Since $(X - A) \cap K = K - K \cap A \in \mathscr{A}$ if K is compact and $A \in \mathscr{F}$, $\mathscr{F}$ is also closed under complementation. Thus $\mathscr{F}$ is a σ-algebra. Since every closed set is in $\mathscr{F}$, $\mathscr{F}$ contains all the weakly Borel sets. To prove that μ^* is a measure on $\mathscr{F}$, it is sufficient to show [because of equation (5.30)] that

$$A \in \mathscr{F},\ \mu^*(A) < \infty \Leftrightarrow A \in \mathscr{A}. \tag{5.31}$$

Since $\mathscr{A}$ is a ring containing all the compact sets, $\mathscr{A} \subset \mathscr{F}$. To prove identity (5.31), let $\mu^*(A) < \infty$. Let V be open $\supset A$ such that $\mu^*(V) < \infty$. Then by equation (5.29), let $K \subset V$ with K compact such that

$$\mu(V) < \mu^*(K) + \varepsilon.$$

Since $A \cap K \in \mathscr{A}$, there is compact $K_1 \subset A \cap K$ such that

$$\mu^*(A \cap K) < \mu^*(K_1) + \varepsilon$$

or

$$\mu^*(A) \leq \mu^*(A \cap K) + \mu^*(V - K) < \mu^*(K_1) + 2\varepsilon$$

proving that $A \in \mathscr{A}$, and therefore (5.31) is true. The proof of the theorem is now complete. ∎

Theorem 5.10. *The Riesz Representation Theorem.* Let I be a positive linear functional on $C_c(X)$, where X is a locally compact Hausdorff space. Then there is a *unique* weakly Borel measure μ satisfying (ii) and (iii) of Theorem 5.9 such that $I(f) = \int f\,d\mu$ for all $f \in C_c(X)$. ∎

Proof. Let μ be the weakly Borel measure of Theorem 5.9. First we show that $I(f) \leq \int f\,d\mu$ for all $f \in C_c(X)$. Let $\varepsilon > 0$, $f \in C_c(X)$, and $m = \sup_{x\in X} |f(x)| > 0$. Then if $K = S_f$, we can find a partition of K by weakly Borel sets $B_1, B_2, \ldots, B_n$ and constants $c_1, c_2, \ldots, c_n$ such that $f < \sum_{i=1}^n c_i \chi_{B_i} \leq f + \varepsilon$. Let V_i be an open set $\supset B_i$ such that $x \in V_i \Rightarrow f(x) < c_i$ and $\mu(V_i) < \mu(B_i) + \varepsilon/nm$. Let $\{g_1, g_2, \ldots, g_n\}$ be functions in $C_c(X)$ such that $h(x) = \sum_{i=1}^n g_i(x) = 1$ for all $x \in K$ with $0 \leq g_i \leq 1$, and $S_{g_i} \subset V_i$ with $1 \leq i \leq n$. (See Problem 5.2.1.) Now let $W = \{x: h(x) > 1 - \varepsilon\}$ and $p(x) \in C_c(X)$ with $0 \leq p \leq 1$ and $S_p \subset W$ such that $\mu(W) \leq I(p) + \varepsilon$. Then $\mu(K) \leq [1/(1-\varepsilon)]I(h) + \varepsilon$, implying $\mu(K) \leq \sum_{i=1}^n I(g_i)$. Now we have

$$\begin{aligned}
I(f) &= \sum_{i=1}^n I(fg_i) \leq \sum_{i=1}^n I(c_i g_i) \\
&= \sum_{i=1}^n c_i I(g_i) \\
&\leq \sum_{i=1}^n (c_i + m) I(g_i) - m\mu(K) \\
&\leq \sum_{i=1}^n (c_i + m)\left[\mu(B_i) + \frac{\varepsilon}{nm}\right] - m\mu(K) \\
&= \sum_{i=1}^n c_i \mu(B_i) + \varepsilon \\
&\leq \int f\,d\mu + \varepsilon \cdot \mu(K) + \varepsilon.
\end{aligned}$$

This proves that $I(f) \leq \int f\,d\mu$ for all $f \in C_c(X)$. Changing f to $-f$, we have $I(f) = \int f\,d\mu$.

To prove uniqueness, let μ_1 and μ_2 be two weakly Borel measures satisfying (ii) and (iii) of Theorem 5.9, and let $I(f) = \int f\,d\mu_1 = \int f\,d\mu_2$ for all $f \in C_c(X)$. If K is a compact set and V open such that $V \supset K$ and

$$\mu_2(V) < \mu_2(K) + \varepsilon,$$

then taking $f \in C_c(X)$ such that $\chi_K(x) \leq f(x) \leq \chi_V(x)$ we have

$$\mu_1(K) \leq \int f\,d\mu_1 = \int f\,d\mu_2 \leq \mu_2(V) < \mu_2(K) + \varepsilon.$$

Hence $\mu_1(K) \leq \mu_2(K)$. By symmetry, $\mu_1 = \mu_2$ on compact sets, and they are equal on all weakly Borel sets by (ii) and (iii) of Theorem 5.9. ∎

Our final consideration in this section is a representation theorem for bounded linear functionals on $C_c(X)$ which is *not necessarily* positive. First we need a lemma.

Lemma 5.8. For each bounded linear functional I on $C_c(X)$, where X is a locally compact Hausdorff space, there exist two positive bounded linear functionals I_+ and I_- such that $I = I_+ - I_-$ and $\| I \| = \| I_+ \| + \| I_- \|$. ∎

Proof. For $f \geq 0$ and $f \in C_c(X)$ let

$$I_+(f) = \sup\{I(g)\colon 0 \leq g \leq f,\ g \in C_c(X)\}.$$

For $f_1, f_2 \in C_c(X)$ with $f_1 \geq 0$, $f_2 \geq 0$ and for any $c \geq 0$, we have

(a) $I_+(f_1) \geq 0$,

(b) $I_+(f_1) \geq I(f_1)$,

(c) $I_+(cf_1) = cI_+(f_1)$,

(d) $I_+(f_1 + f_2) = I_+(f_1) + I_+(f_2)$.

We establish only (d). If $0 \leq g_1 \leq f_1$ and $0 \leq g_2 \leq f_2$, then $0 \leq g_1 + g_2 \leq f_1 + f_2$; and so

$$I_+(f_1 + f_2) \geq I(g_1 + g_2) = I(g_1) + I(g_2).$$

Taking the supremum over all such g_1 and g_2,

$$I_+(f_1 + f_2) \geq I_+(f_1) + I_+(f_2).$$

Conversely, if $0 \leq g \leq f_1 + f_2$, $g_1 = g \wedge f_1$, and $g_2 = g - g_1$, then $0 \leq$

$g_1 \leq f_1$ and $0 \leq g_2 \leq f_2$ and so

$$I(g) = I(g_1) + I(g_2) \leq I_+(f_1) + I_+(f_2),$$

which implies (d).

To extend I_+ to all $f \in C_c(X)$, let $f = f_1 - f_2$, where $f_1 \geq 0$ and $f_2 \geq 0$ are in $C_c(X)$. By taking $I_+(f) = I_+(f_1) - I_+(f_2)$, it is obvious that I_+ is *well defined* because of (d). I_+ is also linear. We define $I_- = I_+ - I$. Then I_+ and I_- are both positive linear and $I = I_+ - I_-$.

Now we will prove $\| I \| = \| I_+ \| + \| I_- \|$. Since $| I_+(f) | \leq \| I \| \| f \|$ for $f \geq 0$, I_+ is obviously bounded and so is I_-. Also, since $| I(f) | = | I_+(f) - I_-(f) | \leq \| I_+ \| \| f \| + \| I_- \| \| f \|$, $\| I \| \leq \| I_+ \| + \| I_- \|$. For the converse inequality, let $\varepsilon > 0$; then we can find $f_1 \in C_c(X)$ with $0 \leq f_1 \leq 1$ such that

$$I(f_1) \geq \| I_+ \| - \varepsilon.$$

Also we can find $f_2 \, \varepsilon \, C_c(X)$ with $0 \leq f_2 \leq 1$ and $f_2 \geq f_1$ such that

$$I_-(f_2) \geq \| I_- \| - \varepsilon.$$

Let $f = 2f_1 - f_2$. Then $| f | \leq 1$ and

$$\begin{aligned} I(f) &= 2I(f_1) - I(f_2) \\ &\geq \| I_+ \| + I_+(f_1) - I(f_2) - \varepsilon \\ &\geq \| I_+ \| + I_+(f_2) - I(f_2) - 2\varepsilon \\ &= \| I_+ \| + I_-(f_2) - 2\varepsilon \\ &\geq \| I_+ \| + \| I_- \| - 3\varepsilon, \end{aligned}$$

thus proving that $\| I \| = \| I_+ \| + \| I_- \|$. ∎

Theorem 5.11. To each bounded linear functional I on $C_c(X)$, where X is a locally compact Hausdorff space, there corresponds a *unique* finite signed regular weakly Borel measure (a signed measure is regular if μ^+ and μ^- are both regular) μ on X such that

$$I(f) = \int f \, d\mu, \qquad \text{for all } f \in C_c(X).$$

Also $\| I \| = | \mu | (X)$, where $| \mu |$ is the total variation of μ. ∎

Proof. By Lemma 5.8 and Theorem 5.10, $I = I_+ - I_-$, where I_+, I_- are positive linear functionals on $C_c(X)$ with $\| I \| = \| I_+ \| + \| I_- \|$;

and

$$I_+(f) = \int f \, d\mu_1$$

and

$$I_-(f) = \int f \, d\mu_2, \qquad \text{for all } f \in C_c(X),$$

where μ_1, μ_2 are unique *finite* regular weakly Borel measures on X. Letting $\mu = \mu_1 - \mu_2$,

$$I(f) = \int f \, d\mu, \qquad \text{for } f \in C_c(X).$$

Clearly, $| I(f) | \leq \int | f | \, d | \mu | \leq \| f \| \cdot | \mu | (X)$ or $\| I \| \leq | \mu | (X)$. Conversely,

$$| \mu | (X) \leq \mu_1(X) + \mu_2(X) = \| I_+ \| + \| I_- \| = \| I \|.$$

Hence $\| I \| = | \mu | (X)$.

To prove uniqueness, let μ, ν be two finite regular signed measures such that for all $f \in C_c(X)$,

$$I(f) = \int f \, d\mu = \int f \, d\nu.$$

Then $\lambda = \mu - \nu$ is also a finite regular signed measure and for all $f \in C_c(X)$,

$$\int f \, d\lambda = 0.$$

This means that if $\lambda = \lambda^+ - \lambda^-$ is the Jordan decomposition of λ, then $\int f \, d\lambda^+ = \int f \, d\lambda^-$ for all $f \in C_c(X)$ so that λ^+ and λ^- induce the same positive linear functional on $C_c(X)$. By the uniqueness part of Theorem 5.9, $\lambda^+ = \lambda^-$ since λ^+ and λ^- are also regular. Therefore $\lambda = 0$ and $\mu = \nu$. ∎

Problems

5.4.1. (i) Let X be a metricizable locally compact noncompact space. For a given nonnegative extended real-valued number α let (α_n) be a sequence of nonnegative real numbers such that $\alpha = \sum_n \alpha_n$. Let (x_n) be a sequence in X with no cluster point. Define μ on $\text{Ba}(X)$ by

$$\mu(E) = \sum \{\alpha_n \colon x_n \in E\}.$$

Show μ is regular on $\text{Ba}(X)$ and $\mu(X) = \alpha$. (Use Theorem 5.3.)

(ii) If X is a σ-compact locally compact Hausdorff space that is not compact, let (K_n) be a sequence of compact G_δ sets such that $K_n{}^\circ$ is a Baire set and $K_n \subsetneqq K_{n+1}^\circ$, and $X = \bigcup_{n=1}^\infty K_n$. Let $C_n = K_{2n} - K_{2n-1}^\circ$ for $n = 1, 2, \ldots$. Let μ_n be any Baire measure on C_n such that $\mu_n(C_n) = \alpha_n$. [α and (α_n) are defined in (i).] Define μ on $\text{Ba}(X)$ by

$$\mu(E) = \sum \mu_n(E \cap C_n).$$

Show that μ is a regular Baire measure such that $\mu(X) = \alpha$.

(iii) If X is a metrizable or σ-compact locally compact Hausdorff space and X is not compact, then show that there exists a discontinuous positive linear functional on $C_c(X)$.

5.4.2. Let $\bar{X}$ be the space of Example 5.11—the space of all ordinals less than or equal to Ω, the first uncountable ordinal.

(i) Prove that to each f in $C(\bar{X})$ there corresponds an $\alpha \neq \Omega$ such that f is constant on $\{\gamma \in \bar{X}: \alpha < \gamma\}$. (Compare Problem 5.3.13.)

(ii) If μ is the measure of Example 5.11, show

$$f(\Omega) = \int f \, d\mu$$

for each $f \in C(\bar{X})$.

(iii) What is the measure $\bar{\mu}$ of Theorem 5.8 corresponding to this functional?

5.4.3. *The Monotone Convergence Theorem for Nets.* Let μ be a regular Borel measure on a compact Hausdorff space X. Let $(f_\alpha)_{\alpha \in A}$ be an increasing net of continuous functions. Show that $f = \lim_\alpha f_\alpha \in L_1(\mu)$ if and only if $\sup_\alpha \| f_\alpha \|_1 < \infty$ and in this case $\lim_\alpha \| f - f_\alpha \|_1 = 0$.

• 5.5. Product Measures and Integration

Recall that if $\mathscr{A}$ and $\mathscr{B}$ are two σ-algebras on which σ-finite measures μ and ν, respectively, are defined, then the product $\mu \times \nu$ of μ and ν is the complete measure on a σ-algebra $\mathscr{M}$ containing $\mathscr{A} \times \mathscr{B}$ extending the function λ on the semialgebra of measurable rectangles given by $\lambda(A \times B) = \mu(A)\nu(B)$ (see Chapter 3, Section 3.4). *Throughout this section the product $\mu \times \nu$ will be assumed to be this product measure restricted to $\mathscr{A} \times \mathscr{B}$.* If μ and ν are σ-finite measures then $\mu \times \nu$ is the unique measure on $\mathscr{A} \times \mathscr{B}$ satisfying

$$\mu \times \nu(A \times B) = \mu(A)\nu(B)$$

for all A in $\mathscr{A}$ and B in $\mathscr{B}$. We will study, in particular, the product of measures μ and ν on σ-algebras of Borel sets. Furthermore, this section will continue the study of the Fubini–Tonelli Theorems on product integration introduced in Chapter 3, Section 3.4, for Borel-measurable functions on the product of two locally compact Hausdorff spaces. The main theorem in this section is Theorem 5.20.

Throughout this section the topological spaces X and Y under consideration will be assumed to be σ-compact locally compact Hausdorff spaces. In this case the σ-ring σ_r(compact sets in X) is identical to the σ-algebra of Borel sets $B(X)$. The results of this section could be proved without the assumption of σ-compactness for σ-rings of Borel sets as in [26], but the flavor is not lost and ease of notation is gained by assuming that all spaces are σ-compact.

Recall that any Borel measure on a σ-compact locally compact Hausdorff space is σ-finite.

If μ_0 and ν_0 are Baire measures on $B_a(X)$ and $B_a(Y)$, then $\mu_0 \times \nu_0$ is a Baire measure since

$$B_a(X) \times B_a(Y) = B_a(X \times Y). \tag{5.32}$$

In contrast, if μ and ν are Borel measures their product is not necessarily a Borel measure. The reason in simple: The domain of $\mu \times \nu$ is not necessarily a σ-algebra of Borel sets as sometimes

$$B(X) \times B(Y) \subsetneqq B(X \times Y).$$

(see Example 5.10). Our first goal in this section is to define a Borel measure on $B(X \times Y)$ which extends $\mu \times \nu$. The next result gives a criterion for when this can be done.

Theorem 5.12. Suppose μ and ν are nonzero Borel measures on $B(X)$ and $B(Y)$, respectively. There exists a unique regular Borel measure $\mu \otimes \nu$ (called the *tensor product* of μ and ν) on $B(X \times Y)$ which extends $\mu \times \nu$ if and only if μ and ν are regular Borel measures.† ∎

Proof. For the sufficiency proof, let μ_0 and ν_0 be the Baire measures which are the restrictions of μ and ν to $B_a(X)$ and $B_a(Y)$. We know from equation (5.32) that $\mu_0 \times \nu_0$ is a Baire measure on $B_a(X \times Y)$. Let $\mu \otimes \nu$ be the unique regular extension of $\mu_0 \times \nu_0$ to $B(X \times Y)$. (See Theorem 5.4.

† Recently D. H. Fremlin in *Canad. Math. Bull.* **19** (1976) has shown that a compact set in $X \times X$ need not be in the domain of the completion of $\mu \times \mu$. The reader may also consult for other related results Godfrey and Sion's paper in *Canad. Math. Bull.* **12** (1969).

Keep in mind that X and Y are σ-compact.) To show that $\mu \otimes \nu$ extends $\mu \times \nu$, it is sufficient to show (see Problem 5.5.2) that

$$\mu \otimes \nu(C \times D) = \mu \times \nu(C \times D)$$

for compact sets C and D. By the regularity of μ and ν, there exist compact G_δ sets (Proposition 5.11) G_C, G_D, and $G_{C\times D}$ such that

$$C \subset G_C \quad \text{and} \quad \mu(C) = \mu(G_C),$$
$$D \subset G_D \quad \text{and} \quad \nu(D) = \nu(G_D),$$

and

$$C \times D \subset G_{C\times D} \quad \text{and} \quad \mu \otimes \nu(C \times D) = \mu \otimes \nu(G_{C\times D}).$$

Since $\mu \otimes \nu$ and $\mu \times \nu$ agree on Baire sets in $X \times Y$, we have

$$\mu \times \nu(C \times D) \leq \mu \times \nu(G_{C\times D}) = \mu \otimes \nu(G_{C\times D}) = \mu \otimes \nu(C \times D),$$

and

$$\begin{aligned} \mu \otimes \nu(C \times D) &\leq \mu \otimes \nu(G_C \times G_D) = \mu \times \nu(G_C \times G_D) \\ &= \mu(G_C)\nu(G_D) = \mu(C)\nu(D) = \mu \times \nu(C \times D). \end{aligned}$$

Uniqueness is clear since any regular extension of $\mu \times \nu$ is a regular extension of $\mu_0 \times \nu_0$ on $B_a(X \times Y)$.

To prove the necessity, let ϱ be a regular extension of $\mu \times \nu$ to $B(X \times Y)$. It is then a regular extension of $\mu_0 \times \nu_0$ and hence by the sufficiency just proved it is the unique regular extension of $\mu' \times \nu'$, where μ' and ν' are the unique *regular* extensions of μ_0 and ν_0 to the σ-algebras $B(X)$ and $B(Y)$. Hence $\mu' \times \nu' = \mu \times \nu$ so that for all Borel sets A in X and B in Y we have

$$\mu'(A)\nu'(B) = \mu' \times \nu'(A \times B) = \mu \times \nu(A \times B) = \mu(A)\nu(B).$$

Since μ and ν are nonzero, we have $\mu = \mu'$ and $\nu = \nu'$. ∎

What happens in the case that μ and ν are arbitrary Borel measures? Can $\mu \times \nu$ be extended to a Borel measure? An answer is known in the case that one of μ or ν is regular. The Theorem 5.13 below shows that in this case $\mu \times \nu$ can be extended to two measures ϱ_1 and ϱ_2 on $B(X \times Y)$. First we prove some useful preliminary results.

If M is any subset of $X \times Y$, then M_x is the set

$$M_x = \{y\colon (x, y) \in M\} = \mathrm{Pr}_Y[M \cap (\{x\} \times Y)]$$

and M^y is the set

$$M^y = \{x\colon (x, y) \in M\} = \mathrm{Pr}_X[M \cap (X \times \{y\})],$$

where Pr_Y and Pr_X are the projection mappings. If $M \in B(X) \times B(Y)$, then $M_x \in B(Y)$ and $M^y \in B(X)$. Moreover, if $M \in B(X \times Y)$, then $M_x \in B(Y)$ and $M^y \in B(X)$ for all x in X and $y \in Y$. (See Problem 5.5.1.)

If μ and ν are Borel measures on $B(X)$ and $B(Y)$, respectively, and $M \in B(X \times Y)$, we define the extended real-valued functions Γ_M on X and Γ^M on Y, respectively, by

$$\Gamma_M(x) = \nu(M_x) \qquad \text{and} \qquad \Gamma^M(y) = \mu(M^y). \tag{5.33}$$

Proposition 5.15. Let C be compact in $X \times Y$. Then

(i) If ν is regular, $\Gamma_C\colon X \to R$ is Borel measurable.[†]
(ii) If μ is regular, $\Gamma^C\colon Y \to R$ is Borel measurable. ▮

Proof. (i) It must be shown that for each $a \in R$ the set $\{x\colon \Gamma_C(x) \geq a\}$ is in $B(X)$. Since Γ_C is a nonnegative function, it suffices to consider only positive numbers a. Denote $\{x\colon \Gamma_C(x) \geq a\}$ by A_a. Clearly, $A_a \subset \text{Pr}_X(C)$. As Pr_X is continuous, $\text{Pr}_X(C)$ is compact in X and A_a is contained in a compact set. To show that A_a is a Borel set, it suffices to show that A_a is closed, since then A_a will itself be compact as a subset of a compact set. To this end, let $z \in A_a{}^c$. This means $\nu(C_z) < a$, and by the regularity of ν there is an open Borel set V in Y such that $C_z \subset V$ and $\nu(V) < a$. Letting U be the open set

$$U = X - \text{Pr}_X[C - (X \times V)],$$

clearly $x \in U$ if and only if $C_x \subset V$. Applying this criterion to z, this means U is a neighborhood of z disjoint from A_a.

(ii) The proof of (ii) is similar to that of (i). ▮

The impetus of Proposition 5.15 is that the hypotheses of the next two results are satisfied when μ and ν are regular.

Lemma 5.9. (i) If Γ_C is Borel measurable for each compact set C in $X \times Y$, then Γ_M is Borel measurable for each bounded Borel set M in $X \times Y$.

(ii) If Γ^C is Borel measurable for each compact set C in $X \times Y$, then Γ^M is Borel measurable for each bounded Borel set M in $X \times Y$. ▮

Proof. Again we prove (i) and note that the proof of (ii) is similar. Let $\mathscr{R}$ be the ring of all finite disjoint unions of proper differences of com-

[†] Problem 5.6.13 gives an example in which Γ_C need not be Borel measurable for nonregular ν even with respect to the completion of μ.

pact subsets of $X \times Y$ and let $\mathscr{R}_b$ be the ring of bounded Borel sets in $X \times Y$. Since

$$\mathscr{R} \subset \mathscr{R}_b \subset B(X \times Y)$$

and the σ-ring generated by $\mathscr{R}$ is $B(X \times Y)$, the σ-ring generated by $\mathscr{R}_b$ is also $B(X \times Y)$.

Let $M \in \mathscr{R}_b$. Choose compact sets C and K such that $M \subset C \times K$. Then M belongs to the σ-ring $B(X \times Y) \cap (C \times K)$ which is generated by the ring $\mathscr{R} \cap (C \times K)$. Let $\mathscr{M}$ be the class of all Borel sets N in $X \times Y$ with $N \subset C \times K$ and Γ_N Borel measurable. As Γ_N is measurable for each N in $\mathscr{R}$ (see Problem 5.5.3) and $\mathscr{R} \cap (C \times K) \subset \mathscr{R}$, we have

$$\mathscr{R} \cap (C \times K) \subset \mathscr{M}.$$

However, $\mathscr{M}$ is a monotone class. Indeed, if (N_n) is a nondecreasing sequence in $\mathscr{M}$, then with $N = \lim_n N_n$ we have

$$\begin{aligned}\{x \in X: \nu(N_x) > a\} &= \{x \in X: \lim_n \nu(N_{n_x}) > a\} \\ &= \bigcup_n \{x \in X: \nu(N_{n_x}) > a\}\end{aligned}$$

as $N_x = \lim_n N_{n_x}$ and (N_{n_x}) is a nondecreasing sequence for each x in X. This means Γ_N is measurable and $\mathrm{N} \in \mathscr{M}$. Similarly, $\mathscr{M}$ can be shown to be closed with respect to nonincreasing sequences. Therefore $\mathscr{M}$ is a monotone class containing $\mathscr{R} \cap (C \times K)$, whereby $\mathscr{M}$ contains the σ-ring generated by $\mathscr{R} \cap (C \times K)$, namely, $B(X \times Y) \cap (C \times K)$. In particular $\mathscr{M}$ contains M, so Γ_M is Borel measurable. ∎

Theorem 5.13. (i) If Γ^C is Borel measurable for each compact set C in $X \times Y$, then there exists a unique Borel measure ϱ_1 on $B(X \times Y)$ such that

$$\varrho_1(M) = \int \Gamma_M \, d\mu \tag{5.34}$$

for all bounded Borel sets M in $X \times Y$. ϱ_1 is an extension of $\mu \times \nu$.

(ii) If Γ^C is Borel measurable for each compact set C in $X \times Y$, then there exists a unique Borel measure ϱ_2 on $B(X \times Y)$ such that

$$\varrho_2(M) = \int \Gamma^M \, d\nu \tag{5.35}$$

for all bounded Borel sets M in $X \times Y$. ϱ_2 is an extension of $\mu \times \nu$. ∎

Proof. (i) From Lemma 5.9 we know Γ_M is measurable for each bounded Borel set M. Inasmuch as $M \subset C \times K$ for compact sets C and K, and

$$0 \leq \Gamma_M \leq \nu(K)\chi_C,$$

we can conclude that Γ_M is μ-integrable. Define ϱ_1' on the ring of bounded Borel sets $\mathscr{R}_b$ by

$$\varrho_1'(M) = \int \Gamma_M \, d\mu.$$

Clearly ϱ_1' is a nonnegative additive real-valued set function on $\mathscr{R}_b$. To show that ϱ_1' is a measure on $\mathscr{R}_b$ it suffices to show that if (M_n) is a nondecreasing sequence of sets in $\mathscr{R}_b$ converging to M in $\mathscr{R}_b$, then $\varrho_1'(M_n) \uparrow \varrho_1'(M)$. This is immediate from the Monotone Convergence Theorem.

Now let ϱ_1 be the unique extension of ϱ_1' to the σ-algebra $B(X \times Y)$ generated by $\mathscr{R}_b$. (See Theorem 2.4 in Chapter 2, giving special attention to the footnote there.) All that remains to be shown is that ϱ_1 is an extension of $\mu \times \nu$. As in Theorem 5.12, it suffices to show that ϱ_1 and $\mu \times \nu$ agree on sets $C \times K$, where C and K are compact. Note that $\Gamma_{C \times K} = \nu(K)\chi_C$. Hence

$$\varrho_1(C \times K) = \int \Gamma_{C \times K} \, d\mu = \nu(K)\mu(C) = \mu \times \nu(C \times K).$$

The proof of (ii) is similar. ∎

In light of Theorem 5.13 we have extensions ϱ_1 and ϱ_2 of $\mu \times \nu$ to $B(X \times Y)$. The next question is whether $\varrho_1 = \varrho_2$.

Example 5.19. (Where $\varrho_1 \neq \varrho_2$). Let $\bar{X}$ again be the set of ordinals less than or equal to Ω, the first uncountable ordinal. Let $X = \bar{X} - \{\Omega\}$. Let μ be the measure on $\bar{X}$ of Example 5.11 and let C be a compact set in $\bar{X} \times \bar{X}$. Is Γ_C Borel measurable? It must be shown that, for any positive real number a, the set A_a equal to $\{x \in \bar{X} : \Gamma_C(x) \geq a\}$ is a Borel set. It suffices to show that $A_a \cap X$ is closed in X. Since X is first countable it suffices to show that whenever (x_n) is a sequence in $A_a \cap X$ converging to x in X, then $x \in A_a \cap X$. By Problem 5.5.4, $\Gamma_C(x) \geq a$ so that $x \in A_a$.

Thus Γ_C is measurable and similarly Γ^C is measurable so that ϱ_1 and ϱ_2 are defined. To show $\varrho_1 \neq \varrho_2$ consider the set Z given by

$$Z = \{(x, y) : x < y < \Omega \text{ or } x = \Omega\}.$$

For any x,

$$\Gamma_Z(x) = \mu(Z_x) = 1$$

and for any y,

$$\Gamma^Z(y) = \mu(Z^y) = 0.$$

Hence

$$\varrho_1(Z) = \int \Gamma_Z \, d\mu = 1 \quad \text{and} \quad \varrho_2(Z) = \int \Gamma^Z \, d\mu = 0.$$

Theorem 5.14. If μ and ν are both regular, then so are ϱ_1 and ϱ_2; and indeed (by Theorem 5.12) $\varrho_1 = \varrho_2 = \mu \otimes \nu$. ∎

Proof. We show ϱ_1 is regular. It suffices to show by Proposition 5.11 that each bounded open measurable set U is inner regular. Given $\varepsilon > 0$, we must find a compact set C in $X \times Y$ with $C \subset U$ and $\varrho_1(U - C) < 2\varepsilon$. Now since U is bounded, $U \subset G \times H$, where G and H are compact sets in X and Y, respectively. If $\nu(H) = 0$, then $\varrho_1(U) = 0$ as

$$\varrho_1(U) \leq \varrho_1(G \times H) = \mu \times \nu(G \times H) = \mu(G)\nu(H) = 0.$$

However, if $\varrho_1(U) = 0$ we can simply take C to be the empty set and be finished. We assume therefore that $\nu(H) > 0$.

Since Γ_U is a measurable function with respect to $B(X)$, using Lusin's Theorem (Problem 3.1.13, in Chapter 3) and the regularity of μ, we may choose a compact set K in X so that $K \subset G$, $\mu(G - K) < \varepsilon/\nu(H)$ and Γ_U is continuous on K. Let $U' = U \cap (K \times Y)$. It is easy to see that

$$U - U' \subset (G - K) \times H,$$

so that

$$\varrho_1(U - U') \leq \varrho_1\big((G - K) \times H\big) = \mu(G - K)\nu(H) < \varepsilon.$$

If we can find a compact set C in $X \times Y$ such that $C \subset U$ and $\varrho_1(U' - C) \leq \varepsilon$, then the relation

$$\varrho_1(U - C) \leq \varrho_1(U - U') + \varrho_1(U' - C) < 2\varepsilon$$

will complete the proof.

Now if $\mu(K) = 0$, then clearly from the definition of U', $\varrho_1(U') = 0$, in which case we can take C to be $\varnothing$ and the proof is complete. Let us therefore assume $\mu(K) > 0$.

Let $x \in K$. The x-section U_x of U is a Borel set and $U_x \subset H$ so that $\nu(U_x) < \infty$. By virtue of the regularity of ν, there exists a compact set $E(x)$ in Y with $E(x) \subset U_x$ and $\nu(U_x) < \nu\big(E(x)\big) + \varepsilon/\mu(K)$. This means

$\Gamma_U(x) < \nu(E(x)) + \varepsilon/\mu(K)$. Since Γ_U is continuous on K, there exists an open neighborhood $V(x)$ of x, relative to the subspace K of X, such that

$$\Gamma_U(t) < \nu(E(x)) + \varepsilon/\mu(K), \qquad \text{for all } t \in V(x). \tag{5.36}$$

Since $[K \times E(x)] - U$ is compact, so is $\text{Pr}_X([K \times E(x)] - U)$. Thus

$$V'(x) \equiv V(x) - \text{Pr}_X([K \times E(x)] - U)$$

is open relative to K. If $t \in V(x)$, then $t \in V'(x)$ if and only if $E(x) \subset U_t$. Hence $V'(x)$ is an open neighborhood of x, relative to K, such that $V'(x) \times E(x) \subset U$. Let $D(x)$ be a compact neighborhood of x, relative to K, such that $D(x) \subset V'(x)$. Then $D(x) \times E(x) \subset U$ and $D(x) \subset V(x)$. By inequality (5.36)

$$\Gamma_U(t) < \nu(E(x)) + \varepsilon/\mu(K), \qquad \text{for all } t \in D(x).$$

It follows that

$$\Gamma_{U-[D(x)\times E(x)]}(t) < \varepsilon/\mu(K), \qquad \text{for all } t \in D(x). \tag{5.37}$$

We repeat the above argument for each x in K. As above we get compact sets $D(x)$ and $E(x)$ such that $D(x)$ is a neighborhood of x relative to K, $D(x) \times E(x) \subset U$, and inequality (5.37) is satisfied.

Since the interiors of the sets in the collection $[D(x)]_{x\in K}$ cover K, there exists by the compactness of K a finite subcovering, say

$$K \subset D(x_1) \cup \cdots \cup D(x_n).$$

Define C by

$$C \equiv \bigcup_{i=1}^{n} D(x_i) \times E(x_i).$$

Clearly C is compact and $C \subset U$. We assert that

$$\Gamma_{U'-C} \leq [\varepsilon/\mu(K)]\chi_K. \tag{5.38}$$

Since $\Gamma_{U'-C} \leq \Gamma_{U'}$ and $\Gamma_{U'}$ vanishes on $X - K$, inequality (5.38) is surely true at points of $X - K$. On the other hand, if $x \in K$ then $x \in D(x_i)$ for some x_i, and so

$$\Gamma_{U-D(x_i)\times E(x_i)}(x) < \varepsilon/\mu(K)$$

by inequality (5.37). Since $U' - C \subset U - D(x_i) \times E(x_i)$, we have $\Gamma_{U'-C}(x) < \varepsilon/\mu(K)$.

Using inequality (5.38) and the definition of ϱ_1 we have

$$\varrho_1(U' - C) = \int \Gamma_{U'-C}\, d\mu \leq [\varepsilon/\mu(K)]\mu(K) = \varepsilon. \qquad \blacksquare$$

Summarizing the results (Theorems 5.12–5.14), we may say ϱ_1 exists if ν is regular, ϱ_2 exists if μ is regular, $\mu \otimes \nu$ exists if μ and ν are regular, and $\varrho_1 = \varrho_2 = \mu \otimes \nu$ if μ and ν are regular.

Our final goal in this section is to give another criterion for ϱ_1 to equal ϱ_2 and to prove a Fubini–Tonelli theorem on iterated integration for the measure $\mu \otimes \nu$.

If $(X, \mathscr{A}, \mu)$ and $(Y, \mathscr{B}, \nu)$ are measure spaces and h is an extended real-valued measurable function on $X \times Y$, then we say the *iterated integral* $\iint h\, d\nu\, d\mu$ exists if there exists a null set E in $\mathscr{A}$ and a μ-integrable function f on X such that h_x is ν-integrable for each x in $X - E$ and $\int h_x\, d\nu = f(x)$. By definition $\iint h\, d\nu\, d\mu$ is then given to be $\int f(x)\, d\mu$. Similarly the iterated integral $\iint h\, d\mu\, d\nu$ is defined.

Let us recall the Fubini and Tonelli Theorems of Chapter 3 (see Theorem 3.10).

Theorem 5.15. *Fubini's Theorem.* If $(X, \mathscr{A}, \mu)$ and $(Y, \mathscr{B}, \nu)$ are σ-finite measure spaces and if h is a $(\mu \times \nu)$-integrable function on $X \times Y$, then both iterated integrals of h exist and

$$\iint h\, d\nu\, d\mu = \int h\, d(\mu \times \nu) = \iint h\, d\mu\, d\nu. \qquad \blacksquare$$

Theorem 5.16. *Tonelli's Theorem.* If $(X, \mathscr{A}, \mu)$ and $(Y, \mathscr{B}, \nu)$ are σ-finite measure spaces and if h is a nonnegative measurable function on $X \times Y$ such that at least one of the iterated integrals of h exist, then h is $(\mu \times \nu)$-integrable. $\blacksquare$

Observe that in both theorems h is assumed to be measurable with respect to the measurable space $(X \times Y, \mathscr{A} \times \mathscr{B})$. In the case that X and Y are σ-compact locally compact Hausdorff spaces and we consider the measurable spaces $(X, B(X))$ and $(Y, B(Y))$, it may happen that a function h is measurable with respect to $B(X \times Y)$ but not with respect to $B(X) \times B(Y)$. It is natural therefore to ask whether the analogs of Theorems 5.15 and 5.16 hold for such functions h when μ and ν are Borel measures on $B(X)$ and $B(Y)$, respectively.

The measures ϱ_1 and ϱ_2 furnish an answer. Of course ϱ_1 exists if Γ_C is Borel measurable for each compact set in $X \times Y$. A similar statement applies to ϱ_2. For this reason we make the assumptions on Γ_C and Γ^C in the following analog of the Fubini Theorem.

Theorem 5.17. Suppose X and Y are σ-compact locally compact Hausdorff spaces.

(i) Suppose Γ_C is Borel measurable for each compact set C in $X \times Y$. If h is a ϱ_1-integrable function on $X \times Y$ [measurable with respect to $B(X \times Y)$], then $\int\int h \, d\nu \, d\mu$ exists and equals $\int h \, d\varrho_1$.

(ii) Suppose Γ^C is Borel measurable for each compact set C in $X \times Y$. If h is a ϱ_2-integrable function on $X \times Y$, then the iterated integral $\int\int h \, d\mu \, d\nu$ exists and is equal to $\int h \, d\varrho_2$. ∎

Proof. (i) Writing h as $h = h^+ - h^-$, where $h^+ = h \vee 0$ and $h^- = -(h \wedge 0)$, we may assume that $h \geq 0$.

First assume h is a simple Borel-measurable function with compact support. Then $h = \sum_{i=1}^n \alpha_i \chi_{E_i}$, where each E_i is a bounded Borel set. For each $i \in \{1, 2, \ldots, n\}$ according to equation (5.34),

$$\varrho_1(E_i) = \int \Gamma_{E_i} \, d\mu,$$

where again $\Gamma_{E_i}(x) = \nu(E_{i_x})$. For each $x \in X$,

$$\int \chi_{E_{i_x}} \, d\nu = \nu(E_{i_x}) < \infty,$$

and by linearity

$$\int h_x \, d\nu = \sum_{i=1}^n \alpha_i \nu(E_{i_x}) < \infty. \tag{5.39}$$

Hence

$$\begin{aligned} \int \left[\int h_x \, d\nu \right] d\mu &= \int \left[\sum_{i=1}^n \alpha_i \nu(E_{i_x}) \right] d\mu = \int \left[\sum_{i=1}^n \alpha_i \Gamma_{E_i}(x) \right] d\mu \\ &= \sum_{i=1}^n \alpha_i \int \Gamma_{E_i}(x) \, d\mu = \sum_{i=1}^n \alpha_i \varrho_1(E_i) = \int h \, d\varrho_1. \end{aligned} \tag{5.40}$$

Now assume h is any nonnegative ϱ_1-integrable function. There is a sequence of simple functions h_n such that $0 \leq h_n \uparrow h$. Since each Borel set in $B(X \times Y)$ (recall that $X \times Y$ is σ-compact) is the limit of an increasing sequence of bounded Borel sets, we may assume that each h_n has compact support. Hence using equations (5.39) and (5.40) we have

$$\int h \, d\varrho_1 = \lim_n \int h_n \, d\varrho_1 = \lim_n \int \left(\int (h_n)_x \, d\nu \right) d\mu < \infty. \tag{5.41}$$

Now for each $x \in X$, $((h_n)_x)_N$ is a nondecreasing sequence so that, if we let

$$f_n(x) = \int (h_n)_x \, d\nu,$$

then $(f_n(x))_N$ is a nondecreasing sequence of μ-integrable functions. For each $x \in X$, let $f(x) = \lim_n f_n(x)$. By the Monotone Convergence Theorem

$$\int f(x) \, d\mu = \lim_n \int f_n(x) \, d\mu = \lim_n \int \left[\int (h_n)_x \, d\nu \right] d\mu < \infty, \tag{5.42}$$

whence f is a μ-integrable function with

$$\int (h_n)_x \, d\nu = f_n(x) \uparrow f(x)$$

for each $x \in X$. Now for each $x \in X$, $(h_n)_x \uparrow h_x$, so that by the Monotone Convergence Theorem again

$$\int h_x \, d\nu = \lim_{n \to \infty} \int (h_n)_x \, d\nu = f(x). \tag{5.43}$$

Since f is μ-integrable there exists a set E with $\mu(E) = 0$ such that f is real-valued on $X - E$. Hence for $x \in X - E$, h_x is ν-integrable by equation (5.43) and we can conclude thereby that the iterated integral $\iint h \, d\mu \, d\nu$ exists. Also combining equations (5.42) and (5.41) we have

$$\iint h \, d\mu \, d\nu = \int f(x) \, d\mu = \int h \, d\varrho_1.$$

The proof of statement (ii) is analogous. ∎

The next theorem is the analog of the Tonelli Theorem (Theorem 5.16) for the measurable space $(X \times Y, B(X \times Y))$.

Theorem 5.18. Suppose X and Y are σ-compact locally compact Hausdorff spaces.

(i) Suppose Γ_C is Borel measurable for each compact set C in $X \times Y$. If h is a nonnegative Borel function on $X \times Y$ [measurable with respect to $B(X \times Y)$] such that $\iint h \, d\nu \, d\mu$ exists, then h is ϱ_1-integrable.

(ii) Suppose Γ^C is Borel measurable for each compact set C in $X \times Y$. If h is a nonnegative Borel function on $X \times Y$ such that $\iint h \, d\mu \, d\nu$ exists, then h is ϱ_2-integrable. ∎

Proof. (i) Since $\int\int h\, d\nu\, d\mu$ exists, there exists a μ-integrable function f and a μ-null set E such that for $x \in X - E$, h_x is ν-integrable and $\int h_x\, d\nu = f(x)$.

There exists a sequence of simple functions h_n with compact support such that $0 \leq h_n \uparrow h$. As shown by equations (5.39) and (5.40), if $f_n(x)$ is defined to be $\int (h_n)_x\, d\nu$, then f_n is μ-integrable and

$$\int f_n\, d\mu = \int h_n\, d\varrho_1. \tag{5.44}$$

If $x \in X - E$, then as $(h_n)_x \leq h_x$ we have

$$f_n(x) = \int (h_n)_x\, d\nu \leq \int h_x\, d\nu = f(x).$$

Hence

$$\int f_n(x)\, d\mu \leq \int f(x)\, d\mu < \infty$$

or by equation (5.44)

$$\int h_n\, d\varrho_1 \leq \int f(x)\, d\mu.$$

By the Monotone Convergence Theorem

$$\int h\, d\varrho_1 = \lim_n \int h_n\, d\varrho_1 \leq \int f\, d\mu < \infty.$$

The proof of (ii) is similar. ∎

Using Theorem 5.17 and 5.18, we now have the following conditions for ϱ_1 to equal ϱ_2.

Theorem 5.19. Suppose X and Y are σ-compact locally compact Hausdorff spaces, and Γ_C and Γ^C are Borel measurable for each compact set in $X \times Y$. The following statements are equivalent:

(i) $\varrho_1 = \varrho_2$.

(ii) For any nonnegative Borel-measurable function h on $X \times Y$, whenever $\int\int h\, d\mu\, d\nu$ and $\int\int h\, d\nu\, d\mu$ exist, they are equal. ∎

Proof. (i) implies (ii) by using Theorems 5.17 and 5.18. Conversely, to show $\varrho_1 = \varrho_2$ when (ii) holds, it is sufficient to show $\varrho_1(C) = \varrho_2(C)$ for each compact set C in $X \times Y$. If C is compact in $X \times Y$, then χ_C is ϱ_1 and ϱ_2 integrable so that by Theorem 5.17, $\int\int \chi_C\, d\mu\, d\nu$ and $\int\int \chi_C\, d\nu\, d\mu$

exist and equal $\varrho_2(C)$ and $\varrho_1(C)$, respectively. By assumption, these iterated integrals are equal, so that $\varrho_1(C) = \varrho_2(C)$.

Combining Theorem 5.19 with Theorems 5.14, 5.17, and 5.18 we obtain a beautiful Fubini–Tonelli theorem for the measure $\mu \otimes \nu$ on $B(X \times Y)$.

Theorem 5.20. Assume μ and ν are regular Borel measures on the σ-compact locally compact Hausdorff spaces X and Y. Let $\mu \otimes \nu$ be the regular tensor product of μ and ν on $B(X \times Y)$. Let h be measurable with respect to $B(X \times Y)$.

(i) If h is $\mu \otimes \nu$-integrable, then both iterated integrals of h exist, and

$$\iint h \, d\nu \, d\mu = \iint h \, d\mu \, d\nu = \int h \, d(\mu \otimes \nu).$$

(ii) If h is nonnegative and one of the iterated integrals of h exists and is finite, then h is $\mu \otimes \nu$-integrable. ∎

Problems

5.5.1. Prove that the class of all sets $M \subset X \times Y$ such that $M_x \in B(Y)$ and $M^y \in B(X)$ is a σ-algebra containing all compact sets in $X \times Y$. Conclude that if $M \in B(X \times Y)$, then $M_x \in B(Y)$ and $M^y \in B(X)$.

5.5.2. In Theorem 5.12, prove that it is sufficient to show $\mu \otimes \nu(C \times D) = \mu \times \nu(C \times D)$ for compact sets C and D. [Hint: Show that $\mu \otimes \nu$ and $\mu \times \nu$ agree on the ring $\mathscr{R}$ generated by such rectangles. ($\mathscr{R}$ is the set of finite disjoint unions of rectangles of the form $(C_1 - C_2) \times (D_1 - D_2)$, where $C_2 \subset C_1$ and $D_2 \subset D_1$). Since the measures are σ-finite on $\mathscr{R}$, show they have a unique extension to the σ-ring generated by $\mathscr{R}$.]

5.5.3. Assume Γ_C is Borel measurable for each compact set C in $X \times Y$. Prove the following:

(i) Γ_M is measurable if $M = C - D$, where C and D are compact with $D \subset C$.

(ii) If M and N are disjoint Borel sets such that Γ_M and Γ_N are measurable, then $\Gamma_{M \cup N}$ is measurable.

5.5.4. Suppose C is a compact set in $X \times Y$ and (x_n) is a sequence in X converging to x in X. If $\Gamma_C(x_n) \geq a$ for all n, show that $\Gamma_C(x) \geq a$. [Hint: Let $E = \lim_n \sup C_{x_n}$. Show $\nu(E) \geq a$. Show $E \subset C_x$ and conclude $\nu(C_x) \geq a$.]

Appendix

A. Differentiation of Borel Measures: The Change-of-Variable Formula

In this section, we will discuss the differentiation of Borel measures on Euclidean n-space R^n with respect to the n-dimensional Lebesgue measure m_n defined in Chapter 2. This discussion will lead to a change-of-variable formula for the integration of real functions on R^n whose domains are transformed by differentiable transformations on R^n. This change-of-variable formula is analogous to that of Section 4.4 for general measure spaces.

Before beginning we make two observations. First, the reader should recall from Chapter 5 that any Borel measure on R^n is regular. This follows from Corollaries 5.1 and 5.2 and Theorem 5.2. Secondly, if A is a Lebesgue-measurable set in R^n and $cA = \{cx \mid x \in A\}$ for some $c > 0$, then $m_n(cA) = c^n m_n(A)$. In particular if B is the open ball $\{x \mid \| x - x_0 \| < r\}$ with center x_0 and radius r in R^n, then $m_n(B) = r^n m_n(U)$, where U is the unit open ball $\{x \mid \| x \| < 1\}$. Verification of this observation follows readily from properties of m_n given in Chapter 2.

We begin with the following definition.

Definition A.1. Let μ be a signed Borel measure on the σ-algebra of Borel sets in R^n. Let $x \in R^n$. μ is said to be *differentiable* at x if there exists a real number $D\mu(x)$ such that for each $\varepsilon > 0$ there is some $\delta > 0$ such that

$$\left| \frac{\mu(B)}{m_n(B)} - D\mu(x) \right| < \varepsilon$$

for each open ball B in R^n with $x \in B$ and radius less than δ. $D\mu(x)$ is called the *derivative* of μ at x. ∎

Remark A.1. The following statements are readily verifiable.

(i) If μ and μ' are signed Borel measures such that $\mu + \mu'$ is a signed measure and $D\mu(x)$ and $D\mu'(x)$ exist, then $D(\mu + \mu')(x)$ exists and $D(\mu + \mu')(x) = D\mu(x) + D\mu'(x)$.

(ii) If $D\mu(x)$ exists, then

$$D\mu(x) = \lim_{r \to 0} D_r\mu(x),$$

where

$$D_r\mu(x) = \sup\{\mu(B)/m_n(B) \mid B \text{ is an open } ball \text{ of radius } \leq r \text{ containing } x\}.$$

By $D\mu$ we will denote the real function [defined in the set of all those x in R^n where $D\mu(x)$ exists] whose value at x is $D\mu(x)$.

Our initial question is "How does $D\mu$ compare with the Radon–Nikodym derivative $d\mu/dm_n$ in the case $\mu << m_n$?" Our first objective is to show that when μ is a signed Borel measure and $\mu << m_n$ then $D\mu$ coincides with $d\mu/dm_n$ almost everywhere with respect to the Lebesgue measure m_n. To this end we use two lemmas.

Lemma A.1. Let $(B_i)_{i \in I}$ be a collection of open balls in R^n. For any real $\alpha < m_n(\bigcup_{i \in I} B_i)$ there exists a finite pairwise disjoint subcollection $B_{k_1}, \ldots, B_{k_p}$ such that

$$\alpha < 3^n \sum_{i=1}^{p} m_n(B_{k_i}).$$ ∎

Proof. Let $\alpha < m_n(\bigcup_{i \in 1} B_i)$. Using the regularity of m_n (see Section 2.3), there exists a compact set $K \subset \bigcup_{i \in I} B_i$ so that $m_n(K) > \alpha$. K being compact implies there exists a finite collection $B_1, B_2, \ldots, B_s$ from $(B_i)_{i \in N}$ which cover K. Let $r_1, r_2, \ldots, r_s$ be the radii of the s given balls. By reordering we may assume $r_1 \geq r_2 \geq \cdots \geq r_s$.

Let $B_{k_1} = B_1$. Let k_2 be the smallest integer among $2, \ldots, s$ such that $B_{k_2} \cap B_{k_1} = \varnothing$ if k_2 exists. Likewise let k_3 be the smallest integer among $k_2 + 1, \ldots, s$ such that $B_{k_3}, B_{k_2}, B_{k_1}$ are pairwise disjoint. Continue in this fashion until $\{B_1, B_2, \ldots, B_s\}$ has been exhausted.

The resulting subcollection $\{B_{k_1}, \ldots, B_{k_p}\}$ is pairwise disjoint. Moreover each B_i for $i = 1, \ldots, s$ must intersect some B_{k_j} for $j = 1, \ldots, p$.

If $B_i \cap B_{k_j} \neq \emptyset$, then $B_i \subset U_{k_j}$, where U_{k_j} is the ball with center identical to that of B_{k_j} but with radius three times that of B_{k_j}. Thus

$$\bigcup_{i=1}^{s} B_i \subset \bigcup_{i=1}^{p} U_{k_i}$$

and

$$\alpha < m_n(K) \leq m_n\left(\bigcup_{i=1}^{s} B_i\right) \leq m_n\left(\bigcup_{i=1}^{p} U_{k_i}\right) \leq \sum_{i=1}^{p} m_n(U_{k_i}) = 3^n \sum_{i=1}^{p} m_n(B_{k_i}). \quad \blacksquare$$

Lemma A.2. Let μ be a Borel measure on R^n and let A be a Borel set such that $\mu(A) = 0$. Then $D\mu(x) = 0$ a.e. (m_n) on A. ∎

Proof. Let $B = \{x \in A \mid \lim_{r\to 0} D_r\mu(x) = 0\}$. Clearly $D\mu(x) = 0$ for $x \in B$. We will show $m_n(A - B) = 0$.

For each $y \in N$ let

$$A_j = \{x \in A \mid \lim_{r\to 0} D_r\mu(x) > 1/j\}.$$

Clearly $A - B = \bigcup_{j=1}^{\infty} A_j$. We will show A_j is measurable and $m_n(A_j) = 0$ for each j.

Let j be fixed and let $\varepsilon > 0$ be arbitrary. By the regularity of μ, there exists an open set $U \supset A$ such that $\mu(U) < \varepsilon$. For each $x \in A_j$, there exists an open ball $B_x \subset U$ containing x such that

$$\mu(B_x) > (1/j)m_n(B_x).$$

Using the preceding lemma, for any $\alpha < m_n(\bigcup_{x\in A_j} B_x)$

$$\alpha \leq 3^n \sum_{i=1}^{p} m_n(B_{k_i}) \leq 3^n j \sum_{i=1}^{p} \mu(B_{k_i})$$

$$= 3^n j\mu\left(\bigcup_{i=1}^{p} B_{k_i}\right) \leq 3^n j\mu(U) < 3^n j\varepsilon.$$

Since $A_j \subset \bigcup_{x\in A_j} B_x$, A_j lies in a set of arbitrarily small Lebesgue measure and thus is measurable with measure zero. ∎

Theorem A.1. If μ is a signed Borel measure on R^n and $\mu << m_n$, then μ is differentiable a.e. (m_n). Moreover $D\mu = d\mu/dm_n$ a.e. (m_n). ∎

Proof. Denote $d\mu/dm_n$ by f. Then f is measurable and $\mu(E) = \int_E f\, dm_n$ for each Borel set E. (Note that since μ is a Borel measure, f is integrable over compact sets.)

Associate with each rational number r the sets

$$A_r = \{x : f(x) \le r\} \quad \text{and} \quad B_r = \{x : f(x) > r\},$$

and define nonnegative Borel measures by

$$\lambda_r(E) = \int_{E \cap B_r} [f(x) - r] \, dm_n.$$

Clearly $\lambda_r(A_r) = 0$. By the preceding lemma $D\lambda_r = 0$ a.e. (m_n) on A_r. Let $C_r = \{x \in A_r \mid D\lambda_r(x) \neq 0\}$ and let $V = \bigcup_{r \in Q} C_r$. Then $m_n(V) = 0$.

Fix $x \in V^c$. Let $\{B_{r_k}(x)\}$ be a sequence of open balls containing x such that the radii $r_k \to 0$. Assuming $f(x) \neq +\infty$ choose a rational s so that $s > f(x)$. Then $x \in A_s$. Since

$$\mu(B_{r_k}(x)) - s m_n(B_{r_k}(x)) = \int_{B_{r_k}(x)} (f - s) \, dm_n \le \lambda_s(B_{r_k}(x)),$$

we have

$$\frac{\mu(B_{r_k}(x))}{m_n(B_{r_k}(x))} \le s + \frac{\lambda_s(B_{r_k}(x))}{m_n(B_{r_k}(x))}.$$

Since $x \in V^c$, we have $x \in A_s - C_s$. Hence $D\lambda_s(x) = 0$ so

$$\limsup_{k \to \infty} \frac{\mu(B_{r_k}(x))}{m_n(B_{r_k}(x))} \le s.$$

Since s is an arbitrary rational $> f(x)$, we have

$$\limsup_{k \to \infty} \frac{\mu(B_{r_k}(x))}{m_n(B_{r_k}(x))} \le f(x) \text{ a.e. } (m_n). \tag{A.1}$$

Similarly, if μ is replaced by $-\mu$ and f by $-f$,

$$\liminf_{k \to \infty} \frac{\mu(B_{r_k}(x))}{m_n(B_{r_k}(x))} \ge f(x), \text{ a.e. } (m_n), \tag{A.2}$$

for every sequence B_{r_k} of open balls containing x with $r_k \to 0$. (A.1) and (A.2) give the desired conclusion. ∎

The following result follows from Theorem A.1.

Theorem A.2. Every signed Borel measure on R^n is differentiable almost everywhere. ∎

Proof. Assume first that μ is a Borel measure. Then μ is a σ-finite measure on R^n. Let $\mu = \mu_1 + \mu_2$ be the Lebesgue decomposition of μ with $\mu_1 << m_n$ and $\mu_2 \perp m_n$. By Theorem A.1 μ_1 is differentiable a.e. (m_n) and $D\mu_1 = d\mu_1/dm_n$ a.e.

Since $\mu_2 \perp m_n$, there is a Borel set B such that $\mu_2(B) = m_n(B^c) = 0$. By Lemma A.2, $D\mu_2 = 0$ a.e. (m_n) on B and therefore $D\mu_2 = 0$ a.e. (m_n) on R^n since $m_n(B^c) = 0$. Therefore

$$D\mu(x) = D(\mu_1 + \mu_2)(x) = D\mu_1(x) + D\mu_2(x) = D\mu_1(x)$$

almost everywhere on R^n.

Secondly, if μ is a signed Borel measure, then write μ as the difference of two Borel measures and proceed as above. ∎

We now turn our attention to transformations on R^n. A transformation $T: U \subset R^n \to R^n$, where U is an open domain, is said to be *differentiable* at $x_0 \in U$ if there exists a linear operator $L: R^n \to R^n$ such that

$$\lim_{x \to x_0} \frac{\| T(x) - T(x_0) - L(x - x_0) \|}{\| x - x_0 \|} = 0.$$

The linear operator is dependent on x_0 and is unique. It will be denoted by $T'(x_0)$ and will be called the *derivative* of T at x_0.

If $T: U \to R^n$ is differentiable at each point x in U then T is said to be *differentiable on U*. If T is differentiable then T is continuous and each partial

$$\frac{\partial T_i}{\partial x_j}(x) \equiv \lim_{h \to 0} \frac{T_i(x + he_j) - T_i(x)}{h}$$

exists at each point x of U. [Here $T(x) = \big(T_1(x), \ldots, T_n(x)\big)$ and e_j is the n-tuple $(0, 0, \ldots, 0, 1, 0, \ldots, 0)$ where 1 is in the jth position.]

The matrix

$$J_T(x) = \begin{bmatrix} \dfrac{\partial T_1}{\partial x_1}(x) \cdots \dfrac{\partial T_1}{\partial x_n}(x) \\ \vdots \\ \dfrac{\partial T_n}{\partial x_1}(x) \cdots \dfrac{\partial T_n}{\partial x_n}(x) \end{bmatrix}$$

is called the *Jacobian matrix* and its determinant $\det J_T(x)$ is called the *Jacobian*. The matrix $J_T(x)$ is the matrix representation of $T'(x)$ in the sense that $T'(x)(h) = J_T(x)[h]$.

A transformation T is said to be C^1 if T is differentiable and each $\partial T_i/\partial x_j$ is continuous.

Lemma A.3. Let $T: U \to R^n$ be C^1 with U open in R^n. If $A \subset U$ and $m_n(A) = 0$, then $T(A)$ is Lebesgue measurable and $m_n(T(A)) = 0$. ∎

Proof. We first observe that we may assume each continuous partial derivative $\partial T_i/\partial x_j$ is bounded on U. Indeed, U can be written as the countable union $\bigcup_{k=1}^{\infty} B_k$ of open balls B_k with closures $\bar{B}_k \subset U$, whose centers consist of rational coordinates and whose radii are rational values. Then $\partial T_i/\partial x_j$: $\bar{B}_k \to R^n$ is bounded since $\bar{B}_k$ is compact. Since $T(A) = T(A \cap (\bigcup_k B_k)) = \bigcup_k T(A \cap B_k)$ and $m_n(A \cap B_k) = 0$, it would be sufficient to restrict T to B_k if $\partial T_i/\partial x_j$ were not bounded.

Thus we assume that there exists a number $M > 0$ such that for all $i, j = 1, \ldots, n$, $| \partial T_i/\partial x_j(x) | < M$ for all x in U. This means that for each $x_0 \in U$, $\| T'(s_0)h \| \leq nM \| h \|$ for each h in R^n. Additionally, since T is differentiable, for each $x_0 \in U$ there exists an open ball B_{x_0} centered at x_0 and contained in U such that for all x in B_{x_0}

$$\| T(x) - T(x_0) \| < nM \| x - x_0 \| + \| x - x_0 \| = K \| x - x_0 \|, \quad (A.3)$$

where $K \equiv nM + 1$.

Let $\varepsilon > 0$. Choose an open set O such that $A \subset O \subset U$ and $m_n(O) < \varepsilon$. We will show $T(O)$ is a Borel set and show $m_n(T(O)) \leq \varepsilon(3K)^n$ by showing $m_n(C) < \varepsilon(3K)^n$ for any compact subset of $T(O)$ and using the regularity of m_n.

The set O can be written as $\bigcup_{i=1}^{\infty} C_i$ where each C_i is compact. This means $T(O) = \bigcup_{i=1}^{\infty} T(C_i)$ and $T(O)$ is Borel since each $T(C_i)$ is also compact as T is continuous.

Let C be a compact subset of $T(O)$. Let $y \in C$. Then $y = T(x_0)$ for some $x_0 \in O$. By (A.3)

$$\| T(x) - y \| < K \| x - x_0 \| \quad (A.4)$$

for all x in some open ball $B_{x_0} \subset O$ centered at x_0. Let B'_{x_0} be the ball centered at y with radius K times the radius of B_{x_0}. Then from (A.4), $T(B_{x_0}) \subset B'_{x_0}$. Since C is compact, $C \subset \bigcup_{i=1}^{q} B'_{y_i}$ for q points $y_1, \ldots, y_q$ in C.

By Lemma A.1., there exist a pairwise disjoint subcollection $B'_{k_1}, \ldots, B'_{k_p}$ from among the collection $B'_{y_1}, \ldots, B'_{y_q}$ such that

$$m_n\left(\bigcup_{i=1}^{q} B'_{y_i} \right) \leq 3^n m_n\left(\bigcup_{i=1}^{p} B'_{k_i} \right) = 3^n K^n \sum_{i=1}^{p} m_n(B_{k_i}).$$

Therefore

$$m_n(C) \leq m_n\Big(\bigcup_{i=1}^{q} B'_{y_i}\Big) \leq (3K)^n \sum_{i=1}^{p} m_n(B_{k_i}) = (3K)^n m_n\Big(\bigcup_{i=1}^{p} B_{k_i}\Big)$$
$$\leq (3K)^n m_n(\mathrm{O}) < (3K)^n \varepsilon.$$

This implies $m_n(T(\mathrm{O})) \leq \varepsilon(3K)^n$ and $T(A)$ is contained in a Borel set of arbitrarily small measure. Consequently $m_n(T(A)) = 0$. ∎

Proposition A.1. Suppose T is a continuous transformation on an open set A of R^n onto an open set B of R^n. Suppose also that T is one-to-one and that its inverse from B onto A is continuous. Then

(i) $T(E)$ is a Borel set in B if and only if E is a Borel set in A.
(ii) If T is C^1, then $T(E)$ is a Lebesgue-measurable set in B whenever E is a Lebesgue-measurable set in A. If also the Jacobian $\det J_T$ of T is nonzero on A, then the converse also holds.

Proof. (i) Let $\mathscr{A}$ and $\mathscr{B}$ denote the Borel subsets of A and B, respectively. The class $T(\mathscr{A}) = \{T(E) : E \in \mathscr{A}\}$ is a σ-algebra of sets from $\mathscr{B}$ containing all open sets from $\mathscr{B}$. Consequently $T(\mathscr{A}) \supset \mathscr{B}$. In the same manner $T^{-1}(\mathscr{B}) \supset \mathscr{A}$, so that $T(\mathscr{A}) = \mathscr{B}$.

(ii) Let E be a Lebesgue-measurable set such that $m_n(E) < \infty$. There exists a Borel set F such that $E \subset F \subset A$ and $m_n(F - E) = 0$. Consequently by Lemma A.3, $T(F - E)$ is Lebesgue measurable. Therefore

$$T(E) = T(F) - T(F - E)$$

is Lebesgue measurable.

The converse holds when $\det J_T \neq 0$ on A since then T^{-1} is also C^1 by the Inverse Function Theorem.[†] ∎

Let now T be a C^1 transformation satisfying the hypothesis of the preceding proposition. Let μ be the Borel measure defined on the Borel-measurable subsets of A by

$$\mu(E) = m_n(T(E)) \tag{A.5}$$

From Lemma A.3 it follows that $\mu << m_n$. The next theorem together with Theorem A.1 shows that $d\mu/dm_n = |\det J_T|$ a.e. (m_n).

Recall from Section 2.3 that if $L : R^n \to R^n$ is a linear operator then for all Lebesgue-measurable sets E, $m_n(L(E)) = |\det L|\, m_n(E)$.

[†] The inverse function theorem can be found in texts on advanced calculus.

Theorem A.3. Let $T : V \to W$ be a bijective differentiable transformation from the open set V in R^n to the open set W in R^n. Let μ be defined on the Borel sets as in (A.5). Then for all x in V

$$D\mu(x) = |\det(J_T(x))|.$$

Proof. Let $x \in V$. We may assume $x = 0$ and $T(\mathrm{O}) = 0$. Indeed, otherwise we could replace T by S given by $S(t) = T(x + t) - T(x)$, mapping the open set $V - x$ onto $W - T(x)$.

Let us first consider the case when $\det J_T(\mathrm{O}) \neq 0$. Then we may also assume that $L \equiv J_T(\mathrm{O})$ is the $n \times n$ identity matrix I. If not we replace T by F given by $F(x) = L^{-1}(T(x))$. Then the ith component $F_i(x)$ of $F(x)$ equals $\sum_{k=1}^{n}(L^{-1})_{ik}T_k(x)$ so that $(\partial F_i/\partial x_j)(\mathrm{O}) = \sum_{k=1}^{n}(L^{-1})_{ik}\partial T_k/\partial x_j(\mathrm{O})$ and so $J_F(\mathrm{O}) = L^{-1}J_T(\mathrm{O}) = I$. Moreover the measure $\nu(E) \equiv m_n(F(E)) = m_n(L^{-1}(T(E)))$ satisfies

$$\nu(E) = |\det L^{-1}|\, m_n(T(E)) = |\det J_T(\mathrm{O})^{-1}|\, \mu(E),$$

so that $D\mu(\mathrm{O}) = |\det(J_T(\mathrm{O}))|\, D\nu(\mathrm{O})$. Hence if the theorem is proved for F, that is, $D\nu(\mathrm{O}) = |\det J_F(\mathrm{O})| = 1$, then the result follows for T.

Let $\varepsilon > 0$ be arbitrary. Assuming that $J_T(\mathrm{O}) = I$, it must then be shown that there is a $\delta > 0$ such that

$$\left| \frac{\mu(B)}{m_n(B)} - 1 \right| < \varepsilon,$$

whenever B is an open ball containing O and with radius less than δ. Fix $\eta > 0$ so that $\eta < 1/4$ and $1 - \varepsilon < (1 - 2\eta)^n < (1 + 2\eta)^n < 1 + \varepsilon$. Since $J_T(\mathrm{O}) = I$ so that $T'(\mathrm{O})$ is the identity operator, there exists a $\delta > 0$ such that

$$\| T(x) - x \| < \eta \, \| x \|, \tag{A.6}$$

whenever $\| x \| < \delta$ and $x \in V$.

Let B be an open ball in V with center at x_0 and containing 0 with radius $r < \delta/2$. Let B_1 and B_2 be the open balls concentric with B but with radii $(1 - 2\eta)r$ and $(1 + 2\eta)r$, respectively. Notice first that if $x \in B$, then $\| x \| < 2r < \delta$ and so by (A.6)

$$\| T(x) - x_0 \| \leq \| Tx - x \| + \| x - x_0 \| < \eta \, \| x \| + r < 2\eta r + r = (2\eta + 1)r.$$

Hence $T(B) \subset B_2$.

We claim also that $B_1 \subset T(B)$.

To this end let $E_1 = B_1 \cap T(B)$ and $E_2 = B_1 - T(B)$. Clearly E_1 is open. Moreover $E_1 \neq \varnothing$ since $\eta < 1/4$ and

$$\| T(x_0) - x_0 \| < \eta \| x_0 \| < \eta r < (1 - 2\eta)r.$$

If $\| x - x_0 \| = r$, then $\| x \| < 2r < \delta$ and

$$r = \| x - x_0 \| \leq \| x - T(x) \| + \| T(x) - x_0 \| < 2r\eta + \| T(x) - x_0 \|,$$

so that $\| T(x) - x_0 \| > (1 - 2\eta)r$. This means the boundary of B does not map into B_1 so that $B_1 - T(B) = B_1 - T(\bar{B})$ and E_2 is also open. Inasmuch as $B_1 = E_1 \cup E_2$ with E_1 and E_2 open and disjoint and $E_1 \neq \varnothing$, E_2 must be empty since the convex set B_1 in R^n cannot be disconnected. Consequently $B_1 \subset T(B)$.

We have proved $B_1 \subset T(B) \subset B_2$. Therefore

$$\begin{aligned}(1 - \varepsilon)m_n(B) &< (1 - 2\eta)^n m_n(B) = m_n(B_1) \leq m_n\big(T(B)\big) \\ &= \mu(B) \leq m_n(B_2) = (1 + 2\eta)^n m_n(B) < (1 + \varepsilon)m_n(B).\end{aligned}$$

Hence

$$1 - \varepsilon < \frac{\mu(B)}{m_n(B)} < 1 + \varepsilon.$$

We now consider the case that $\det J_T(\mathrm{O}) = 0$, i.e., $L = J_T(\mathrm{O})$ is not invertible. If U is the closed unit ball of R^k, then $m_n\big(L(U)\big) = 0$ (see Section 2.3). Let $\varepsilon > 0$. Since $L(U) = \bigcap_{m=1}^{\infty} E_{1/m}$, where $E_{1/m}$ is the set of all points in R^n whose distance from $L(U)$ is less than $1/m$, there exists $\eta > 0$ such that $m_n(E_\eta) < \varepsilon$. Also as in (A.6), since $T'(\mathrm{O}) = L$, by definition there exists a $\delta > 0$ such that when $\| x \| < \delta$ and $x \in V$

$$\| Tx - L(x) \| < \eta \| x \|. \tag{A.7}$$

Let B be as above with $r < \delta/2$. Let $B_0 = \{x \in R^n \mid \| x \| < 2r\}$. Then $B \subset B_0$. Let $E = \{y \varepsilon R^k \mid \text{the distance of } y \text{ from } L(B_0) \text{ is less than } 2r\eta\}$. By (A.7), $T(B) \subset E$. Clearly $y \in E_\eta$ if and only if $2ry \in E$ so that $m_n(E) = (2r)^n m_n(E_\eta) < (2r)^n \varepsilon$. Therefore

$$\frac{\mu(B)}{m_n(B)} = \frac{m_n\big(T(B)\big)}{m_n(B)} \leq \frac{m_n(E)}{m_n(B)} = \frac{m_n(E)}{2^n m_n(B_0)} = \frac{m_n(E)}{2^n (2r)^n m_n(U)} < \frac{\varepsilon}{m_n(U)}.$$

Consequently $D\mu(\mathrm{O}) = 0 = | \det J_T(\mathrm{O}) |$. ∎

We now come to the concluding result of this section—the so-called "change of variable formula."

Theorem A.4. Let $T : V \to W$ be a bijective C^1 differentiable transformation from the open set V in R^n onto the open set W in R^n. Then

(i) for each Lebesgue-measurable subset E of V

$$m_n(T(E)) = \int_E |\det J_T(x)|\, dm_n; \text{ and}$$

(ii) for each Lebesgue-integrable function f in $L_1(W)$ the function $(f \circ T) \cdot \det |J_T|$ is integrable over V and

$$\int_W f(x)\, dm_n = \int_V (f \circ T)(x)\, |\det J_T(x)|\, dm_n. \tag{A.8}$$

Proof. (i) Let $\mu(E) = m_n(T(E))$ for each Lebesgue-measurable set E in V. By Lemma A.3 and Proposition A.1, $\mu << m_n$ so that for each Lebesgue-measurable set E

$$\mu(E) = \int_E d\mu/dm_n\, dm_n.$$

However by Theorems A.1 and A.3, $d\mu/dm_n = |\det J_T|$ a.e. (m_n). The conclusion is immediate.

(ii) Let us first consider the case where $\det J_T \neq 0$ on V. With this assumption suppose first that $f = \chi_E$ for a Lebesgue-measurable set E in W. Then $T^{-1}(E)$ is Lebesgue measurable and by (i) we have

$$\int_E \chi_E\, dm_n = m_n(E) = m_n[T(T^{-1}(E))] = \int_{T^{-1}(E)} |\det J_T|\, dm_n$$

$$= \int_V (\chi_E \circ T)\, |\det J_T|\, dm_n = \int_V (f \circ T)\, |\det J_T|\, dm_n.$$

By the usual process we can extend this argument to simple functions f and then to general integrable functions and (A.8) is proved in this case.

If $\det J_T(x) = 0$ for some x in V, let $C = \{x \in V \mid \det J_T(x) = 0\}$. Then C is closed in V and $T(C)$ is closed in W. Consequently if $f \in L_1(W)$ and T is considered as a linear transformation from $V - C$ onto $W - T(C)$ we have

$$\int_{W-T(C)} f(x)\, dm_n = \int_{V-C} (f \circ T)(x)\, |\det J_T(x)|\, dm_n \tag{A.9}$$

by the above argument. However from (i), $m_n(T(C)) = 0$ so that

$$\int_{T(C)} f(x)\, dm_n = 0 = \int_C (f \circ T)(x) \mid \det J_T(x) \mid dm_n. \tag{A.10}$$

The result follows by combining (A.9) and (A.10).

B. Product Measures

In this section, we will discuss how a product measure can be constructed on the cartesian product space $\prod_{i\in A} X_i$, where $(X_i, \mathscr{A}_i, \mu_i)$, $i \in A$, is a given family of measure spaces.

When A is a finite set, the construction given in Chapter 3 can be easily extended. Indeed, in this case, we can again consider the algebra $\mathscr{A}$ of all finite disjoint unions of measurable rectangles of the form $\prod_{i\in A} E_i$, $E_i \in \mathscr{A}_i$, and define μ on $\mathscr{A}$ by

$$\mu\Big(\prod_{i\in A} E_i\Big) = \prod_{i\in A} \mu_i(E_i).$$

(Here, of course, we take $0 \cdot \infty = 0$.) For a finite disjoint union $\bigcup_{i=1}^n Y_i$ of measurable rectangles Y_i, we take $\mu(\bigcup_{i=1}^n Y_i) = \sum_{i=1}^n \mu(Y_i)$. Thus, if $A = \{1, 2, \ldots, m\}$, then we can write for Y in $\mathscr{A}$,

$$\mu(Y) = \int_{X_m} \Big[\cdots \Big[\int_{X_1} \chi_Y(x_1, x_2, \ldots, x_m)\, d\mu_1(x_1) \Big] \cdots \Big] d\mu_m(x_m). \tag{A.11}$$

It follows easily from (A.11), using one-dimensional integration theory, that μ is a measure on $\mathscr{A}$. We can now state the following theorem, using Theorem 2.4.

Theorem B.1. Suppose that the measure spaces $(X_i, \mathscr{A}_i, \mu_i)$, $i = 1, 2, \ldots, m$, are σ-finite. Then there is a unique product measure $\mu = \prod_{i=1}^m \mu_i$ defined on the σ-algebra generated by the measurable rectangles $E = \prod_{i=1}^m E_i$, $E_i \in \mathscr{A}_i$, such that $\mu(E) = \prod_{i=1}^m \mu_i(E_i)$. ∎

Remark B.1. If the finite set A is the disjoint union of two nonempty finite subsets A_1 and A_2, then for σ-finite measure spaces $(X_i, \mathscr{A}_i, \mu_i)$, $i \in A$, we may identify the product measure space $(\prod_{i\in A} X_i, \prod_{i\in A} \mathscr{A}_i,$

$\prod_{i\in A}\mu_i$) with the product of the measure spaces

$$\Big(\prod_{i\in A_1} X_i, \prod_{i\in A_1} \mathscr{A}_i, \prod_{i\in A_1} \mu_i\Big) \quad \text{and} \quad \Big(\prod_{i\in A_2} X_i, \prod_{i\in A_2} \mathscr{A}_i, \prod_{i\in A_2} \mu_i\Big).$$

Thus, it follows that the Fubini and Tonelli theorems of Chapter 3 are also valid in this more general situation. In other words, the integral of a nonnegative function measurable on the product measure space of a finite number of σ-finite measure spaces or of a function that is integrable in this product space can be evaluated by iterated integration over the individual measure spaces in any order.

We will now consider products of an infinite number of measure spaces $(X_i, \mathscr{A}_i, \mu_i)$, $i \in A$ and A infinite. To avoid difficulties in handling infinite products of the form $\prod_{i\in A}\mu_i(E_i)$, where $E_i = X_i$ for all but finitely many i, we assume that $\mu_i(X_i) = 1$ for all $i \in A$. For a finite subset $F \subset A$, let $\mathscr{A}_F$ be the class of all finite disjoint union of measurable rectangles of the form $\prod_{i\in A}E_i$, where $E_i \in \mathscr{A}_i$ for each $i \in A$ and $E_i = X_i$ for $i \notin F$. Then $\mathscr{A}_F$ is an algebra. Let $\mathscr{F}$ be the class of all finite subsets of A. Then $\mathscr{A}_0 = \bigcup_{F\in\mathscr{F}}\mathscr{A}_F$ is also an algebra. We define μ on $\mathscr{A}_0$ in the usual way:

$$\mu\Big(\prod_{i\in A} E_i\Big) = \prod_{i\in A} \mu_i(E_i),$$

where $\prod_{i\in A}E_i$ ($\in \mathscr{A}_F$) is a measurable rectangle. [As usual, for disjoint measurable rectangles Y and Z in $\mathscr{A}_0$, we take $\mu(Y \cup Z) = \mu(Y) + \mu(Z)$.] It can be easily verified that μ so defined is well defined and finitely additive. Thus, once we show that μ is a measure on $\mathscr{A}_0$, the extension theorem 2.4 is immediately applicable to extend μ uniquely to the σ-algebra generated by $\mathscr{A}_0$.

To show that μ is a measure on $\mathscr{A}_0$, let (Y_n) be a sequence of disjoint sets in $\mathscr{A}_0$ whose union is also in $\mathscr{A}_0$. Notice that it is enough to show that $\lim_{n\to\infty}\mu(Z_n) = 0$, where $Z_n = \bigcup_{i=n+1}^{\infty} Y_i$, since then

$$\mu\Big(\bigcup_{i=1}^{\infty} Y_i\Big) = \mu\Big(\bigcup_{i=1}^{n} Y_i\Big) + \mu(Z_n) = \sum_{i=1}^{n} \mu(Y_i) + \mu(Z_n) \to \sum_{i=1}^{\infty} \mu(Y_i)$$

as $n \to \infty$. Note that $\bigcap_{n=1}^{\infty} Z_n = \varnothing$. Let us now assume that $\mu(Z_n) \not\to 0$ as $n \to \infty$. We show below that this assumption will imply that $\bigcap_{n=1}^{\infty} Z_n$ is nonempty.

Let us first observe that there are finite subsets $F_n \in \mathscr{F}$ such that $Z_n \in \mathscr{A}_{F_n}$ and we can assume, with no loss of generality, that $F_n \subset F_{n+1}$

and $F_n = \{a_1, a_2, \ldots, a_{i_n}\}$. Using Fubini's theorem, we can write

$$\mu(Z_n) = \int f_n(x)\, d\mu_{a_1}(x),$$

where the function f_n is given by

$$f_n(x_1) = \int_{X_{a_2}} \left[\cdots \left[\int_{X_{a_{i_n}}} \chi_{Z_n}(x_1, x_2, \ldots, x_{i_n})\, d\mu_{a_{i_n}}(x_{i_n}) \right] \cdots \right] d\mu_{a_2}(x_2).$$

Since $\mu(Z_n) \nrightarrow 0$ as $n \to \infty$, there is at least one $x \in X_{a_1}$, say y_1, such that $f_n(y_1)$ is defined for all n and $f_n(y_1) \nrightarrow 0$ as $n \to \infty$. Repeating this argument and using induction, it is easily seen that there is a sequence (y_n), $y_n \in X_{a_n}$, such that for each n and for each $k < i_n$, the integral

$$\int_{X_{a_k}} \left[\cdots \left[\int_{X_{a_{i_n}}} \chi_{Z_n}(y_1, \ldots, y_{k-1}, x_k, \ldots, x_{i_n})\, d\mu_{a_{i_n}}(x_{i_n}) \right] \cdots \right] d\mu_{a_k}(x_k)$$

exists and does not converge to zero as $n \to \infty$. Thus, given any m, there is some $n > i_m$ such that

$$(y_1, \ldots, y_{i_m}, x_{i_m+1}, \ldots, x_{i_n}) \in Z_n$$

for some $(x_{i_m+1}, \ldots, x_{i_n}) \in X_{a_{i_m+1}} \times \cdots \times X_{a_{i_n}}$. Since $Z_n \subset Z_m$, $Z_m \in \mathscr{A}_{F_m}$ and $F_m = \{a_1, a_2, \ldots, a_{i_m}\}$, it is clear that $(y_1, y_2, \ldots, y_{i_m}, p) \in Z_m$, whenever $p \in \prod_{i \notin F_m} X_i$. Let $u \in \prod_{i \notin \cup_{m=1}^{\infty} F_m} X_i$. Let us define $x \in \prod_{i \in A} X_i$ by $x_i = y_k$ when $i = a_k$, and $x_i = u_i$ when $i \notin \bigcup_{m=1}^{\infty} F_m$. Then $x \in \bigcap_{m=1}^{\infty} Z_m$. This is a contradiction.

Thus, we can now state the following theorem.

Theorem B.2. For each $i \in A$, let $(X_i, \mathscr{A}_i, \mu_i)$ be a measure space such that $\mu_i(X_i) = 1$. Then there is a unique measure μ on the σ-algebra $\mathscr{A}$ generated by the measurable rectangles in $X = \prod_{i \in A} X_i$ such that $\mu(\prod_{i \in A} E_i) = \prod_{i \in A} \mu_i(E_i)$ for every measurable rectangle $\prod_{i \in A} E_i$ in $\mathscr{A}$. ∎

The measure space $(X, \mathscr{A}, \mu)$ in the above theorem is called the product of the measure spaces $(X_i, \mathscr{A}_i, \mu_i)$, $i \in A$. We write $(X, \mathscr{A}, \mu) = \prod_{i \in A} (X_i, \mathscr{A}_i, \mu_i)$ and $\mu = \prod_{i \in A} \mu_i$. There are several analogs of Fubini's theorem for infinite products, which we will not discuss here. The interested reader can consult [16] for such results. We also leave to the reader the verification of the following associative property for infinite products.

Theorem B.3. For each $i \in A$, let $(X_i, \mathscr{A}_i, \mu_i)$ be a measure space such that $\mu_i(X_i) = 1$. Let the sets $\{A_\beta : \beta \in B\}$ form a partition of the set A. Then we can identify the measure spaces $\prod_{i \in A}(X_i, \mathscr{A}_i, \mu_i)$ and $\prod_{\beta \in B}[\prod_{i \in A_\beta}(X_i, \mathscr{A}_i, \mu_i)]$. ∎

We conclude by showing how we can utilize the product measure (for infinite products) to construct the Lebesgue measure on $[0, 1)$, and then on the reals.

For each positive integer i, let $X_i = \{0, 1\}$, $\mathscr{A}_i = \{\varnothing, \{0\}, \{1\}, \{0, 1\}\}$, and μ_i defined by $\mu_i(\{0\}) = \mu_i(\{1\}) = 1/2$. Let $(X, \mathscr{A}, \mu)$ be the product $\prod_{i=1}^\infty (X_i, \mathscr{A}_i, \mu_i)$. Then the set X_0 defined by

$$X_0 = \{x = (x_i): x_i = 1 \quad \text{for all but finitely many } i\}$$

is countable, measurable, and has μ-measure zero. Consider the measure space $(Y, \mathscr{A}', \mu)$, where $Y = X - X_0$ and $\mathscr{A}' = \mathscr{A} \cap Y$. Let us define $h: Y \to [0, 1)$ by $h(x) = \sum_{i=1}^\infty (x_i/2^i)$. Noting that each real number in $[0, 1)$ has a binary representation (which is unique when we consider only the terminating representation for a number which has more than one binary representation), we see that h is a one-to-one correspondence. Let $0 \le a < b \le 1$ and $E = h^{-1}(a, b)$. We claim that $\mu(h^{-1}(a, b)) = b - a$. It is sufficient to establish this when a and b both have terminating binary representations, since the numbers with such representations are dense in $[0, 1)$.

Let $a = .a_1 a_2 \ldots a_n$ and $b = .b_1 b_2 \ldots b_n$ be the binary representations of a and b. Let the sets K and L be defined by

$$K = \{i: 1 \le i \le n, \ a_i = 0 \quad \text{and} \quad b_i = 1\}$$

and

$$L = \{i: 1 \le i \le n, \ a_i = 1 \quad \text{and} \quad b_i = 0\}.$$

Then, since $a < b$, $\min K < \min L$. For each $i \in K$, consider the set U_i defined by

$$U_i = \{x = (x_k): x_k = b_k \quad \text{for} \quad 1 \le k < i, \ x_i = b_i - 1, \quad \text{and} \\ x_k \in \{0, 1\} \quad \text{for} \quad k > i\}.$$

Then the U_i's (for $i \in K$) are disjoint measurable rectangles and $\mu(U_i) = 1/2^i$. For each $j \in L$, let $p(j) = \max\{i \in K: i < j\}$. Define the set $V_{p(j)}$ by

$$V_{p(j)} = \{x = (x_k): x_k = b_k \quad \text{for} \quad 1 \le k < p(j), x_{p(j)} = 0, \\ x_k = a_k \quad \text{for} \quad p(j) < k \le j - 1, x_j = 0, x_k \in \{0, 1\} \quad \text{for} \quad k > j\}.$$

Then, $V_{p(j)} \subset U_{p(j)}$, $\mu(V_{p(j)}) = 1/2^j$, and when j's are different, the $V_{p(j)}$'s are disjoint. It is easily verified that

$$h^{-1}(a, b) = \bigcup_{i \in K} \left[U_i - \bigcup_{\substack{j \in L \\ p(j)=i}} V_{p(j)} \right].$$

It follows that

$$\mu\big(h^{-1}(a, b)\big) = b - a.$$

Since $\Sigma = \{B \subset [0, 1): h^{-1}(B)$ is product measurable$\}$ is a σ-algebra containing all open subintervals (a, b), Σ contains all Borel subsets of $[0, 1)$. It is now clear that the completion of the measure λ defined on the Borel subsets of $[0, 1)$ by $\lambda(B) = \mu\big(h^{-1}(B)\big)$ is the Lebesgue measure on $[0, 1)$. The measure can be extended to the reals by

$$\lambda(B) = \sum_{n=-\infty}^{\infty} \lambda[B \cap (n, n + 1) - n].$$

Bibliography

1. Aliprantis, C. D., and Burkinshaw, O., *Principles of Real Analysis*, North-Holland, New York (1981).
2. Ash, R. B., *Measure, Integration, and Functional Analysis*, Academic Press, New York (1972).
3. Asplund, E., and Bubgart, L., *A First Course in Integration*, Holt, Rinehart, and Winston, New York (1966).
4. Bagchi, S. C., and Sitaram, A., Determining sets for measures on R^n, *Illinois J. Math.* **26**, 419–422 (1982).
5. Bauer, H., *Probability Theory and Elements of Measure Theory*, Holt, Rinehart, and Winston, New York (1972).
6. Berberian, S. K., *Measure and Integration*, MacMillan, New York (1965).
7. Berberian, S. K., On the extension of Borel measures, *Prov. Amer. Math. Soc.* **16**, 415–418 (1965).
8. Bochner, S., Additive set functions on groups, *Ann. Math.* **40**, 852–855 (1939).
9. Bourbaki, N., *Topologie Générale*, Herman, Paris (1951).
10. Briggs, J. M., and Schaffter, T., Measure and Cardinality, *Amer. Math. Monthly* **86**, 852–855 (1979).
11. Broughton, A., and Huff, B. W., A Comment on unions of sigma-fields, *Amer. Math. Monthly* **84**, 553–554 (1977).
12. Burrill, C. W., *Measure, Integration, and Probability*, McGraw-Hill, New York (1972).
13. Cohen, P. J., *Set Theory and the Continuum Hypothesis*, W. A. Benjamin, Inc., New York (1966).
14. Doss, R., The Hahn decomposition theorem, *Proc. Amer. Math. Soc.* **80**, 377 (1980).
15. Dubins, L. E., An elementary proof of Bochner's finitely additive Radon–Nikodym theorem, *Amer. Math. Monthly* **76**, 520–523 (1969).
16. Dunford, N., and Schwartz, J., *Linear Operators, Part I: General Theory*, Wiley-Interscience, New York (1958).
17. Feldman, M. B., A Proof of Lusin's Theorem, *Amer. Math. Monthly* **88**, 191–192 (1981).
18. Graves, L. M., *The Theory of Functions of Real Variables* (2nd ed.), McGraw-Hill, New York (1956).

19. Halmos, P. R., *Lectures on Ergodic Theory*, Chelsea, New York (1956).
20. Halmos, P. R., *Measure Theory*, Van Nostrand, Princeton, New Jersey (1950).
21. Halmos, P. R., *Naive Set Theory*, Van Nostrand, Princeton, New Jersey (1960).
22. Hardy, G. H., Weierstrass's non-differentiable function, *Trans. Amer. Math. Soc.* **17**, 301–325 (1916).
23. Hewitt, E., and Stromberg, K., *Real and Abstract Analysis*, Springer-Verlag, Berlin (1965).
24. Huggins, F. N., Some interesting properties of the variation function, *Amer. Math. Monthly* **83**, 538–546 (1976).
25. Johnson, R. A., A compact non-metrizable space such that every closed subset is a G-delta, *Amer. Math. Monthly* **77**, 172–176 (1970).
26. Johnson, R. A., On product measures and Fubini's theorem in locally compact spaces, *Trans. Amer. Math. Soc.* **123**, 112–129 (1966).
27. Johnson, R. A., On the Lebesgue decomposition theorem, *Proc. Amer. Math. Soc.* **18**, 628–632 (1967).
28. Johnson, R. A., Some types of Borel measures, *Proc. Amer. Math. Soc.* **22**, 94–99 (1969).
29. Kelley, J. L., *General Topology*, Van Nostrand, Princeton, New Jersey (1955).
30. Levin, M., and Stiles, W., On the regularity of measures on locally compact spaces, *Proc. Amer. Math. Soc.* **16**, 201–206 (1972).
31. Luther, N. Y., Lebesgue decomposition and weakly Borel measures, *Duke Math. J.* **35**, 601–615 (1968).
32. Magill, K. D. Jr., N-point compactifications, *Amer. Math. Monthly*, 1075–1081 (1965).
33. McShane, E. J., *Integration*, Princeton University Press, Princeton, New Jersey (1944).
34. Monroe, M. E., *Measure and Integration* (2nd ed.), Addison-Wesley, Reading, Massachusetts (1971).
35. Mukherjea, A., Remark on Tonelli's theorem on integration in product spaces, *Indiana Univ. Math. J.* **23**, 679–684 (1974).
36. Ostrowski, A. M., Note on the Bernoulli–L'Hospital rule, *Amer. Math. Monthly* **83**, 239–242 (1976).
37. Oxtoby, J. C., *Measure and Category*, Springer-Verlag, New York (1971).
38. Parthasarathy, K. R., *Probability Measures on Metric Spaces*, Academic Press, New York (1967).
39. Penot, J. P., Fixed point theorems without convexity, *Bull. Soc. Math. France, Memoire* **60**, 129–152 (1979).
40. Phillips, E., *An Introduction to Analysis and Integration Theory*, Intext Educational Pub., Scranton, Toronto, London (1971).
41. Reed, M., and Simon, B., *Functional Analysis*, Academic Press, New York (1972).
42. Rogers, C. A., A less strange version of Milner's proof of Brouwer's fixed point theorem, *Amer. Math. Monthly* **87**, 525–527 (1980).
43. Ross, K. A., and Stromberg, K., Baire sets and Baire measures, *Ark. Mat.* **6**, 151–160 (1965).
44. Royden, H. L. *Real Analysis* (2nd ed.), MacMillan, New York (1968).
45. Rudin, W., *Real and Complex Analysis*, McGraw-Hill, New York (1966).
46. Sabharwal, C. L., and Alexiades, V., On the extension of Lebesgue measure, unpublished.
47. Saks, S., *Theory of the Integral*, Warszawa-Lwów (1937).

48. Segal, I. E., and Kunze, R. A., *Integrals and Operators*, McGraw-Hill, New York (1968).
49. Sehgal, V. M., A fixed point theorem for mappings with a contractive iterate, *Proc. Amer. Math. Soc.* **23** (3), 631–634 (1969).
50. Sehgal, V. M., On fixed and periodic points for a class of mappings, *J. London Math. Soc.* **5** (2), 571–576 (1972).
51. Serrin, J., and Varberg, D., A general chain rule for derivatives and the change of variables formula for the Lebesgue integral, *Amer. Math. Monthly* **76**, 514–520 (1969).
52. Simoson, A., An "Archimedean" paradox, *Amer. Math. Monthly* **89**, 114–125 (1982).
53. Stone, M. H., *Linear Transformations in Hilbert Space*, American Mathematical Society, Providence, Rhode Island (1964).
54. Sucheston, L., Banach Limits, *Amer. Math. Monthly* **74** (3), 308–311 (1967).
55. Suppes, P., *Axiomatic Set Theory*, Van Nostrand, Princeton, New Jersey (1960).
56. Titchmarsh, E. C., *The Theory of Functions* (2nd ed.), Oxford University Press, London and New York (1939).
57. Vaidyanathaswamy, R., *Set Topology*, Chelsea, New York (1960).
58. Varberg, D. E., On absolutely continuous functions, *Amer. Math. Monthly* **72**, 831–841 (1965).
59. Walker, P. L., On Lebesgue integrable derivatives, *Amer. Math. Monthly* **84**, 287–288 (1977).
60. Willard, S., *General Topology*, Addison-Wesley, Reading, Massachusetts (1970).
61. Zaanen, A. C., *Integration*, North-Holland, Amsterdam (1967).
62. Zaanen, A. C., *Linear Analysis*, North-Holland, Amsterdam (1953).

Definition, Theorem, Proposition, Lemma, and Corollary Index

Definitions

1.1 2
1.2 3
1.3 4
1.4 4
1.5 5
1.6 6
1.7 10
1.8 10
1.9 11
1.10 11
1.11 12
1.12 16
1.13 17
1.14 22
1.15 23
1.16 27
1.17 27
1.18 28
1.19 28
1.20 28
1.21 28
1.22 29
1.23 31
1.24 33
1.25 33
1.26 34
1.27 35
1.28 37
1.29 37
1.30 38
1.31 40
1.32 45
1.33 48
1.34 48
1.35 49
1.36 49
1.37 51
1.38 52
1.39 52
1.40 54
1.41 78
1.42 82
2.1 87
2.2 98
2.3 100
2.4 103
2.5 120
2.7 124
3.1 133
3.2 136
3.3 137
3.4 138
3.5 139
3.6 144
3.7 144
3.8 148
3.9 162
3.10 163
3.11 164
3.12 188
3.13 189
4.1 202
4.2 205
4.3 209
4.4 220
4.5 234
4.6 237
4.7 239
4.8 249
4.9 251
4.10 252
5.1 269
5.2 269
5.3 281
5.4 291
5.5 291
A.1 332

Theorems

1.1 3
1.2 12
1.3 13
1.4 22
1.5 30
1.6 31
1.7 36
1.8 47
1.9 47
1.10 53
1.11 53
1.12 54
1.13 54
1.14 54
1.15 55

1.16 56
1.17 56
1.18 57
1.19 58
1.20 58
1.21 60
1.22 62
1.23 63
1.24 64
1.25 65
1.26 66
1.27 67
1.28 78
1.29 80
1.30 82
2.1 94
2.2 100
2.3 103
2.4 106
2.5 118
2.6 120
2.8 126
3.1 138
3.2 146
3.3 151
3.4 152
3.5 164
3.6 166
3.7 175
3.8 176
3.9 176
3.10 177
3.11 179
3.12 191
3.13 193
4.1 205
4.2 207
4.3 223
4.4 224
4.5 238
4.6 240
4.7 247
4.8 253
4.9 263
4.10 265
5.1 274
5.2 295
5.3 295
5.4 298
5.5 300
5.6 303
5.7 308
5.8 310
5.9 312
5.10 315
5.11 317
5.12 320
5.13 323
5.14 325
5.15 327
5.16 327
5.17 328
5.18 329
5.19 330
5.20 331
A.1 334
A.2 335
A.3 339
A.4 341
B.1 342
B.2 344
B.3 345

Propositions

1.1 4
1.2 5
1.3 6
1.4 7
1.5 7
1.6 13
1.7 19
1.8 19
1.9 20
1.10 20
1.11 23
1.12 23
1.13 29
1.14 33
1.15 33
1.16 33
1.17 37
1.18 37
1.19 40
1.20 47
1.21 50
1.22 51
1.23 52
2.1 88
2.2 88
2.3 90
2.4 98
2.5 102
3.1 134
3.2 135
3.3 136
3.4 136
3.5 136
3.6 137
3.7 139
3.8 145
3.9 160
3.10 175
3.11 176
3.12 189
3.13 190
3.14 191
4.1 202
4.2 204
4.3 210
4.4 214
4.5 217
4.6 218
4.7 221
4.8 222
4.9 235
4.10 239
4.11 240
4.12 251
4.13 252
4.14 262
4.16 264
5.1 271
5.2 271
5.3 272
5.4 276
5.5 282
5.6 282
5.7 284
5.8 285
5.9 286
5.10 288
5.11 293
5.12 297
5.13 307
5.14 308
5.15 322
A.1 338

Lemmas

1.1 35
2.1 94
2.2 94
2.3 125
3.1 143
3.2 147
3.3 170
3.4 171
3.5 172
3.6 173
3.7 173
3.8 174
4.1 220
4.2 221
4.3 223
4.4 238
4.5 241
5.1 272
5.2 273
5.3 279
5.4 279

5.5 283
5.6 293
5.7 293
5.8 316
5.9 322
A.1 333
A.2 334
A.3 337

Corollaries

1.1 79
1.2 80
3.1 147
3.2 147
3.3 153
3.4 165
4.1 210
4.2 221
4.3 246
4.4 255
4.5 256
4.6 256
4.7 263
5.1 283
5.2 285
5.3 287
5.5 302
5.6 302

Symbol and Notation Index

Set and element notations

A, B, C, X, Y, or Z	2
a, b, c, x, y, or z	2
α, β, γ	2
$\mathscr{A}, \mathscr{B}, \mathscr{C}$	2
ϕ	2
N, Q, R, Z, Z^+	2
$\overline{R}$	14
F	249
$[x]$	10
Ω	13
$\aleph_0, c$	6

Notation used in set operations and relations

$\subset, \subsetneqq$	2
$\cap, \cup$	2
$A - B$	2
$A \Delta B$	3
$A \times B$	3
$\lim_n \sup A_n, \lim_n \inf A_n$	8
A^c	3
$\sup A, \inf A$	11
card A	6
2^U	6

Function notations

D_f, R_f	4
$g \circ f, f \subset g$	4
$f(A), f^{-1}(B)$	4
χ_A	5
δ	5
1_A	5
f^+, f^-	133
$f \wedge g, f \vee g$	79, 133
$D^+f(a), D_+f(a), D^-f(a), D_-f(a)$	202
$V_a^b f$	209
$f'_+(a), f'_-(a), f'(a)$	202
$V_{-\infty}^{\infty} f$	224

Measure theoretic notations

$\mathscr{R}(\mathscr{E}), \mathscr{A}(\mathscr{E})$	18
$\sigma_r(\mathscr{E}), \sigma(\mathscr{E})$	18
$\mathscr{M}(\mathscr{E})$	22
$\mathscr{D}(\mathscr{E})$	22
μ^*	98
l, m_0	94
m^*	105
m_F	109
μ_*	120

Measure theoretic notations (*cont.*)

a.e. 136
$\int f\,d\mu$ 144, 145
μ_f 162
$\mathscr{A} \times \mathscr{B}$ 172
$\mu \times \nu$ 172, 319
E_x, E^y 172, 321
f_x, f^y 173
$\mathscr{R}_{\sigma\delta}$ 173
Γ_E, Γ^E 173, 322
$\nu \ll \mu$ 234, 237
$\nu^+, \nu^-, |\nu|$ 239
$d\nu/d\mu$ 246
$\mu_1 \perp \mu_2$ 239
$\sigma(\mathscr{G})$, σ(VL) 272
$\sigma(\mathscr{L})$ 280
$B_w(X)$, $B(X)$, $\mathrm{Ba}_w(X)$
  $\mathrm{Ba}(X)$ 281
$\mu \otimes \nu$ 320

Topological notations

$\overline{A}$ 27
A°, A_b 28
$(X, \mathscr{E})$ 28
$\beta(X)$ 44
(X, d) 49
$S_x(\epsilon)$ 50
S_f 278

Function space notations

$C_b(X)$ 52
$C(X)$ 78
$C_1(X)$ 80
$C_0(X)$ 81
$C_1[0, 1]$ 188
L_p 186
L_∞ 192
$\mathscr{L}^1(\mu)$ 268
$C_c(X)$ 270
VL, VLU 269, 270

Normed space notations

l_p 188
$L(X, Y)$ 252
$X^*[=L(X, F)]$ 252

Norm notations

$\|x\|$ 228
$\|f\|_p$ 235
ess sup $|f|$ 239
$\|f\|_\infty$ 239
$\|T\|$ 247

Miscellaneous notation

lim sup x_n or $\overline{\lim}_n\, x_n$ 14
lim inf x_n or $\underline{\lim}_n\, x_n$ 14
$x_n \uparrow x, x_n \downarrow x$ 15

Subject Index

Absolutely continuous functions, 220, 224, 230
Absolutely continuous measures, 234, 237
Absolutely summable, 189
Algebra
  generated by a class of sets, 18
  of sets, 17
  subspace of $C_b(X)$, 78
Almost everywhere (a.e.), 136
Almost uniform convergence, 137
Arzela–Ascoli Theorem, 82
Atom, 92
Axiom of Choice, 12

Baire-Category Theorem, 53
  an application, 117, 259
Baire measure, 291
Baire sets, 281
Baire space, 40
Banach Fixed Point Theorem, 55
  a converse of, 76
  an extension of, 76
Banach indicatrix, 228
Banach space, 188
Base (for a topology), 28
Bijection, 5
Bing Metrization Theorem, 54
Borel–Cantelli Lemma, 91
Borel measurable function, 133
Borel measure, 291
Borel sets (in a topological space), 281
Borel sets (of R), 26, 103
Boundary point, 28
Bounded linear functional, 252
Bounded linear operator, 251
Bounded metric space, 51
Bounded variation, function of, 210, 224
Brouwer Fixed Point Theorem, 47

Cantor function, 71, 232
Cantor set, 71
Cantor–Lebesgue Theorem, 156
Carathéodory Extension Theorem, 106
Cardinalities of σ-algebras, 26
Cardinality of a set, 6
Cartesian product, 3, 12
Cauchy sequence, 16, 52
  in measure, 139
$C_c(X)$-open, 283
Chain, 11
Choice function, 11
Closed function, 28
Cluster point, 37
Compact space, 29
Compactification
  one-point, 38, 278
  Stone–Čech, 44
Complement, 2
Complete measure, 102
Complete metric space, 52
Completely regular space, 32
Component, 48
Connected space, 45
Continuous
  absolutely, 220, 224, 230
  uniformly, 71
Continuous function, 28
Continuum hypothesis, 119

Contraction mapping, 55
Contractive mapping, 55
Convergence
  almost uniform, 137
  in L_p, 197
  in measure, 138
Convex function, 72, 142
Convex set, 72
Convolution (of integrable functions), 260
Countable additivity, 87
Countable ordinals, set of, 13
Countable subadditivity, 88
Countably compact space, 37
Countably infinite set, 6
Cover, 29
  open, 29

Daniell integral, 269
Daniell–Stone Representation Theorem, 274
DeMorgan's laws, 3
Dense set, 35
Derivate, 202
Derivative (left, right), 202
  of a measure, 333
  of a transformation, 336
Differentiable function (left, right), 202
Differentiable transformation, 336
Dini's Theorem, 71
Directed system, 43
Discrete topology, 27
Distribution function, 97
Domain of a function, 4
Dominated Convergence Theorem, 151, 158
Dual (conjugate) space, 252
Dynkin system, 23

Egoroff's Theorem, 138
  for families of functions,128
Equicontinuity, 82
Equivalence class, 10
Equivalence relation, 10
Extended real numbers, 13

Fatou's Lemma, 147
Finite intersection property (f.i.p.), 29
First category, 40
First countable, 33
First uncountable ordinal, 13
F_σ-set, 113, 278
Fubini's Theorem, 175, 327
Fubini–Tonelli Theorem, 177
Function, 5
  absolutely continuous, 220, 224, 230
  Borel measurable, 133
  of bounded variation, 210, 224
  choice, 11
  closed, 28
  continuous, 28
  distribution, 97
  injection, 5
  integrable, 145
  integrable nonnegative, 144
  integrable simple, 144
  Kronecker's delta (δ), 5
  Lebesgue measurable, 133
  measurable, 133
  onto, 5
  open, 28
  sequence, 5
  simple, 136
  surjection, 5

Greatest lower bound, 11
G_δ-set, 41, 113, 278

Hahn Decomposition Theorem, 238
Half-open interval topology, 41
Hausdorff Maximal Principle, 12
Hausdorff space, 31
Heine–Borel–Bolzano–Weierstrass Theorem, 31
Hölder Inequality, 190
Homeomorphic spaces, 46
Homeomorphism, 41

Image of a set, 4
Index set, 6
Indiscrete topology, 27
Individual Ergodic Theorem, 193
Induction
  principle of mathematical, 11
  Principle of Transfinite Induction, 13
Injection (*see* function)
Inner measure, 120
Inner regular, 291
Integrable function, 145
Integral, 144, 145, 148
  Daniell, 269

Lebesgue, 148
Lebesgue–Stieltjes, 163
Riemann–Stieltjes, 164
Integration
of complex-valued functions, 157
by parts, 167, 182
Interior of a set, 28
Interior point, 28
Intersection of sets, 2
Inverse image of a set, 4
Isolated point, 41
Isometric metric spaces, 52
Isometry, 52

Jacobian, 336
Jensen's inequality, 158
Jordan decomposition, 239

Kronecker's delta (δ) function, 5

Least upper bound, 11
Lebesgue Convergence Theorem
a generalization, 196
see also Dominated Convergence Theorem
Lebesgue Covering Lenma, 70
Lebesgue Decomposition Theorem, 247
an extension, 258
Lebesgue Density Theorem, 231
Lebesgue integral, 148
Lebesgue-measurable function, 133
Lebesgue-measurable set, 106
Lebesgue measure, 94, 106, 108
Lebesgue number, 70
Lebesgue outer measure, 100
Lebesgue–Stieltjes integral, 163
Lebesgue–Stieltjes measure, 97, 162
L'Hospital Rule, 230
Limit inferior, 8, 14
Limit point, 27
Limit superior, 8, 14
$\lim_n$ inf, 8, 14
$\lim_n$ sup, 8, 14
Lindelöf space, 34
Linear functional, bounded, 252
Linear operator, 249
bounded, 251
Linear space, 186
Lipschitz condition, 54, 213, 220
Local connectedness, 48
Locally compact, 38, 278
Locally connected, 48
Locally finite, 54
Locally measurable, 112, 289
Lower bound, 11
greatest, 11
Lower semicontinuous, 33
l_p, l_∞, L_∞ space, 192
L_p space, 186
L_p^*, l_p^* spaces, 256
Lusin's Theorem, 141

Mapping, 5
open (*see* Function, open)
Maximal element, 11
Mean-square-continuous mapping (m.s.c.), 196
Measurable function, 133
Measurable kernel, 121
Measurable modification, 304
Measurable, μ^*-, 100
Measurable transformation, 264
Measure, 87
Baire, 291
Borel, 291
complete, 102
counting, 87
differentiable, 332
finite, 89
finitely additive, 87
inner, 120
Lebesgue, 94, 106, 108
Lebesgue outer, 100
Lebesgue–Stieltjes, 97, 162
metric outer, 103
outer, 98
regular, 292
semifinite, 90
signed, 234
σ-finite, 89
weakly Baire, 291
weakly Borel, 291
Measure-preserving mapping, 193
Measure space, 132
Metric, 49
Lèvy, 70
Metric space, completion of, 52
Metrizable, 54
Minimal element, 11
Minkowski Inequality, 191
Monotone class, 22

Monotone Convergence Theorem, 146
Monotone Convergence Theorem for Nets, 319
Mutually singular, 239

Nagata–Smirnov Metrization Theorem, 54
Negative set, 237
Negative variation, 239
Net, 43
Nondecreasing sequence, 15
Nonexpansive mapping, 55
Norm, 188
  ess sup, 192
Normal space, 31
Normed linear space, 188
Nowhere dense, 40
Null set (with respect to a measure), 237

One-point compactification, 38, 278
Open function, 28
Open set, 27
Operator
  bounded linear, 251
  linear, 249
Order complete, 11
Order topology, 38
Ordering
  partial, 10
  total, 10
  well, 10
Ordinal
  first uncountable, 13
  set of countable, 13
Outer measure, 98
Outer regular, 292

Pairwise disjoint, 3
Paracompact, 42, 295
Partial ordering, 10
Path-connected, 49
Perfect set, 115
Periods of measurable functions, 232
Picard Existence Theorem, 67
Point
  boundary, 28
  cluster, 37
  interior, 28
  isolated, 41
  limit, 27
Point at infinity, 278
Positive linear functional (on a vector lattice), 269
Positive set, 237
Positive variation, 239
Power set, 6
Principle of mathematical induction, 11
Principle of Transfinite Induction, 13
Product
  Cartesian, 3, 72
  topological, 29
Product measure, 172, 319
Pseudocompact, 42
Pseudometric, 73
Pseudonorm, 187
Purely atomic, 258

Quotient space, 73

Radon–Nikodym derivative, 246
Radon–Nikodym Theorem, 240
Range, 4
Rectangle, measurable, 171
Recurrence Theorem, 143
Regular measure, 292
Regular space, 31
Relation, 4
  composition of, 4
  equivalence, 10
  extension of, 4
  inverse of, 4
  restriction of, 4
Relative topology, 27
Retract of a topological space, 74
Riemann–Lebesgue Theorem, 156, 157
Riemann–Stieltjes integral, 164
Riesz–Fischer Theorem, 191
Riesz Representation Theorem, 253, 315
Ring (generated by a class of sets), 18
Ring of sets, 16

Scheffe's Theorem, 158
Schroeder–Bernstein Theorem, 9
Second category, 40
Second countable, 33
Semifinite measure, 90
Sequence, 5
  Cauchy, 16, 52
  Cauchy in measure, 139
  nondecreasing, 15
Semialgebra, 170

Separable, 35
Separate points, 78
Sequentially compact, 37
Set
  cardinality of, 6
  compact, 29
  countable, 6
  of countable ordinals, 13
  countably infinite, 6
  finite, 6
  index, 6
  μ^*-measurable, 100
  null (with respect to a measure), 237
  open, 27
  partially ordered, 10
  positive (negative), 237
  power, 6
  subset, 2
  totally ordered, 10
  uncountable, 6
  well-ordered, 10
σ-algebra, 18
  separable, 27
σ-algebra generated by a class of sets, 18
σ-bounded, 278
σ-class, 25
σ-compact, 278
σ-finite measure, 89
σ-ring, 16
σ-ring of Baire sets, 281
σ-ring of Borel sets, 281
Signed measure, 234
Simple function, 136
Space
  Baire, 40
  Banach, 188
  Borel measure, 132
  compact topological, 29
  completely regular, 32
  connected, 45
  dual, 252
  Hausdorff, 31
  Lebesgue measure, 132
  Lindelöf, 34
  linear (or vector), 186
  l_p, l_∞, L_∞, 192
  L_p, 186
  L_p^*, l_p^*, 256
  measure, 132
  metric, 49
  normal, 31
  normed linear, 188
Space (*cont.*)
  regular, 31
  T_1, T_2, T_3, T_4, 31
  topological, 28
Stone–Čech compactification, 44
Stone–Weierstrass Theorem, 78
Stone–Weierstrass Theorem for Lattices, 84
Stone–Weierstrass Theorem for L_2-Valued Real Functions, 196
Subbase, 28
Summable series (absolutely), 189
Support, 278
Surjection, 5
Symmetric difference, 3

T_1, T_2, T_3, T_4 spaces, 31
Tensor product of measures, 320
Thick set, 128
Tietze Extension Theorem, 36
Tonelli's Theorem, 176, 327
Topological product, 29
Topology, 27
  discrete, 27
  half-open interval, 41
  indiscrete, 27
  order, 38
  of pointwise convergence, 73
  quotient, 73
  relative, 27
  usual, 27
Total ordering, 10
Total variation function, 221
Total variation (of a function), 210
Total variation (of a measure), 239
Totally bounded, 51
Totally disconnected, 48
Transfinite Induction, 13
Transformation, 5
  differentiable, 336
  measurable, 264
Transformation, measure preserving (*see* Measure-preserving mapping)
Translation invariance of the Lebesgue integral, 154
Translation invariant, 96
Tychonoff Theorem, 30

Ulam Theorem, 120
Uncountable set, 6

Uniform integrability, 154
Uniformly continuous, 71
Union, 2
Upper bound, 11
  least, 11
Urysohn Metrization Theorem, 54
Urysohn's Lemma, 35

Validity of Tonelli's Theorem, 179
Vanish at infinity, 81
Vector lattice, 269
Vitali–Caratheodory Theorem, 159
Vitali covering, 205
Vitali's Covering Theorem, 205
VL-open, 271

Weakly Baire measure, 291
Weakly Baire sets, 281
Weakly Borel measure, 291
Weakly Borel sets, 281
Weakly regular, 306
Well-ordering, 10
Well-Ordering Principle, 12

Zero set, 278
Zorn's Lemma, 12